PROGRESS IN SMART MATERIALS AND STRUCTURES

PROGRESS IN SMART MATERIALS AND STRUCTURES

PETER L. REECE
EDITOR

Nova Science Publishers, Inc.
New York

Library of Congress Cataloging-in-Publication Data

Progress in smart materials and structures / Peter L. Reece (editor).
    p. cm.
  Includes index.
  ISBN 13: 978-1-60021-106-5
  ISBN 10: 1-60021-106-2
  1. Smart materials. 2. Smart structures. I. Reece, Peter L.
  TA418.9.S62P76   2006
  620.1'1--dc22   2006007417

Published by Nova Science Publishers, Inc. ✛*New York*

CONTENTS

PREFACE

"Smart" materials respond to environmental stimuli with particular changes in some variables. For that reason they are often also called responsive materials. Depending on changes in some external conditions, "smart" materials change either their properties (mechanical, electrical, appearance), their structure or composition, or their functions. Mostly, "smart" materials are embedded in systems whose inherent properties can be favorably changed to meet performance needs. Smart materials and structures have widespread applications in ; 1. Materials science: composites, ceramics, processing science, interface science, sensor/actuator materials, chiral materials, conducting and chiral polymers, electrochromic materials, liquid crystals, molecular-level smart materials, biomaterials. 2. Sensing and actuation: electromagnetic, acoustic, chemical and mechanical sensing and actuation, single-measurand sensors, multiplexed multimeasurand distributed sensors and actuators, sensor/actuator signal processing, compatibility of sensors and actuators with conventional and advanced materials, smart sensors for materials and composites processing. 3. Optics and electromagnetics: optical fibre technology, active and adaptive optical systems and components, tunable high-dielectric phase shifters, tunable surface control. 4. Structures: smart skins for drag and turbulence control, other applications in aerospace/hydrospace structures, civil infrastructures, transportation vehicles, manufacturing equipment, repairability and maintainability. 5. Control: structural acoustic control, distributed control, analogue and digital feedback control, real-time implementation, adaptive structure stability, damage implications for structural control. 6. Information processing: neural networks, data processing, data visualization and reliability. This new book presents leading new research from around the globe in this field.

The shape-memory composite belt with a TiNi-SMA wire fiber and a polyurethane-SMP sheet matrix was fabricated. The bending actuation characteristics of the belt were investigated by the thermomechanical tests. The results obtained in Chapter 1 can be summarized as follows. (1) Residual deflection close to the maximum deflection is obtained by cooling under constant maximum deflection. The residual deflection disappears by heating under no load. Both the rate of shape fixity and the rate of shape recovery are close to 100%. (2) Recovery force appears by heating under constant residual deflection. The recovery force is 93-94% of the maximum force. The development of high functionality of shape-memory composite elements is expected by various combinations of SMAs and SMPs.

The creation of an effective two-way shape memory alloy (TWSMA) requires particular heat treatment and training considerations. Chapter 2 investigates methods of training TWSMA from the viewpoint of creating an optimal actuator for engineering applications. A review of recommended heat treatment conditions and previously reported training methods is followed by findings from an experimental study that compared different training methods for producing NiTi TWSMA wires with a hot shape of an arc and a cold shape of a straight line. These methods are: shape memory cycling, constrained cycling of deformed martensite, pseudoelastic cycling and combined shape memory & pseudoelastic cycling. In order to give a meaningful evaluation of their performance that is relevant to training for practical applications, these training methods are assessed in terms of maximum two-way strain, changes in the original hot shape and the transformation temperatures after the training process, and the effective production of the cold shape. Only the combined shape memory & pseudoelastic cycling provides an effective training method for creating NiTi TWSM with a non-uniaxial two-way shape change. The undesirable side effects of training were that the NiTi TWSMA wire lost partial memory of the original hot shape and its transformation temperatures shifted to the lower values. There also exists an optimal number of training cycles and possibly an optimal training load for obtaining the best cold shape memory and the greatest two-way recoverable strain. These findings point to future directions in research for advancing TWSMA training technology.

Detwinning in crystalline solids is a unique deformation mechanism partially responsible for the shape memory effect. Owing to an insignificant dislocation process during detwinning, the residual strain can be recovered through a reverse phase transformation. The maximum shape recovery strain is intrinsically related to the lattice geometry and twinning mode, while the magnitude of shape recovery is related to a competition between detwinning and dislocation generation responsible for the macroscopically observed deformation. The detwinning magnitude is directional, and in the polycrystalline materials it is related to the textures and deformation mode. With textures, the detwinning process is enhanced for certain directions and reduced for other directions and so is the shape recovery strain. The anisotropy in the detwinning process allows the possibility of maximizing the potential of the polycrystalline shape memory alloys. Factors that affect the detwinning process will in turn affect the subsequent shape recovery.

Chapter 3 deals with detwinning and shape recovery in NiTi SMAs and influencing factors. NiTi SMAs in several initial conditions are studied including textured wire and bar, rolled sheet, and less textured ingot. Detwinning anisotropy due to texture and deformation mode is studied and explained based on crystallographic analysis. Various experimental results on the relation between detwinning anisotropy and shape recovery anisotropy due to texture and deformation mode are presented and discussed. The present research has also studied the evolution of the crystallographic texture as a result of martensite deformation. In-depth understanding of the atomistic arrangement along <011> type II twin and its role in detwinning are also presented.

Currently, IPMCs (Ionic Polymer-Metal Composites)--as biomimetic actuators and sensors--are being rigorously studied because of their enormous potential for engineering applications in medical, electrical, mechanical, and aerospace engineering. IPMCs have been considered as suitable, soft actuators/sensors to develop biologically-inspired smart structures and artificial-muscle systems since they can create large deformations under low input voltages and generate sensor signals under proper mechanical deformations. Chapter 4 describes general aspects and recent progress in the application of IPMCs and also deals with several design examples in detail. We introduce fundamentals, manufacturing techniques, actuation principles, and electrical properties of typical IPMCs. Further, the equivalent bimorph beam model is introduced, and several design examples--such as flapping wing, IPMC diaphragm, ZNMF (Zero-Net-Mass-Flux) pump, valveless micropump, and muscle-like linear actuators--are presented.

Smart structures require actuation and this actuation is provided by a control system. Hence, control of structure is an important element in the study of smart structures. A control system is driven by a set of control algorithms called the control law. Traditionally, control law is designed based on the reduced order finite element model. By this approach, one cannot handle multi-modal phenomenon such as wave propagation, wherein the transients may cause steep increase in the responses causing the failure of structures. In Chapter 5, a new method of designing control law for laminated composite structure is addressed using the novel Active Spectral Finite Element for controlling responses caused by high frequency impact type loading. The advantage of the spectral element approach is that, it gives a very small system sizes due to its ability to represent the inertia distribution exactly and hence all modes are contained in this small system size. As a result, one need not resort to modal order reduction that is normally associated with any finite element approach for control design. In this work, the design of a PID controller for a smart laminated composite beam embedded/surface mounted using Piezo-Fiber Composite (PFC) actuators, is addressed. The efficiency of the proposed method is demonstrated on a few problems.

Thrusters are commonly used in satellite attitude control systems to provide controllable external thrust force, and hence maintain or change satellite orbits. Ideally, the thrust vector of a satellite thruster should pass through its mass center. In reality, this thrust vector alignment is not always met during the lifetime of a satellite, and consequently resulting in a disturbance torque that could cause the losses of the satellite orientation and orbit keeping. To eliminate the effects of this disturbance torque in the presence of a large thrust level, a reaction control system (RCS), which spins the satellite in the opposite direction, is currently used. This RCS consists of several small auxiliary thrusters that impose a significant mass penalty and onboard fuel consumption, and hence increase the launch cost and shorten the satellite lifetime. In addition, the firing of thrusters in a satellite generates vibration that resonates throughout the entire satellite structure, and hence renders onboard sensitive devices non-operational.

In Chapter 6, a novel satellite thrust vector control and vibration suppression technology is introduced, including the following five aspects: 1) the concept of the novel technology, 2) the development of a two-degree-of-freedom smart composite platform, 3) the platform control, 4) satellite thrust vector control using the platform, and 5) the satellite structure vibration suppression using the platform.

The core of this novel technology is to place a smart structural interface ---- the UHM smart composite platform, between the satellite thruster and the satellite structure to make the satellite thruster steerable. By steering the thruster in real time the thrust vector of a satellite can always be pointed in the desired direction and a vibration suppression capability can also be provided by the same smart platform simultaneously. To achieve this novel technology, the UHM smart composite platform employing the state-of-the-art smart structures science and technology is developed. This platform consists of smart composite panels, smart composite struts, and advanced control systems, and possesses simultaneous precision positioning and vibration suppression capabilities. The inverse kinematics of the platform is analyzed, and an adaptive nonlinear modeling method is proposed to model the platform kinematics. Advanced control strategies are investigated and applied to specially treat the existing nonlinearities, and two control strategies, namely local control strategy and global control strategy, are presented and compared experimentally.

To investigate the satellite thrust vector control, the platform is assembled onto a satellite structure. The satellite attitude dynamic model for thrust vector control is then built, and the satellite attitude controller and intelligent controller for the smart composite platform are designed. The successful performance of the thrust vector control employing the UHM smart composite platform is proven here. The results indicate that the smart composite platform can precisely achieve the thrust vector control, and the misalignment of the thrust vector of the satellite can be corrected effectively with satisfactory position accuracy of the thrust vector.

The vibration suppression capability of this novel technology is also assessed. The combined system dynamics of the satellite structure and the smart composite platform is analyzed, and the dominant modes of the satellite structure are determined. A MIMO adaptive control scheme is then developed to suppress the satellite structure vibration employing four PZT stack actuators in the three smart composite struts and the central support of the platform as well as three PZT patch actuator pairs in the platform device plate. A convergence factor vector concept is also introduced to ease the multi-channel convergent rate control. This vibration controller is adjusted based on the vibration information of the satellite structure and drives the smart composite platform to isolate the vibration transmission from the firing thruster to the satellite structure. Eleven vibration components of the structure and platform are controlled. The results demonstrate that the entire vibration of the structure at its dominant frequency can be suppressed to 7-10% of its uncontrolled value for various device plate position configurations.

Optical fiber grating sensor technology has attracted considerable interests of research and development in the last decade. The normal optical fiber Bragg grating (FBG) sensor systems, in which the measurand is related to the Bragg (or resonant) wavelength of one or several FBGs, usually require wavelength interrogators (or modulators) and temperature compensators when they are used practically. Therefore the cost may be increased and the construction of the system may become complex. Another FBG sensor prototype based on chirp-tuned FBGs may avoid these problems. In Chapter 7, we study the basic sensing principle of and, based on this study, propose two novel sensor designs for displacement measurement and tilt measurement, respectively. In this sensor prototype, the measurand is related to the bandwidth (or chirp) of the involved FBG(s) so that it can be sensed by direct measurement of the reflected optical power from the FBG(s). Due to this optical power-encoding property, the proposed sensors show great advantages including simple construction (no need of wavelength interrogator) and inherently insensitive to temperature thus eliminating the need for temperature compensation. The proposed sensor designs may have potential applications in the optical fiber sensor area.

Guided Ultrasonic Waves (GUWs) are a useful tool in those structural health monitoring applications that can benefit from built-in transduction, moderately large inspection ranges and high sensitivity to small flaws. Chapter 8 describes two complementary methods, one based on unsupervised learning algorithms, and one based on supervised learning algorithms, for structural damage detection and classification based on GUWs. Both methods combine the advantages of GUW inspection with the outcomes of the Discrete Wavelet Transform (DWT), that is used for extracting robust defect-sensitive features that can be combined to perform a multivariate diagnosis of damage. In particular, the DWT is exploited to de-noise and compress the ultrasonic signals in real-time and generate a set of relevant wavelet coefficients to construct a uni-dimensional or multi-dimensional damage index. The damage index is then fed to an outlier analysis (unsupervised algorithm) to detect anomalous structural states, or to an artificial neural network (supervised algorithm) that classifies the size and the location of the defects.

The general framework proposed in this chapter is applied to the detection of crack-like and notch-like defects in seven-wire steel strands and in railroad tracks. In the first application, the probing hardware consists of narrowband magnetostrictive transducers used for both ultrasound generation and detection. In the second application, the hardware consists of a hybrid laser/air-coupled system for broadband ultrasound generation and detection. These applications demonstrate the effectiveness of the DWT-aided structural diagnosis of defects that are small compared to the waveguide cross-sectional area.

The proposed signal analysis approaches are general, and are extendable to many other structural monitoring applications using GUWs as the main defect diagnosis tool.

By embedding a smart material in a polymeric and/or metallic constituent, a smart composite is obtained which can be utilized in various technical applications. In order to predict the behavior of this composite, a micromechanical analysis is performed which provides the overall (global) response from the known constitutive relations of the individual phases, their material properties, volume ratios, and by considering their detailed interaction. In Chapter 9, micromechanical analyses of multiphase materials are presented which are capable of establishing the global constitutive relations of various types of smart composites which possess periodic microstructure. These include composites with piezoelectric, piezomagnetic, electrostrictive, magnetostrictive and shape-memory alloy phases. Both linear, nonlinear and large deformation analyses are presented. In addition, polymeric and metallic matrix composites are considered, where in the latter case, the inelastic behavior of the elastoplastic or viscoplastic constituent must be taken into account. Furthermore, bounded composites with arbitrarily distributed embedded smart phases (piezoelectric, shape-memory alloy, electrorheological, magnetorheological and fiber optic) are analyzed. Finally, the resulting micromechanically established global constitutive relations are employed to predict the dynamic behavior and buckling of smart composite plates which form a type of smart composite structure.

In: Progress in Smart Materials and Structures
Editor: Peter L. Reece, pp. 1-13
ISBN: 1-60021-106-2
© 2007 Nova Science Publishers, Inc.

Chapter 1

THERMOMECHANICAL PROPERTIES OF SHAPE MEMORY COMPOSITE WITH SMA AND SMP

Hisaaki Tobushi

Aichi Institute of Technology, Japan

Abstract

The shape-memory composite belt with a TiNi-SMA wire fiber and a polyurethane-SMP sheet matrix was fabricated. The bending actuation characteristics of the belt were investigated by the thermomechanical tests. The results obtained can be summarized as follows. (1) Residual deflection close to the maximum deflection is obtained by cooling under constant maximum deflection. The residual deflection disappears by heating under no load. Both the rate of shape fixity and the rate of shape recovery are close to 100%. (2) Recovery force appears by heating under constant residual deflection. The recovery force is 93-94% of the maximum force. The development of high functionality of shape-memory composite elements is expected by various combinations of SMAs and SMPs.

Introduction

Shape memory materials (SMMs) have been developed as smart materials. In SMMs, shape memory alloy (SMA) and shape memory polymer (SMP) have been used in practical applications. The shape memory effect (SME), superelasticity (SE) and large recovery stress appear in SMA [Funakubo, 1987; Otsuka and Wayman, 1998]. However, both elastic modulus and yield stress or transformation stress are low at low temperature and high at high temperature. This means the fact that rigidity of SMA elements is low at low temperature and high at high temperature. In order to obtain the two-way shape memory effect (TWSME), that is, two-way movement of SMA elements, the combination of SMA and bias materials, for example, steel is used in practical applications. On the other hand, the shape fixity (SF) and shape recovery (SR) appear in SMP [Hayashi, 1993; Tobushi et al., 1998]. However, both elastic modulus and yield stress, that is, proportional limit are low at high temperature and high at low temperature. This means the fact that rigidity of SMP elements is low at high temperature and high at low temperature.

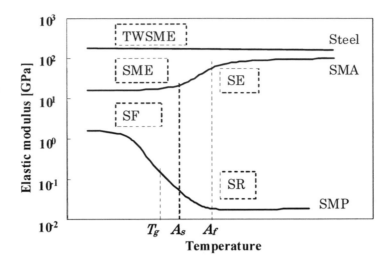

(a) Elastic modulus of SMA, SMP and steel

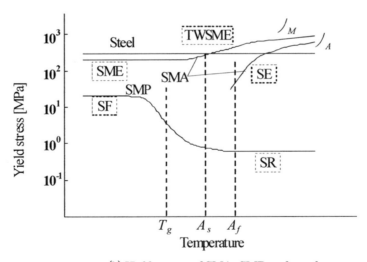

(b) Yield stress of SMA, SMP and steel

Figure 1. Dependence of elastic modulus and yield stress on temperature for SMA, SMP and steel.

The relationships between elastic modulus and temperature and between yield stress and temperature for SMA, SMP and steel are schematically shown in Figure 1. As can be seen, the dependence of elastic modulus and yield stress on temperature are quite different among SMA, SMP and steel. Therefore, if the composite materials with these materials are developed, new properties which can not be obtained by themself can be achieved. For example, if SMA and SMP are combined, the composite material with large recovery strain, high recovery stress, high rigidity, TWSME and light weight can be obtained. Therefore, if the shape memory composites (SMCs) with SMA and SMP are developed, new and higher functionality of the material can be expected [Sterzl et al., 2003]. That is, shape fixity and high rigidity at low temperature and shape recovery and recovery stress at high temperature can be used in the SMC.

In the present chapter, in order to confirm the basic characteristics of the SMC with SMA and SMP, the SMC belt was fabricated and the bending actuation characteristics were investigated. A TiNi SMA wire was used for the fiber of the SMC. A polyurethane SMP sheet was used for the matrix of the SMC. The thermomechanical bending tests were carried out for evaluation of characteristics of the SMC belt. Based on the experimental data, high performance of the shape fixity, shape recovery and recovery force was confirmed. The development of high functionality of the SMC belt is expected.

Experimental Method

Materials and Specimen

Polyurethane SMP (Diary MP5510 produced by Mitsubishi Heavy Industries, Ltd.) was used as the matrix of the SMC belt. The glass transition temperature of the SMP sheet T_g was 328K. A TiNi SMA wire (produced by Furukawa Electric Co.) was used as the fiber of the SMC. The reverse-transformation finish temperature A_f of the SMA wire was 325K. Diameter and length of the wire were 0.75mm and 50mm, respectively. The SMC belt was fabricated as follows. The lower half of the SMP sheet with thickness of 2.1mm was molded by mixing two kinds of liquid in a block at first. In a few minutes after molding, viscosity of the SMP became high enough to support the SMA wire. SMA wires were put on the lower half of the

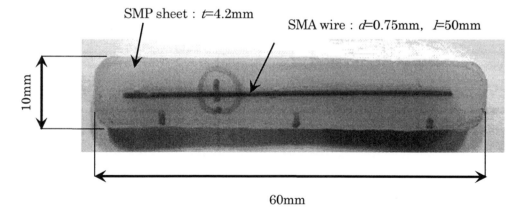

Figure 2. Specimen.

SMP sheet. The SMP was molded again on them for the upper half of the SMP sheet. The SMC sheet was obtained by curing. Specimens of the SMC belt were cut out from the SMC sheet at the same spacing. The photograph of a specimen of the SMC is shown in Figure 2. As can be seen in Figure 2, a SMA wire was located in the central part of the SMC belt. Width, thickness and length of the SMC-belt specimen were 10mm, 4.2mm and 60mm, respectively.

Experimental Apparatus

The SMM-characteristic testing machine [Tobushi, Hayashi and Kojima, 1992] was used for the thermomechanical tests. The machine was composed of a tension-compression machine and a heating-cooling device. The experiments were performed by the three-point bending test. Distance between two supports was 40mm. Radius of an end of the supports was 2.5mm. An upper jig with end-radius of 5mm pressed the center of the specimen between two supports. Two Teflon sheets were used between the specimen and two supports in order to avoid the friction on the contact surface. Temperature was measured by a thermocouple with diameter of 0.1mm. The thermocouple was set in the neighborhood of the end of the upper jig.

Experimental Procedures

The following three kinds of three-point bending test were carried out. Rate of heating and cooling was about 1K/min. Two specimens were used in each test.

Deformation Test at Various Temperatures

Three-point bending tests with loading and unloading were carried out at various constant ambient temperatures. Maximum deflection was 15mm. The ambient temperatures were T_g-25K, T_g and T_g+30K. Rate of deflection was 2mm/min.

Shape Fixity and Recovery Test

The experimental procedure in the shape fixity and recovery test is schematically shown in Figure 3. Rate of heating and cooling was about 1K/min.

①　First, the maximum deflection y_{max} = 10mm was applied at T_g+30K. At the point of y_{max}, force took the maximum value which was observed at T_g+30K in the deformation test at various temperatures.

②　The specimen was cooled down to T_g-30K by keeping y_{max} constant. Force decreased almost to zero during cooling.

③　Force was removed perfectly at T_g-30K. The deflection after unloading was measured. The rate of shape fixity R_f was obtained. The R_f will be defined in Experimental Results and Discussion.

④　The specimen was heated up to T_g+30K under no load. The deflection after heating was measured. The rate of shape recovery R_r was obtained. The R_r will be defined in Experimental Results and Discussion.

Recovery Force Test

The experimental procedure in the recovery force test is schematically shown in Figure 4. The procedures $\overline{}$ –β were the same as those in the shape fixity and recovery test.

χ The specimen was heated up to T_g+30K by keeping the fixed deflection constant. Force during heating was measured and the recovery force F was obtained

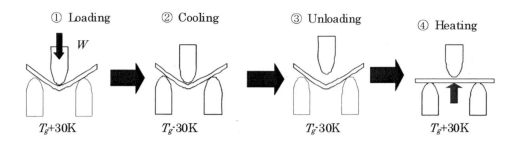

Figure 3. Experimental procedure in the shape fixity and recovery test.

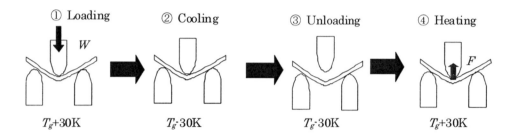

Figure 4. Experimental procedure in the recovery force test.

Experimental Results and Discussion

Deformation Properties at Various Temperatures

The relationship between force and deflection obtained by the deformation test at various temperatures is shown in Figure 5. As can be seen in Figure 5, force increases in proportion to deflection in the initial loading region. After taking the maximum force, force decreases gradually in proportion to deflection. The slope of the initial loading curve denotes the spring constant for bending of the material. Both spring constant and maximum force are high at low temperature but are low at high temperature. The maximum force appears at deflection of 5mm at T_g-25K but at deflection of 10mm at T_g+30K. In the unloading process at T_g+30K, deflection is recovered and residual deflection is small after unloading. On the other hand, in the unloading process at T_g-25K, recovery of defection is small and large residual deflection appears after unloading. The residual deformation appeared after unloading at T_g-25K disappears by heating up to T_g+30K under no load. The differences of deformation properties between two specimens at each temperature are small.

Evaluation of Elastic Modulus and Spring Constant

If a concentrated force W is applied at the center of a simply supported beam, the maximum deflection y is expressed by the following equation

$$y = \frac{Wl^3}{48EI} \tag{1}$$

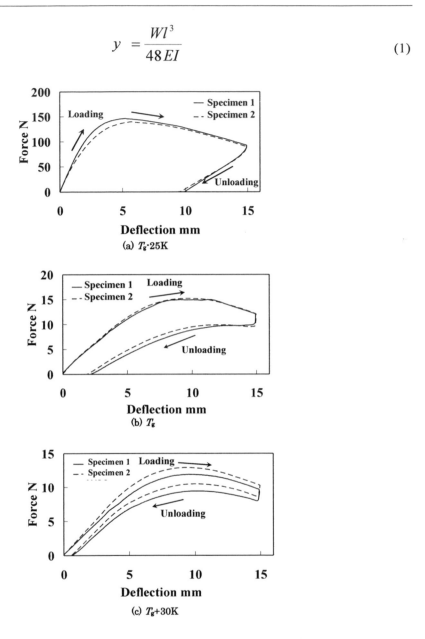

Figure 5. Relationship between force and deflection in deformation test at various temperatures.

where E, I and l denote elastic modulus, second moment of area and distance between two supports. The relationship between W and y is expressed by using the bending-spring constant k as follows

$$W = k \cdot y \tag{2}$$

In the case of a SMA wire of diameter d, by using second moment of area $I_a = \pi d^4/64$, the bending-spring constant k_a is given

$$k_a = \frac{3\pi}{4l^3} E_a \cdot d^4 \tag{3}$$

where E_a denotes elastic modulus of the SMA.

In the case of a SMP beam of a rectangular cross section with width b and high t, by using second moment of area $I_p = bt^3/12$, the bending-spring constant k_p is expressed by the following equation

$$k_p = \frac{4}{l^3} E_p \cdot bt^3 \tag{4}$$

where E_p represents elastic modulus of the SMP.

In the present study, since the volume fraction of the SMA wire in the SMC belt is 1.05%, the bending-spring constant of the SMC belt k_c can be evaluated by a sum of k_a and k_p as follows

$$k_c = k_a + k_p = \frac{1}{l^3}\left(\frac{3}{4}\pi E_a d^4 + 4E_p bt^3\right) \tag{5}$$

With respect to the SMC belt, the bending-spring constant k_c is obtained from a slope of the initial loading curve by using Eq. (2). The SMC belt has a rectangular cross section of width b and height t. Therefore, elastic modulus in bending E_c is given by the following equation

$$E_c = \frac{k_c}{4b}\left(\frac{l}{t}\right)^3 \tag{6}$$

The experimental results and calculated results of k_c and E_c are shown in Table 1. As can be seen in Table 1, the bending-spring constant and elastic modulus of the SMC belt can be evaluated by the proposed model. The value of elastic modulus of the SMC belt E_c at T_g+30K is 48MPa which is larger than that of SMP $E_p=30MPa$ at T_g+30K. This means the fact that higher rigidity of SMC elements than that of SMP elements can be obtained in practical applications. The rigidity of SMC elements depends on the volume fraction and the arrangement of SMA fibers.

Table 1. Values of elastic modulus and spring constant

Temperature	Experimental value				Calculated value		
	E_p [MPa]	E_a [GPa]	k_c [N/m]	k_p [N/m]	k_a [N/m]	k_c [N/m]	E_c [MPa]
T_g-25K	1150	20	55500	53250	233	53483	1155
T_g+30K	30	70	2240	1389	815	2204	48

Shape Fixity and Shape Recovery

The relationship between force and deflection, that between force and temperature and that between deflection and temperature obtained by the shape fixity and recovery test are shown in Figs. 6, 7 and 8, respectively.

As can be seen in Figure 6, the maximum force appears at deflection of y_{max}=10mm. Force decreased almost to zero during the cooling process (2) under constant y_{max}.

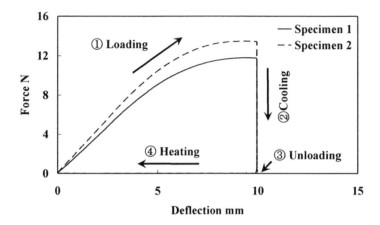

Figure 6. Relationship between force and deflection in shape fixity and recovery test.

As can be seen in Figure 7, force decreases slightly just after the start of cooling (2). In the region at T_g+30K which is above A_f, a SMA fiber is in the superelastic state and therefore stress relaxation of SMP matrix is prevented, resulting in slight decrease in force. In the midway of the cooling process (2), force decreases according to thermal contraction of the material due to decrease in temperature. In the final stage of cooling process (2), force decreases significantly almost to zero because of high elastic modulus of SMP.

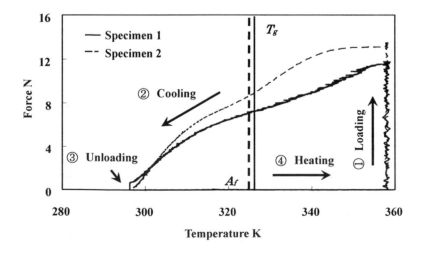

Figure 7. Relationship between force and temperature in shape fixity and recovery test.

As can be seen in Figure 8, deflection decreases at temperatures in the vicinity of A_f and T_g during the heating process (4) and becomes almost to zero at T_g+30K.

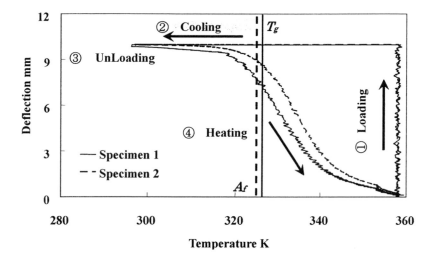

Figure 8. Relationship between deflection and temperature in shape fixity and recovery test.

The rate of shape fixity R_f and the rate of shape recovery R_r are defined by the following equation, respectively

$$R_f = \frac{y_u}{y_{max}} \qquad (7)$$

$$R_r = \frac{y_{max} - y_h}{y_{max}} \qquad (8)$$

where y_u and y_h represent the deflection after unloading at T_g-30K and the deflection after heating up to T_g+30K, respectively.

The values of the rate of shape fixity R_f and the rate of shape recovery R_r are shown in Table 2. As can be seen in Table 2, both R_f and R_r are close to 100%, showing the excellent characteristics of shape fixity and shape recovery.

The difference of shape fixity and shape recovery between two specimens is small.

Table 2. Values of shape fixity and shape recovery rates

	Shape fixity rate R_f [%]	Shape recovery rate R_r [%]
Specimen 1	98.7	98.8
Specimen 2	99.7	99.6

Deformation State of the Material

In order to observe the deformation state of the material, the photographs of an original state of the specimen, a deformed state of y_{max}=10mm at T_g+30K, a shape-fixed state after cooling followed by unloading at T_g-30K and a shape-recovered state after heating up to T_g+30K under no-load are shown in Figure 9 (a), (b), (c) and (d), respectively.

(a) Original state of the specimen

(b) Deformed state at y_{max}

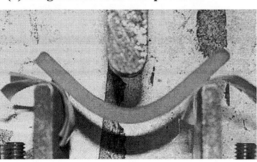

(c) Shape-fixed state after cooling

(d) Shape-recovered state after heating

Figure 9. Photographs of specimen at various states.

As can be seen from the comparison between Figs. 9 (b) and (c), the deformed state of y_{max}=10mm at T_g+30K is almost maintained after cooling down to T_g-30K followed by unloading. From the comparison between Figs. 9 (a) and (d), it can be confirmed that the original shape is almost recovered by heating up to T_g+30K under no-load.

From the photograph of the deformed state at y_{max}=10mm, the radius of curvature of the SMA wire on the neutral plane of the SMC belt can be measured. If the radius of curvature is denoted by r, the maximum bending strains ε on the surface elements of the SMA wire and the SMC belt are given by the following equations, respectively,

$$\varepsilon = \frac{d}{2r} : \qquad \text{for SMA wire}$$

$$\varepsilon = \frac{t}{2r} \qquad \text{for SMC belt}$$

where d and t denote the diameter of the SMA wire and the thickness of the SMC belt, respectively.

The values of the maximum bending strain of the SMA wire and the SMC belt are 3.2% and 18%, respectively. With respect to the SMA wire, the maximum strain is in the region of the martensitic transformation, resulting in high performance of the shape memory effect. With respect to the SMP sheet, the maximum bending strain is in the region of recoverable deformation under cyclic loading. The SMC belt can be therefore used as the reliable bending actuator.

Recovery Force

The relationship between force F and deflection and that between force F and temperature obtained by the recovery force test are shown in Figs. 10 and 11, respectively.

As can be seen in Figure 10, the force-deflection curves during the loading process (1) and cooling process (2) are almost the same as those for the shape fixity and recovery test (2) observed in Figure 6. Recovery force increases during the heating process (4) under constant y_{max}.

As can be seen in Figure 11, the force-temperature curves during the loading process (1), the cooling process (2) and the unloading process (3) are almost the same as those observed in Figure 7. In the heating process (4), force increases significantly at temperatures in the vicinity of A_f and T_g. The value of maximum recovery force obtained by heating up to T_g+30K is close to the maximum force given in the loading process (1). The values of maximum recovery force obtained by heating up to T_g+30K and the rate of recovery force to maximum force given in the loading process (1) are shown is Table 3. The rate of recovery force is 93-94%. The value of recovery force is higher than that of 7N which is obtained in the SMP belt without the SMA wire. Therefore, high recovery force can be obtained by SMC elements in practical applications. The recovery force of SMC elements depends on the volume fraction and the arrangement of SMA fibers.

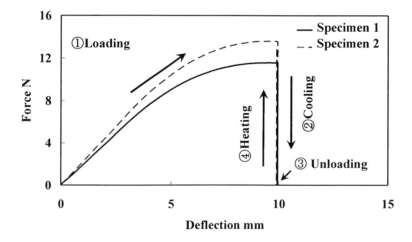

Figure 10. Relationship between force and deflection in recovery force test.

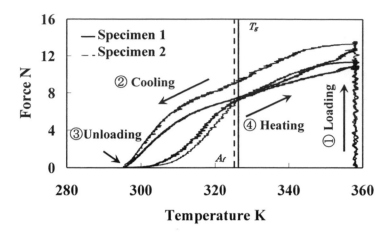

Figure 11. Relationship between force and temperature in recovery force test.

Table 3. Values of maximum force, recovery force and the rate of recovery force to maximum force

	Maximum force [N]	Recovery force [N]	Rate of recovery force [%]
Specimen 1	11.6	10.9	93.2
Specimen 2	13.6	12.8	94.3

Conclusion

The shape memory composite was fabricated by combining SMA and SMP. The SMC belt was composed of the TiNi-SMA wire fiber and the polyurethane-SMP sheet matrix. The bending actuation characteristics of the SMC belt were investigated by the thermomechanical tests with loading-unloading and heating-cooling. The results obtained can be summarized as follows.

(1) Deflection is recovered by unloading at high temperature, but large residual deflection appears after unloading at low temperature. Spring constant is high at low temperature but is low at high temperature. The spring constant of the SMC belt can be evaluated by the model of simple combination of the SMA wire and the SMP sheet.

(2) Force decreases by cooling under constant maximum deflection and large residual deflection close to the maximum deflection is obtained. The residual deflection disappears by heating under no load. The rate of shape fixity and the rate of shape recovery are close to 100%, showing the excellent characteristics of shape fixity and shape recovery.

(3) Recovery force appears by heating under constant residual deflection. The recovery force is 93-94% of the maximum force, showing excellent characteristics of recovery force. The development of high functionality of SMC elements can be expected by various combinations of SMAs and SMPs. That is, if the volume fraction of SMA is large, large recovery stress can be obtained. If A_f of SMA is higher than T_g of SMP,

the two-way property can be used during heating and cooling, and high rigidity of shape-fixed SMC can be obtained at low temperature.

References

Funakubo, H. ed.,1987. "Shape Memory Alloys", *Gordon and Breach Science Pub.*, New York.

Hayashi, S. 1993. "Properties and Applications of Polyurethane Series Shape Memory Polymer", *Int. Progr. Urethanes*, **6**: 90-115.

Otsuka, K. and C. M. Wayman ed., 1998. "Shape Memory Materials", *Cambridge University Press*, Cambridge.

Sterzl, T., B. Winzek, M. Mennicken, R. Nagelsdiek, H. Keul, H. Hocker and E. Quandt. 2003. "Bistable Shape Memory Thin Film Actuators", Smart Struct. Mater., *Proc. SPIE*, Vol.5053: 101-109.

Tobushi, H., T. Hashimoto, N. Ito, S. Hayashi and E. Yamada.1998. "Shape Fixity and Shape Recovery in a Film of Shape Memory Polymer of Polyurethane Series", *J. Intell. Mater. Syst. Struct.*, **9**: 127-136

Tobushi, H., S. Hayashi and S. Kojima. 1992. "Mechanical Properties of Shape Memory Polymer of Polyurethane Series", *JSME Inter. J.*, Ser. I, 35-3: 296-302.

In: Progress in Smart Materials and Structures
Editor: Peter L. Reece, pp. 15-28

ISBN: 1-60021-106-2
© 2007 Nova Science Publishers, Inc.

Chapter 2

ISSUES FOR THE OPTIMAL TRAINING OF TWO-WAY SHAPE MEMORY ALLOY

Eric Abel and Hongyan Luo

Division of Mechanical Enginering & Mechatronics,
University of Dundee, Dundee, UK

Abstract

The creation of an effective two-way shape memory alloy (TWSMA) requires particular heat treatment and training considerations. This chapter investigates methods of training TWSMA from the viewpoint of creating an optimal actuator for engineering applications. A review of recommended heat treatment conditions and previously reported training methods is followed by findings from an experimental study that compared different training methods for producing NiTi TWSMA wires with a hot shape of an arc and a cold shape of a straight line. These methods are: shape memory cycling, constrained cycling of deformed martensite, pseudoelastic cycling and combined shape memory & pseudoelastic cycling. In order to give a meaningful evaluation of their performance that is relevant to training for practical applications, these training methods are assessed in terms of maximum two-way strain, changes in the original hot shape and the transformation temperatures after the training process, and the effective production of the cold shape. Only the combined shape memory & pseudoelastic cycling provides an effective training method for creating NiTi TWSM with a non-uniaxial two-way shape change. The undesirable side effects of training were that the NiTi TWSMA wire lost partial memory of the original hot shape and its transformation temperatures shifted to the lower values. There also exists an optimal number of training cycles and possibly an optimal training load for obtaining the best cold shape memory and the greatest two-way recoverable strain. These findings point to future directions in research for advancing TWSMA training technology.

Introduction

Two-way shape memory alloy (TWSMA) exhibits the ability to produce a reversible change between a high-temperature shape in austenite (hot shape) and a low-temperature shape in martensite (cold shape) by means of a temperature change. This phenomenon, known as the two-way shape memory effect, arises from a solid-solid diffusionless phase transformation between martensite and austenite, which is characterized by four transformation temperatures: martensite start temperature (Ms), martensite finish temperature (Mf), austenite start temperature (As) and austenite finish temperature (Af). In comparison with one-way shape memory alloy (OWSMA), where an additional element such as a spring is needed for providing an external force to reset the cold shape upon cooling, TWSMA can lead to a more compact and simplified configuration of smart actuators and structures. This may be of benefit to industrial and surgical applications where a limited working space is often encountered. It is well known that the memory of both a hot shape and a cold shape is not a natural feature of the alloy, but rather is a 'learned' behaviour. The process of creating TWSMA usually starts with heat treating the raw material to memorize a certain shape as a hot shape, in other words, to form OWSMA. Then a cold shape is induced in the heat treated alloy element to bring a two-way shape memory effect into being by means of training. There are specific requirements for the heat treatment conditions and the optimal training considerations for the effective production of TWSMA. In particular, the training method used plays a key role. Research into producing TWSMA started with copper based alloy, but has focused more recently on the more commonly used NiTi alloy. However, most studies [1-13] have generally been more concerned with demonstrating the two-way shape memory effect, rather than offering a detailed understanding of the training processes associated with creating specific hot and cold shapes. Moreover, the different training methods have been assessed only in terms of the maximum two-way recoverable strain, but the changes in the original hot shape and the transformation temperatures resulting from them were not considered. This means of assessment has very limited relevance to the training of TWSMA for engineering applications, where the final hot and cold shapes and the transformation temperatures of the trained alloy element need to be known. From the viewpoint of practical applications, the optimal training method should be determined by a combination of the maximum two-way recoverable strain, a minimum change in the original hot shape and in the transformation temperatures after the training process, and the effective introduction of the desired cold shape.

This chapter starts with a review of the heat treatment conditions that have been recommended for NiTi alloy. Experimental work is then described, using four typical methods of training NiTi TWSMA, the example used being a wire with a hot shape of an arc and a cold shape of a straight line. The performance of each of these methods is evaluated and recommendations are made for future research into ways of optimizing the training process.

Heat Treatment Conditions for Twsma

NiTi alloy does not exhibit any shape memory effect in its raw material form, which is in a cold-worked state. Appropriate heat treatment can produce OWSMA, which remembers the shape set during the heat treatment process as the hot shape. The actual procedure is accomplished by first constraining the material in the desired shape, then heating it in an oven at an appropriate temperature for some time, followed by water cooling, or rapid air cooling if the element and the shaping fixture are small in size. It has been reported that the heat treated alloy allows an easier introduction of a cold shape during the subsequent training procedure than does the cold-worked alloy, and that the magnitude of two-way recoverable strain achieved in the former is much higher than in the latter [3, 14]. This is attributed to the fact that the high density of "random" dislocations in the alloy originally caused by cold working, which are microstructural defects and impede the mobility of martensite twin boundaries, can be reduced and also rearranged by the heat treatment.

Wang et al further investigated the effect of heat treatment conditions on training NiTi TWSMA and found that, after training by the same means, a sample heat treated at a higher temperature showed a stronger two-way shape memory effect, because the density of dislocation decreases as the heat treatment temperature increases [7]. They obtained a maximal two-way shape recovery with heat treatment at 550° C, which was three times that obtained at 400° C. NiTi alloy heat-treated to between 400°C and 500°C exhibits two transformations, an R-phase transformation and a martensitic transformation, at different temperatures [15]. So the results of their study give support to the conclusion made in another paper [13], that a single stage transformation between the parent phase and martensite is essential for producing a significant two-way shape memory effect in NiTi alloy. A heat treatment temperature of 550° has therefore been chosen for the experimental work in this study. In contrast, the two-way shape memory effect is much less dependent on the heat treatment time [7]. Generally, heat treatment time is determined by the equipment used for heat treatment, the size of the sample and the thermal mass of the shaping fixture.

Training Methods for TWSMA

Several training methods have been proposed in the past and are summarized in Table 1. These methods are typically classified as one-time martensite deformation, thermomechanical cycling treatment and reheat treatment.

The one-time martensite deformation method is accomplished simply, by deforming the heat treated alloy sample in the martensite phase. However, it is seldom used since usually only a small two-way memory effect can be produced [16, 17]. A later study by Liu et al [1] envisaged that the two-way recoverable strain could be increased by imposing a severe martensitic deformation. The maximum strain of 4.1% was achieved in the near-equiatomic NiTi alloy wire sample at a deformation strain of 13.3% beyond the recoverable limit of NiTi OWSMA (about 8%), which is comparable to the best two-way shape memory effect achieved by other training methods. However, the large amount of plastic strain introduced by such a high level of deformation will seriously deteriorate the memory of the hot shape. These findings suggest that the choice of deformation strain in martensite dictates a trade-off between the two-way recoverable strain and the loss of the hot shape, when developing TWSMA by the single-step deformation of martensite. As a result, optimal training results can not be expected with this training method.

Thermomechanical cycling treatment is based on the repetition of a cycle that must include the transformation from austenite to preferentially oriented martensite or from deformed martensite to austenite. This method has been proved to be far more effective than one-time martensite deformation and is therefore widely applied. There are four basic thermomechanical cycling training methods, known as shape memory cycling (a), constrained cycling of deformed martensite (b), pseudoelastic cycling (c) and combined shape memory & pseudoelastic cycling (d) [18]. Other themomechanical cycling training methods are also documented in the literature [19, 20], but they are actually variants or combinations of the above four methods with fewer new aspects than similarities and have shown little improvement in performance, so they are not described here.

The reheat treatment method is a recently reported TWSMA training method involving a second heat treatment procedure on a previously heat treated alloy specimen, but with it fixed in a different shape from that used in the first heat treatment and for a different heat treatment time [21]. The experimental data presented shows that, after a reheat treatment for 30 min, a

Table 1. Main types of training methods for TWSMA.

Type			Training Procedure
One-time martensite deformation			1. Cool the specimen below Mf; 2. Deform it in martensite state; 3. Unload it completely.
Thermomechanical cycling treatment	a	Shape memory cycling	1. Cool the specimen below Mf; 2. Load it in martensite state to a desired cold shape; 3. Unload it completely; 4. Heat it above Af; 5. Repeat the above steps for a number of times.
	b	Constrained cycling of deformed martensite	1. Cool the specimen below Mf; 2. Load it in martensite state to a desired cold shape; 3. Heat and cool it in the constrained condition through Mf and Af; 4. Repeat step 3 for a number of times.
	c	Pseudoelastic cycling	1. Heat the specimen above Af and below Md where pseudoelastic behaviour is expected; 2. Load it in austenite state to a desired cold shape; 3. Unload it completely; 4. Repeat steps 2 and 3 for a number of times.
	d	Combined shape memory and pseudoelastic cycling	1. Heat the specimen above Af and below Md where pseudoelastic behaviour is expected; 2. Load it in austenite state to a desired cold shape; 3. Cool it in the loaded condition below Mf; 4. Unload it completely; 5. Repeat the above steps for a number of times.
Reheat treatment			1. Repeat the heat treatment procedure with the specimen fixed in another shape for about 30 min.

NiTi sample demonstrated the memory of a cold shape close to that of the original shape defined in the first heat treatment and the memory of a hot shape close to that of the new shape defined in the second heat treatment, while the shorter reheat treatment time of 10min and the longer reheat treatment time of 60min were either insufficient to set the new shape in the sample or too long to erase the original shape from it. Compared to the thermomechanical cycling methods, the reheat treatment method involves a much simpler training procedure. However, the initial heat treatment conditions, the reheat treatment temperature and the composition of NiTi material used in their tests were not provided by the authors, so there is difficulty in repeating their experiment exactly. Moreover, there are no other published papers containing these details. Working on the basis that the missing information would not be critical, this training method was tried on NiTi wires in our laboratory, which were initially heat treated at 550°C for 20 min, then water-cooled. Unfortunately, instead of demonstrating the two-way shape memory effect as reported, all the samples tested only remembered the new shape set during the second heat treatment. While the reheat treatment method may eventually be proven to produce good results, there is a need for more published research to assess its effectiveness.

In view of the above, the one-time martensite deformation and the reheat treatment method were not pursued and instead the investigation focused on the four basic thermomechanical cycling training methods to provide information about optimal training parameters.

Experiments for Training NiTi TWSMA Wires

Material and Preliminary Heat Treatment

The material used in this study was a commercial NiTi wire in a cold-worked state from Memory Metals Ltd (Ipswich, UK), with a Ni content of 55.5 wt%, a nominal Af of 80°C and a diameter of 0.8mm. Specimens were 110mm in length. Each was initially held in a steel frame shaped in an arc of 90° with a 5mm straight section at each end, as shown in Figure 1, and heated in an oven at 550°C for 20min, then water-cooled. It is known that the two-way

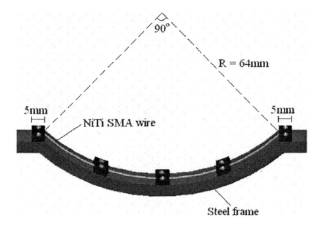

Figure 1. Shape setting of a NiTi wire specimen prior to heat treatment.

recoverable strain that may be achieved is significantly less than the one-way recoverable strain for the same alloy and typically about 2% [18]. According to general beam theory, the maximum strain in the curved wire can be quantified approximately as the ratio of the radius of wire to the radius of its curvature. The target two-way recoverable strain between the desired hot shape of the arc and the desired cold shape of a straight line in this experiment was 0.625%, which is less than the maximum 2% recoverable strain. It is therefore theoretically possible to produce a TWSMA wire through training.

Experimental System

The experimental system is illustrated in Figure 2. A heat treated NiTi wire specimen was placed horizontally with one end held in a clamp that was attached to a steel supporting structure (not shown) and the other end held in a similar clamp attached to the central slider of a low-friction linear variable displacement transducer (LVDT, Model DC25, Solartron Metrology, Steyning, UK), which measured the horizontal displacement of the moveable end during the deformation and shape recovery of the wire. A K-type thermocouple (with a 0.31 5mm bead) was secured to the wire and kept in good thermal contact with it by conductive silicone grease. Its output was converted to a linear voltage using a thermocouple module (A5B47-K-05, Amplicon Ltd, Brighton, UK). A computer-based data acquisition and control system using a PCI multifunction card (PCI 6035E, National Instruments, Newbury, UK) was operated using the Matlab 6.5 Data Acquisition Toolbox v.2.2 (Mathworks, Natick, USA). Heating the wire to the high-temperature austenite was achieved by resistive heating using an electrical current, generated by a computer controlled power operational amplifier (OPA541, Burr-Brown, Tucson, USA), which was capable of delivering up to 5A current. The deformation of the wire was realized by imposing a dead mass on the pulley cord. The horizontal displacement of its moveable end and its temperature during the training procedure were measured by the LVDT and the thermocouple respectively, and acquired to the PC at a sampling rate of 5Hz. The applied current during heating was set as 2.6A, which heated the specimen slowly to allow a gradual training process. Cooling it to martensite was realized through natural convection by switching off the power supply. All the tests were carried out at a room temperature of about 20^{c}C.

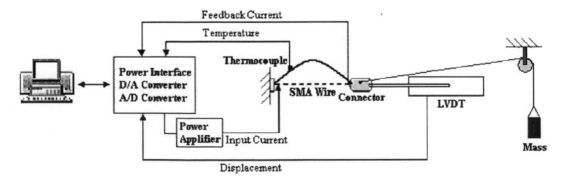

Figure 2. Schematic diagram of the experimental system for training NiTi TWSMA wires.

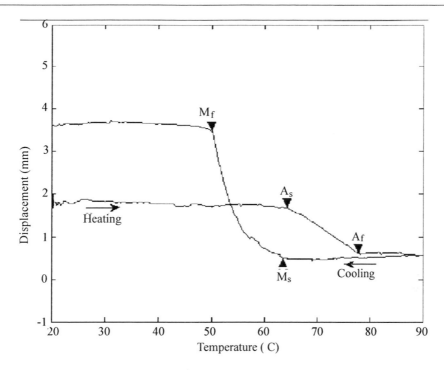

Figure 3. Displacement vs. temperature curve for a heat treated NiTi wire sample measured by the applied loading method.

Transformation Temperatures of the Heat Treated NiTi Wire Samples

The original transformation temperatures of the heat treated samples were measured using the applied loading method before the thermomechanical cycling treatment, since this method has been proved to be more effective for providing practical information about the stress-dependent transformation temperatures than other principal methods, such as differential scanning calorimetry (DSC) and electrical resistance [22]. The underlying principle is that large changes in geometry occur over the transformation temperature range of a OWSMA element as it is heated and cooled under an external load, and that these geometrical changes can be used as a straightforward determination of the transformation temperatures. In order to optimize the sensitivity of detection of these temperatures, a clearly observable shape recovery during the heating procedure and deformation during the cooling procedure is needed. Accordingly, an appropriately large mass is initially used to cause a correspondingly large deformation in the OWSMA element while it is in the martensite and then replaced by a relatively small load which is maintained in the following thermal cycle. In this case, a heat treated NiTi sample was pre-deformed at room temperature by placing a dead mass of 400g on the pulley, removing it and subjecting the sample to a heating and cooling cycle between $20^{c}C$ and $90^{c}C$ under a smaller dead mass of 100g. The deformation and shape recovery behaviour of the sample during this procedure are represented by a displacement vs. temperature curve, illustrated in Figure 3, from which the transformation temperatures were found to be: $Mf = 50°C$, $Ms = 63°C$, $As = 65°C$, $Af = 76°C$.

Experimental Procedure

Each heat treated NiTi wire sample was trained by one of the four thermomechanical cycling methods (see Table 1) over 40 training cycles. It has been shown previously [23] that the deformation behaviour of NiTi OWSMA wire varies with the material state at which loading is imposed and with the heating and cooling conditions. Therefore the training stress required for deforming the specimen into the desired cold shape would be different in each method. The magnitude of this load was determined prior to the training tests, based on a mass of 650g being used to straighten the NiTi wire in martensite for method (a) and (b), 1100g to straighten it in austenite for method (c) and 250g to straighten it upon cooling for method (d). After each training cycle, the specimen experienced a free thermal cycle, by heating it to above Af followed by cooling it to below Mf in the unloaded condition. The hot shape in austenite and the cold shape in martensite were represented by the displacements of the moveable end of the specimen, which were recorded during a free thermal cycle. The transformation temperatures were determined from the displacement vs. temperature curves at each free thermal cycle. The two-way shape memory effect was evaluated from the two-way recoverable shape change, which is defined as the difference between the hot shape and the cold shape, and can be found from the curves.

Results

The hot shape and cold shapes of representative specimens over the range of training cycles for the four training methods are presented in Figure 4 and Figure 5 respectively. The desired

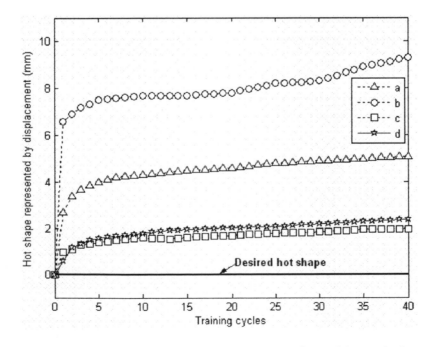

Figure 4. Hot shape vs. training cycles curves for the four training methods.

hot shape of the arc corresponds to zero displacement, while the desired cold shape of the straight line corresponds to a LVDT displacement of 10mm. It is evident that in all the training methods, the specimen has a tendency to remember the desired cold shape at the cost

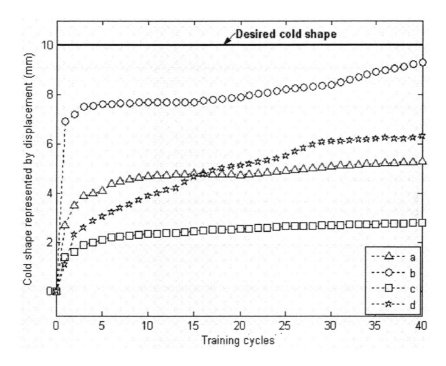

Figure 5. Cold shape vs. training cycles curves for the four training methods.

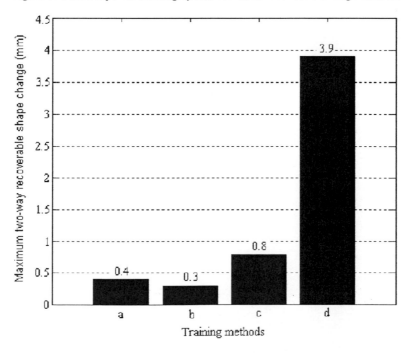

Figure 6. Maximum two-way recoverable shape change obtained by the four training methods.

of "forgetting" the desired hot shape, with increasing effect as the number of training cycles increases. As a result, none of the specimens maintains the original hot shape. Comparing the hot shape curve and the cold shape curve for each training method, it can be seen that only the specimen trained by method (d) shows a significant difference between its hot shape and cold shape, while those trained by the other three methods exhibit a similar hot shape and cold shape. In particular, method (b) gives the final hot and cold shapes of the specimen that are most close to the desired cold shape and method (c) leads to the opposite result. The specimens trained by method (c) and method (d) have a very similar hot shape. The plots of the maximum two-way shape change of each specimen given in Figure 6 show clearly that only method (d) results in a substantial NiTi TWSMA wire.

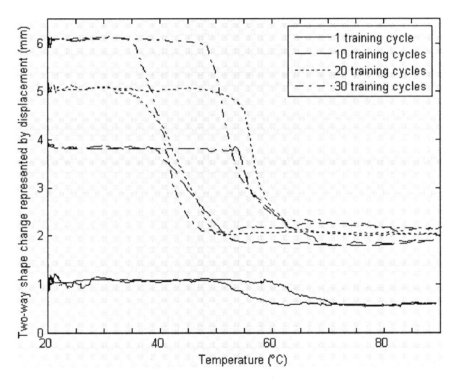

Figure 7. Two-way shape memory behaviours of the specimen trained by method (d) after different numbers of training cycles.

As shown in Figure 7, the two-way shape memory behaviour of the sample trained by method (d) after different numbers of training cycles are demonstrated by the corresponding displacement vs. temperature curves. The gradual establishment of the two-way shape memory effect through training is evident. Figure 8 further illustrates the two-way recoverable shape change as a function of the number of training cycles. Following a rapid initial rise, its value increases gradually at a nearly constant rate and reaches a maximum after about 28 training cycles, after which it remains fairly constant.

The transformation temperatures for all training cycles for method (d) are shown in Figure 9. All the transformation temperatures dropped rapidly during the initial cycling stage and tended towards a steady final value, some 10°C below the initial values.

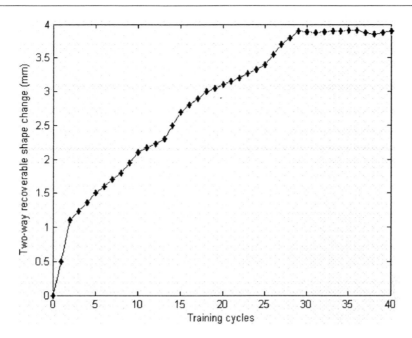

Figure 8. Two-way recoverable shape change of a specimen trained by method (d) over 40 training cycles.

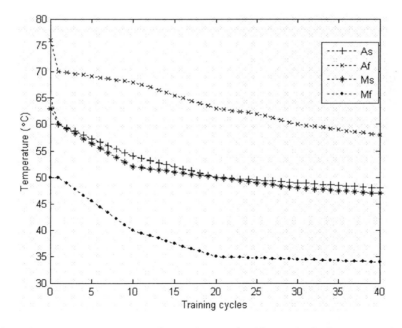

Figure 9. Transformation temperatures of a specimen trained by method (d) over 40 training cycles.

Discussion

The mechanism of thermomechanical cycling treatment to obtain TWSMA lies in developing some 'reminders' of the desired cold shape (known as dislocation arrangements) in the austenite phase of OWSMA by repeating the transformation cycle between the martensite phase and the austenite phase with the application of an external stress [7, 17]. These dislocation arrangements create an anisotropic stress field in the matrix of austenite, which guides the nucleation and growth of the martensite variants towards the preferred orientations in relation to the deformation adopted in the training procedure to form a stress-biased martensite. As a result, a spontaneous shape change towards the trained shape upon cooling will occur. It has been widely accepted that the generation of the dislocation arrangements during training is an essential prerequisite for the exhibition of reversible shape changes on thermal cycling [8]. However, the introduction of the dislocation structure is often accompanied by a permanent strain that would degrade the memory of the hot shape, a negative side effect that should be minimized. According to the findings presented here (see Figures 4-6), it seems that the material state at which the heat treated NiTi wire sample is deformed in the training method is a crucial factor in determining the effective introduction of 'reminders' and the level of the resultant permanent strain. The deformation to the trained shape in full martensite (applying loading below Mf in methods (a) and (b)) originates from the reorientation of martensite, a process in which the resulting dislocation arrangements bring about a large permanent strain, so that the memory of the hot shape is seriously affected. The fact that the sample trained by method (b), which involves heating the sample under a load, "forgot" the hot shape most, further reveals that the dislocation arrangements introduced during the reverse transformation from martensite to austenite yield the highest level of permanent strain. In method (c), where loading and unloading occur in full austenite, it is difficult to introduce 'reminders', which is probably due to the high level of stability of the austenite phase, so the amount of dislocation arrangements produced is too small to remind the trained sample of the cold shape.

By relating the experimental results to the training details in the cases of methods (c) and (d), two points are clear. One is that the permanent strain in the specimen trained by method (d) results mainly from the loading that takes place in full austenite. The other is that the effective dislocation arrangements, those that give rise to the formation of the cold shape upon cooling with little influence on the recovery of the hot shape upon heating, can only be introduced during the forward transformation by inducing the newly nucleated martensite into the particular martensite that is oriented to the applied stress.

In method (d), the final hot shape of the trained NiTi wire was still different from the desired hot shape, because of the permanent strain, which caused the residual deformation in austenite. Referring to Figure 4, the difference between the actual hot shape of the trained sample and the desired hot shape indicates residual deformation, which changes with the number of training cycles. After an initial rapid increase, the residual deformation accumulates very slowly during further training cycles due to the work hardening reaching a saturated level, and accordingly the hot shape changes only slightly. The evolution of the cold shape, the magnitude of the two-way recoverable shape change and the shift in the transformation temperatures of the obtained NiTi TWSMA in method (d) are linked to the progress of the dislocation arrangements. During the initial training cycles, the dislocation arrangements can be introduced readily so increase in number greatly, leading to a rapidly improved level of memory in the cold shape and in the magnitude of the two-way recoverable shape change, as observed in Figure 5 and Figure 8. The rapid decrease in the transformation temperatures (see Figure 9) may be attributed to the internal stress field formed by these dislocations, which has the combined effect of suppressing the martensite transformation upon cooling and promoting the reverse transformation upon heating.

The more dislocation arrangements that are developed over the training cycles, the more difficult it becomes to introduce new ones. After a certain number of training cycles, the dislocation arrangements reach a saturation level. This accounts for the observation (see Figures 5, 8 and 9) that the transformation temperatures approximate some steady values gradually and that the cold shape as well as the two-way recoverable shape change varies only slightly after reaching a maximal value. This suggests that an optimal number of training cycles is associated with the generation of a substantial two-way shape memory effect and that further improvement cannot be made by applying additional training cycles. Noticeably, there is still a difference between the final cold shape of the trained wire and the desired cold shape. This indicates that the effect of the generated dislocation arrangements in the formation of a preferential martensite morphology is less influential than the effect of the external load applied during training. This implies that the training load used in this study is insufficient to introduce enough dislocation arrangements for establishing the desired cold shape in the trained wire. According to the mechanism of TWSMA, in a training cycle, there might exist an optimal force that is able to maximize the dislocation arrangements in the newly generated martensite upon cooling. In theory, this force would be a function of the phase transformation rate and it would vary with temperature. Therefore, the use of a variable force rather than a constant load to train the heat treated NiTi wire in the method (d) may lead to a final cold shape quite similar to the desired one.

Conclusion

In previous studies, NiTi TWSMA wire with uniaxial shape change has been produced using the four thermomechanical cycling training methods referred to in Table 1. Based on the present work, combined shape memory & pseudoelastic cycling is the only effective training method that can generate NiTi TWSMA wire with a non-uniaxial shape change. Although the desired hot shape of an arc of 90° and the desired cold shape of a straight line were not successfully achieved in the trained wire, the new findings from this experimental study point to future directions that could improve TWSMA training. On the one hand, the type of applied training load and the temperature at which the load is imposed could be optimized for achieving the desired cold shape while at the same time reducing the loss of memory of the desired hot shape. For example, a loading profile variable with temperature, rather than a dead mass as the training load might introduce more dislocation arrangements, and loading at Ms rather than above Af might reduce the permanent strain. On the other hand, it is possible to compensate for the loss of the memory of the desired hot shape during training by allowing a margin between the original hot shape set through heat treatment and the desired hot shape after training. For example, in the case of the material used in this study, the NiTi wire could be given an initial shape of a tighter arc than the desired arc of 90° prior to heat treatment. Owing to a large shift of transformation temperatures to lower values, as a concomitant effect of training, a raw material with considerably higher transformation temperatures than the required transformation temperatures for TWSMA should be selected.

References

[1] Liu Y.; Liu Y.; Van Humbeeck J. *Acta Mater.* 1999, Vol. 47, 199-209.

[2] Liu Y.; Liu Y.; Van Humbeeck J. *Scipta Mater.* 1998, Vol. 39, 1047-1055.

[3] Scherngell H.; Kneissl A. C. Mater. *Sci. Eng.*, A. 1999, Vol. 300, 400-403.

[4] BlonK B. J. D.; Lagoudas D. C. Smart Mater. *Struct.* 1998, Vol. 7, 771-783.

[5] Wang Z. G.; Zu X. T.; Feng X. D.; Zhu S.; Dai J. Y.; Lin L. B.; Wang L. M. *Mater. Lett.* 2002, Vol. 56, 284-288.

[6] Hebda D. A.; White S. R. Smart Mater. *Struct.* 1995, Vol. 4, 298-304.

[7] Wang Z. G.; Zu X. T.; Feng X. D.; Lin L. B.; Zhu S.; You L. P.; Wang L. M. *Mater. Sci. Eng.*, A. 2003, Vol. 345, 249-254.

[8] Wang Z. G.; Zu X. T.; Dai J. Y.; Fu P.; Feng X. D. *Mater. Lett.* 2003, Vol. 57, 1501-1507.

[9] Wang Z. G.; Zu X. T.; Feng X. D.; Dai J. Y. *Mater. Lett.* 2002, Vol. 54, 55-61.

[10] Scherngell H.; Kneissl A. C. *Acta Mater.* 2002, Vol.50, 327-341.

[11] Derek A.; Scott R. *Smart Mater. Struct.* 1995, Vol.4, 298-304.

[12] Brett J.; Dimitris C. *Smart Mater. Struc.* 1998, Vol.7, 771-783.

[13] Liu Y.; McCormick P. G. *Acta Metall. Mater.* 1990, Vol. 38, 1321-1326.

[14] Wayman C. M.; Duerig T. W. In Engineering Aspects of Shape Memory Alloys; Duerig T. W.; Melton K. N.; Stцckel D.; Wayman C. M.; Ed.; *Butterworth-Heinemann Ltd:* London, UK, 1990; pp 28-34.

[15] Ostuka K.; In Engineering Aspects of Shape Memory Alloys; Duerig T. W.; Melton K. N.; Stцckel D.; Wayman C. M.; Ed.; *Butterworth-Heinemann Ltd:* London, UK, 1990; pp 36-45.

[16] Contardo L.; Guenin G. *Acta Metall. Mater.* 1990, Vol. 38, 1267-1272.

[17] Van Humbeeck J.; Stalmans R. In Shape Memory Materials; Otsuka K.; Wayman C. M.; Ed.; *Cambridge University Press*: Cambridge, UK, 1998; pp 161-162.

[18] Perkins J.; Hodgson D.; In Engineering Aspects of Shape Memory Alloys; Duerig T. W.; Melton K. N.; Stцckel D.; Wayman C. M.; Ed.; *Butterworth-Heinemann Ltd*: London, UK, 1990; pp 195-206.

[19] Mellor B. G.; Guilemany J. M.; Fernandez J. European Symp. on Martensitic Transformation and Shape Memory Properties J. *Physique Col.* 1991, 457-462.

[20] Sun L.; Wu K. H. *Proc. of SPIE North American Conf. on Smart Structures and Materials* (Orlando, FL). 1994, Vol. 2189, 298-305.

[21] Huang W.; Toh W. J. *Mater. Sci. Lett.* 2000, Vol. 19, 1549-1550.

[22] Abel E.; Luo H. Y. *Smart Mater. Struc.* 2004, Vol. 13, 1111-1117.

[23] Luo H. Y.; Abel E. *Smart Mater. Struc.* (submitted)

In: Progress in Smart Materials and Structures
Editor: Peter L. Reece, pp. 29-65

ISBN: 1-60021-106-2
© 2007 Nova Science Publishers, Inc.

Chapter 3

DETWINNING IN SHAPE MEMORY ALLOY

Yong Liu and Zeliang Xie*
Nanyang Technological University, Singapore

Abstract

Detwinning in crystalline solids is a unique deformation mechanism partially responsible for the shape memory effect. Owing to an insignificant dislocation process during detwinning, the residual strain can be recovered through a reverse phase transformation. The maximum shape recovery strain is intrinsically related to the lattice geometry and twinning mode, while the magnitude of shape recovery is related to a competition between detwinning and dislocation generation responsible for the macroscopically observed deformation. The detwinning magnitude is directional, and in the polycrystalline materials it is related to the textures and deformation mode. With textures, the detwinning process is enhanced for certain directions and reduced for other directions and so is the shape recovery strain. The anisotropy in the detwinning process allows the possibility of maximizing the potential of the polycrystalline shape memory alloys. Factors that affect the detwinning process will in turn affect the subsequent shape recovery.

This chapter deals with detwinning and shape recovery in NiTi SMAs and influencing factors. NiTi SMAs in several initial conditions are studied including textured wire and bar, rolled sheet, and less textured ingot. Detwinning anisotropy due to texture and deformation mode is studied and explained based on crystallographic analysis. Various experimental results on the relation between detwinning anisotropy and shape recovery anisotropy due to texture and deformation mode are presented and discussed. The present research has also studied the evolution of the crystallographic texture as a result of martensite deformation. In-depth understanding of the atomistic arrangement along <011> type II twin and its role in detwinning are also presented.

Keywords: shape memory alloy, twinning, detwinning, dislocation, martensite, deformation, shape recovery, textures.

---

* Correspondence author. E mail adress: mliuy@ntu.edu.sg, Tel./Fax: 65-67904951.

Introduction

Seventy years ago, Swedish physicist Arne Ölander [1932a;b] discovered the rubberlike behavior in an Au-47.5Cd alloy which was later explained by Chang and Read in 1951 as due to "reorientation" of martensite twinned lattices. This rubbery effect was believed to be related to the shape memory effect (SME) discovered 20 years later in AuCd [Chang & Read 1951], InTl [Burkart & Read 1953], CuZn [Suoninen 1954; Hornbogen & Wassermann 1956], and CuAlNi [Chen 1957] among others. In 1961, Buehler and colleagues [Buehler & Wiley 1961] at the US Naval Ordinance Laboratory (NOL) discovered the SME in a binary NiTi alloy containing about 50 at.% Ni which was later named Nitinol (Ni + Ti + NOL). The discovery of the NiTi shape memory alloy quickly stimulated wide interest in both the scientific community and in engineering practices due to its fascinating behavior and relatively lower cost. Research on various aspects of these materials was intensely conducted including phase transformation characteristics, deformation mechanisms, shape memory effect, superelasticity, damping capacity, fatigue property, wear resistance, corrosion resistance, biocompatibility, and development of new SMAs. The applications of various properties of SMAs increased continuously. Meanwhile, the fundamental research to understand SME and its influencing factors has continued ever since.

It was interestingly noted that the material was flexible at low temperature but stiffer when heated. In 1969, Buehler and Cross [1969] wrote:

> … First, at a temperature below their TTR (transition temperature range), the 55-nitinol alloys are highly ductile and may be plastically deformed. Under this condition a relatively low stress in the range of about 10 to 20 ksi will result in about 8% deformation (strain). This initial plastic flow in 55-nitinol is associated with a "martensite shear" of "diffusionless" transformation. This mechanism in a simplified sense is much like applying a shearing force to an aligned deck of playing cards. Under this stimulus each card is made to slide slightly out of alignment with its immediate neighboring card. In the case of a martensitic shear this total atomic movement between adjacent planes of atoms is less than a full interatomic distance…
>
> In 55-nitinol if the total deformation is limited to that which may be completely accounted for by the above "martensitic shear" movement of atoms, then a highly efficient "recovery" or reversed atomic shift is possible through the application of heat…

The above description was perhaps the earliest account on the deformation mechanism of SMA and its relation to the subsequent shape recovery. It is clearly based on careful observations and the main points are still valid even today. For example, the deformation of martensite can be achieved under relatively low stress, approximately 8% strain can be achieved within this low stress amplitude and high shape recovery can be achieved if the martensite deformation is within 8%. It was a very interesting observation that the martensite deformation was proposed to be identical to the shearing of a deck of playing cards. The above observations were, however, empirical.

With the help of an optical microscope, it has been noted [Wayman 1975] that under tensile deformation, some martensite plate groups in a CuZn alloy expand by consuming neighboring groups. By 1975, it was generally agreed that in order to obtain shape memory effect the irreversible deformation associated with slips should be minimized [Owen 1975]. Nevertheless, as written in 1975, Wayman concluded: "… at this stage it is clear that the detailed processes of martensite deformation and consequent shape recovery are many and complex. Consequently, the SME is still incompletely understood". Based on the efforts of various researchers, it is now understood that the deformation mechanism of SMAs is different from the classical deformation mechanism of structural materials. It is associated with domain reorientation rather than dislocation mechanism. Such domain reorientation results in *detwinning* of martensite twins.

SMAs have a unique combination of various novel properties including shape memory effect, superelasticity, high damping capacity, good fatigue and wear resistance, high kinetic output per unit volume, and excellent biocompatibility for NiTi SMAs. As listed in figure 1, most of these properties are associated with the *detwinning* process. Understanding this unique deformation mechanism and influencing factors will significantly contribute to the property optimization and effective fabrication of these materials as well as their performance prediction. This knowledge may also provide reference to understand the deformation mechanism of other types of smart materials having similar domain reorientation in microscopic scale.

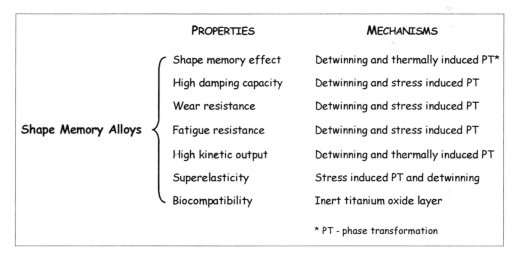

Figure 1. Shape memory alloys, especially NiTi SMAs, possess a unique combination of various novel properties highly attractive for applications in biomedical, MEMS, sensors & actuators, energy dissipation and vibration suppression, aerospace, etc. Most of the properties of SMAs are related to their deformation mechanisms of both detwinning and stress-induced phase transformation.

This chapter will present the recent results on the deformation mechanism of SMAs, major factors affecting the deformation mechanism and their significance in the subsequent shape recovery process in the following order.

1. Twinning and detwinning of <011> type II twin.
2. Deformation of polycrystalline NiTi under tension.
3. Detwinning anisotropy between tension and compression.
4. Detwinning anisotropy due to textures.
5. Prediction of the detwinning anisotropy.
6. Dynamic deformation of SMA.

Twinning and Detwinning of <011> Type II Twin

Formation of lattice twins as a result of phase transformation is an important microstructural prerequisite for a crystalline solid to exhibit shape memory effect. Besides, these lattice twins should be detwinable under stresses. These two conditions are satisfied in many alloy systems having shape memory effect. Several lattice twins are observed in B19′ martensite i.e., <011> type II, $\{11\bar{1}\}$ type I, $\{111\}$ type I, (011) type I, (100) compound and (001) compound twins as summarized in Table 1.

Table 1. Twinning modes observed in NiTi B19′ martensite.

Twinning Mode	K_1	η_1	K_2	η_2	Twinning Shear Strain
$\{11\bar{1}\}$ type I	$(11\bar{1})$	$[0.5404\ 0.45957\ 1]$	$(0.24695\ 0.5061\ 1)$	$[\bar{2}\ \bar{1}\ 1]$	0.3096
$\{111\}$ type I	(111)	$[\overline{1.5117\ 0.5117\ 1}]$	$(\overline{0.66875\ 0.3375}\ 1)$	$[2\ \bar{1}\ 1]$	0.1422
(011) type I	(011)	$[1.5727\ 1\ \bar{1}]$	$(0.7205\ 1\ \bar{1})$	$[0\ 1\ \bar{1}]$	0.2804
<011> type II	$(0.7205\ 1\ \bar{1})$	$[0\ 1\ 1]$	$(0\ \bar{1}\ 1)$	$[1.5727\ 1\ \bar{1}]$	0.2804
(100) compound	$(1\ 0\ 0)$	$[0\ 0\ 1]$	$(0\ 0\ 1)$	$[1\ 0\ 0]$	0.2385
(001) compound	$(0\ 0\ 1)$	$[1\ 0\ 0]$	$(1\ 0\ 0)$	$[0\ 0\ 1]$	

Among the twins observed, the <011> type II twin is generally found to be a major twinning mode. It consists of two martensite variants related to each other by a 180° rotational symmetry about their [011] direction [Knowles & Smith 1981]. The twinning elements of the [011] type II twin reported by Knowles and Smith are the following: twin plane $K_1 = (0.7205\ 1\ \bar{1})$, shear direction $\eta_1 = [011]$, $K_2 = (011)$ and $\eta_2 = [1.5727\ 11]$. Since the twin plane has an irrational index, how the atoms are arranged along the twin boundary is a question during the last two decades. As proposed by Christian [1965; 2002], the irrational twin boundary consists of ledges and steps of rational twin plane. The validity of such proposal has been studied by a number of researchers, including Knowles [1982], Otsuka and colleagues [Onda *et al.* 1992; Otsuka & Ren 2005] and Nishida and colleagues [1995a; 1998]. A strong disagreement on the nature of the type II twin plane in atomic scale exists.

It has been found that various deformation processes of NiTi SMAs are strongly related to the response of type II twins [see, for example, Miyazaki *et al.* 1989a,b; Liu *et al.* 2000]. Thus, a complete understanding on the nature of the irrational type II twin plane is needed in order to understand its detwinning process and the subsequent atomic rearrangement during shape recovery.

Recently, we reported using a rational $(34\bar{4})$ or $(0.75\ 1\ \bar{1})$ plane to approximate the irrational $(0.7205\ 1\ \bar{1})$ twin plane of [011] type II twin [Liu & Xie 2003]. By doing so we were able to construct a crystal structure of type II twin with available crystal visualization software. The advantage of such crystal structural visualization is that we can conveniently rotate and view the atomic arrangement of the lattice twin from any direction of interest. Such visualization provides us a possibility of knowing the details of atomic arrangement before HRTEM observation, thus, helps to interpret the experimental observations. We can also simulate the HRTEM images of the constructed model of twin plane and compare them with that experimentally obtained. Such comparison can help to understand the experimental results and, on the other hand, to provide experimental verifications on the constructed model.

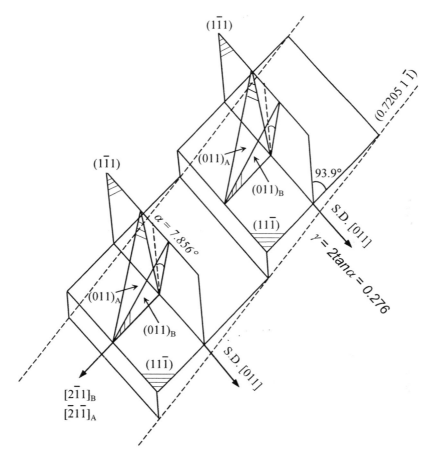

Figure 2. Schematic illustration of <011> type II twin. Relation between irrational twin plane ($0.7205\ 1\ \bar{1}$), rational twin plane ($11\bar{1}$), shear plane ($1\ \bar{1}\ 1$), (011) K_2 plane, and shear direction [011] is highlighted. The <011> type II twin has an overall 180° rotational symmetry about [011] and a pseudo-mirror-plane symmetry with respect to each ledge of ($11\bar{1}$) plane.

Through HRTEM observations [Liu & Xie 2003], HRTEM image simulation [Xie & Liu 2004], and analysis of HRTEM images reported by other researchers [Liu & Xie 2006], we have proved that the <011> type II twin is atomically rational. Its twin plane consists of rational ($11\bar{1}$) ledges that are the atomic-scaled twin plane as shown in Figure 2. According to our analysis, this model is able to unify all the major HRTEM observations reported. Based on this model, we can reasonably explain that the detwinning of <011> type II twin is achieved through atomic shear on the rational ($11\bar{1}$) plane along [011] shear direction, where the shear plane is ($1\bar{1}1$). Such detwinning mechanism can be schematically shown in Figure 3. The atomic shear on the ($1\bar{1}1$) plane results in the advancement of ($11\bar{1}$) ledges along their normal directions (Figure 3a). A synergetic movement of the ($11\bar{1}$) ledges leads to the migration of the irrational twin plane (0.7205 1 $\bar{1}$), leading to the growth of one variant at the expense of the other (Figure 3b). Such mechanism can be used to account for the "macroscopic" observations [Miyazaki *et al.* 1989b].

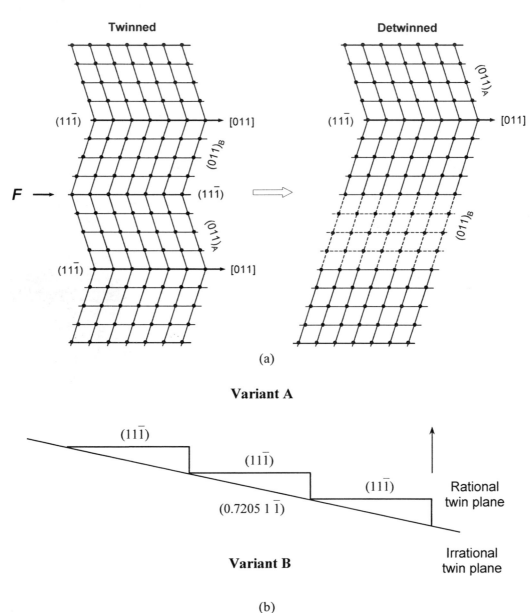

Figure 3. Schematic two-dimensional representation of the detwinning mechanism of <011> type II twin. (a) Atomic shear on (11$\bar{1}$) plane along [011] direction leads to the advancement of the (11$\bar{1}$) ledges (viewed along [2$\bar{1}$1]$_A$//[$\bar{2}$1$\bar{1}$]$_B$.). (b) Synergetic advancement of the (11$\bar{1}$) ledges results in the migration of the irrational (0.7205 1 $\bar{1}$) twin boundary, leading to the growth of one variant at the expense of the other (viewed along [011]$_{A,B}$).

The above detwinning model was further verified experimentally. When a NiTi martensite is deformed under tension to 6%, some of the martensite twins were detwinned, as typically shown in Figure 4. In this observed area, both twinned and detwinned martensite variants coexist. In the twinned region, several <011> type II twin bands are clearly visible. Confirmed by the electron diffraction analysis, the neighboring martensite variants are orientated in $[101]_A//[\bar{1}10]_B$ directions. In the other regions, mainly martensite variant B exists, while variant A became very narrow and was nearly vanished, as indicated by white arrows. As a result of detwinning, variant B grew significantly at the expense of the variant A. In addition, some black traces that are parallel to (001) plane are visible in the detwinned area as indicated by the black arrows. They are believed to be planar defects formed during detwinning process [Liu & Xie 2003].

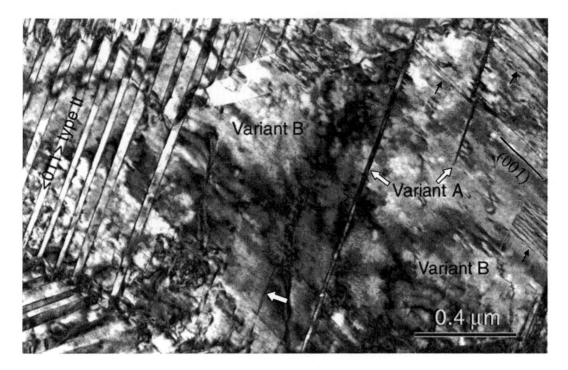

Figure 4. Detwinning of <011> type II twin in NiTi sheet after 6% tension deformation along rolling direction. $\mathbf{B}//[101]_A//[\bar{1}10]_B$ in twinned area, and $\mathbf{B}//[\bar{1}10]$ in detwinned area.

The atomic arrangement of a nearly detwinned area was further studied by using HRTEM and is shown in Figure 5a. The observation direction is the same as that of Figure 4, i.e., electron beam was parallel to $[101]_A//[\bar{1}10]_B$. The variant A in between the two variants B nearly disappeared after the deformation. The migration of twin boundary results in a strain contrast along the boundary, and the boundary is still locally parallel to $(11\bar{1})$ plane. The detailed atomic configuration of detwinning area can be seen in Figure 5b and 5c, which are the IFFT images corresponding to the two framed regions in Figure 5a. As is visible, in the whole detwinning area, all the $(11\bar{1})$ planes are still parallel to each other and no distortion of $(11\bar{1})$ plane is visible. This suggests that the atomic shear was in the $(11\bar{1})$ plane. Between the two variants B, the (010) plane of variant A is reoriented, on the way to be aligned with

$(001)_B$ via atomic shear on $(11\bar{1})$ planes. This observation confirms the detwinning mechanism of $<011>$ type II twin proposed in Figure 3.

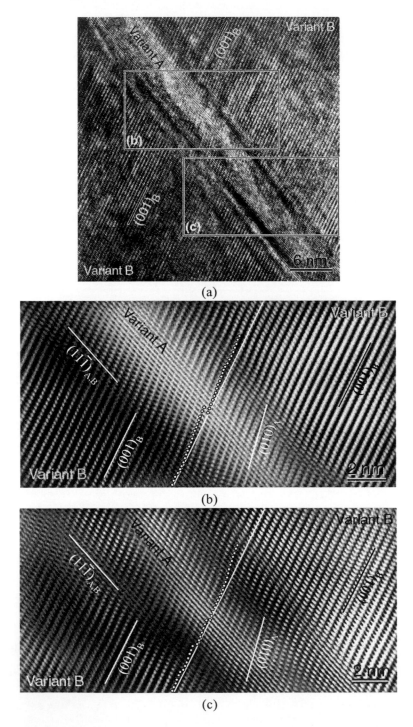

(a)

(b)

(c)

Figure 5. (a) HRTEM micrograph of the detwinned region of $<011>$ type II twin. IFFT images (b, c) corresponding to the framed areas (b, c) in (a).

Deformation of Polycrystalline NiTi under Tension

The deformation of a twinned martensitic polycrystalline NiTi SMA under tension has been recognized to consist of several macroscopic steps [Mohamed *et al.* 1977; Miyazaki *et al.* 1989b; Wayman *et al.* 1990]; 1) at the beginning of the deformation, a monotonic increase in the stress with increasing strain amplitude till about 1.2% strain, 2) a slight stress-drop followed by a stress-plateau till about 6% strain, 3) a further increase in the stress with increasing deformation, and 4) a final plastic deformation leading to fracture. Starting from the end of the stress-plateau, the deformation behavior of SMAs becomes similar to that of traditional structural materials. The deformation before the stress-drop has been attributed to the elastic accommodation of twin bands. The stress-plateau is related to the martensite reorientation process. The deformation starting from the end of the stress-plateau was proposed to be the elastic deformation of the fully re-oriented martensite, and a further plastic deformation of the re-oriented martensite. It was later found that most of the "secret" of SMAs lies in the region of stress-plateau. Through a careful study of the microstructures under deformation to various strains, the above deformation procedures were refined [Liu *et al.* 1999a;b; 2000].

A specially designed experiment by using three extensometers to record the deformation strain of both a NiTi wire and a NiTi drawn bar (Liu *et al.* 1998a; 2000) was performed as schematically illustrated in figure 6. The sample length between the two clamps was 100 mm, while three extensometers with 25 mm gauge length were fixed onto the upper, middle and the lower portion of the specimen.

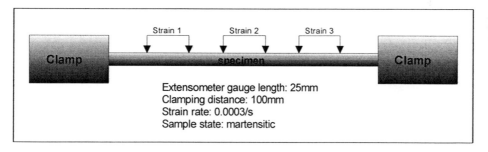

Figure 6. Schematic illustration of the tensile test performed on a NiTi bar with three extensometers.

Figure 7a shows the stress-strain curves recorded by the three extensometers during tension, where strain 1, 2 and 3 are recorded by the upper, middle and lower extensometers, respectively. It can be seen that the onset of the stress-plateau begins at different strains for different parts of the specimen. Extensometer 2 recorded that the stress-drop begins at about 2% strain while extensometer 3 recorded a stress-drop at about 1.7% strain. The stress-drop is in fact related only with time but not with strain (figure 7b). It occurs only once throughout the specimen and it may be originated from a microstructural change took place very locally. This result may suggest that the detwinning process took place initially through "nucleation" of detwinned area within twins and propagated through "growth" of the detwinned region. Such deformation characteristic was proposed to be a Lüders-like deformation [Liu *et al.* 1998a] through propagation of deformation bands in macroscopic scale. A fundamental understanding on the relation between detwinning and deformation band propagation is yet to be established.

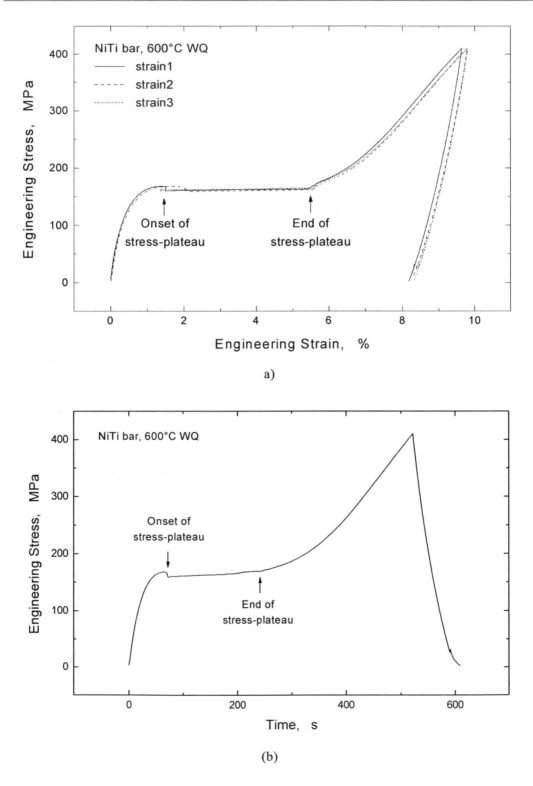

a)

(b)

Figure 7. Continued on next page.

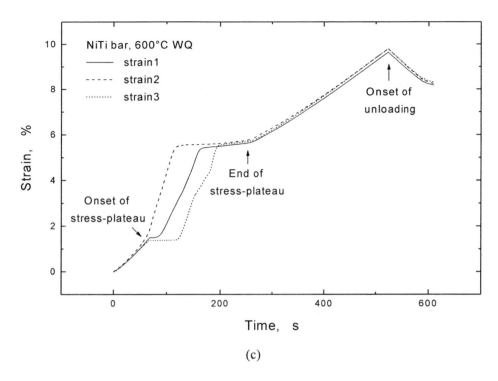

(c)

Figure 7. Localized deformation behavior of martensitic NiTi SMA as determined with 3 extensometers fixed at different parts of the NiTi bar specimen.

Figure 7c shows the three strains as a function of time. It shows that the deformation path of the specimen can be divided into 3 regions depending on the local deformation rate. (1) Before the stress-drop, the deformation rates of the material within 3 extensometers are about the same. (2) Following the stress-drop, the localized deformation process starts and proceeds within the whole range of the stress-plateau, represented by the different deformation amplitudes within different extensometers. (3) As soon as the deformation amplitudes of all parts of the specimen reach the end of the stress-plateau (~6% strain), the deformation rate within different extensometers is again the same. The stress-plateau is ended at about 250 seconds for the present experiment and is marked in both figure 7b and figure 7c. The transition of the strain at the end of the plateau-strain can be clearly seen in figure 7c. While the extra part of the strain-plateau in figure 7c beyond the junction point of three strains should be related to the further deformation of the portion of the specimen outside the extensometers.

The onset of the stress-drop has been generally recognized to be the onset of the martensite reorientation process [Wayman *et al.* 1990]. However, the loading-unloading experiment within the strain region I (Figure 8) shows that a residual strain exits before the onset of stress-plateau. This clearly shows that the initial deformation of the martensitic SMA was not due to elastic deformation of the accommodation twins. Result [Liu *et al.* 1999c] has also shown that no measurable two-way memory effect has been developed after deformation within this region. This suggests that no significant dislocation rearrangement has taken place. Unloading within this region seems not to affect the following stress-drop and the further deformation behavior into region II.

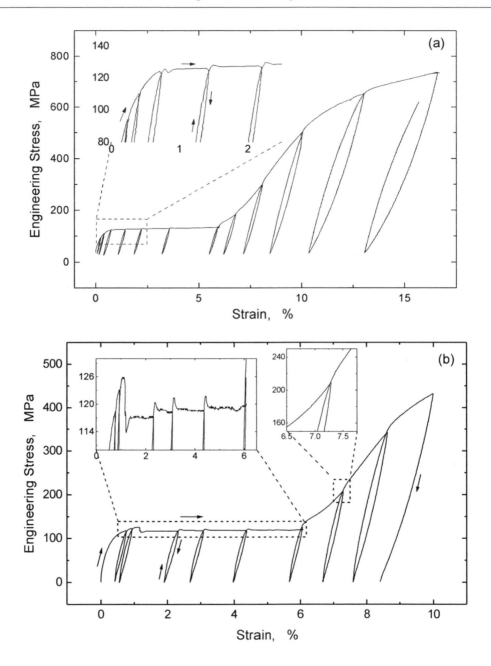

Figure 8. Stress-strain curves under tensile loading and unloading for a martensitic NiTi wire (a) and rolled sheet (b) along the rolling direction.

A phenomenon frequently observed in the samples tensioned to 1.2% strain is the formation of parallel plates (Figure 9). The twin bands between two neighboring plates are well coherent to each other at the plate boundaries. In addition, a (100) compound twin relation between neighboring martensite plates is frequently observed. This shows that some martensite twins have already started to re-arrange even at 1.2% strain. The frequently observed formation of parallel bands within this strain region may be partially responsible for its inelastic nature in Region I [Liu *et al.* 2000].

Figure 9. BF micrograph of martensite tensioned to 1.2% strain. Parallel coherent plates consisting of <011> type II martensite twins have been frequently observed. **B** // $[101]_M$ // $[\bar{1}10]_T$. A (100) compound twin relation between neighboring martensite plates is frequently observed.

It is known that the (100) compound twin does not fit in the phenomenological crystallographic theory and hence it has been recognized as a deformation twin [Knowles & Smith 1981; Onda *et al.* 1992; Nishida *et al.* 1995b]. Since the (100) compound twins are seldom observed in the undeformed samples in the present material while martensite plates having (100) compound twin plane as junction planes are frequently observed in the 1.2% and 4% deformed samples (Figure 9) [Liu *et al.* 2000], it can be concluded that the formation of the (100) compound twins is a result of the tensile deformation. Hence the formation of an atomic coherent plate boundary seems to be the first step for the martensite reorientation and detwinning process. The formation of (100) compound twins may allow the plate boundaries to move more readily under stress.

Summarizing the experimental observations, the martensite deformation process as a function of deformation strain amplitude can be refined. Figure 10a summarizes the microstructural changes of martensite twins especially <011> type II twins during tension. Before the onset of the stress-plateau, under a constant movement of the cross-heads, the deformation rate is the same over the length of the specimen (figure 7c), showing a uniform deformation feature within strain region I. Some martensite twins are inelastically rearranged under the externally applied stress. Formation of coherent and parallel bands within martensite plates has been observed after deformation to 1.2% strain. In addition, generation of dislocations inside (11$\bar{1}$) type I twins has also been observed. It should be mentioned that there are still a large amount of martensite plates that do not show a significant change at 1.2% strain.

The second region starts from the onset till the end of the stress-plateau. The major characteristic of this deformation process is marked by a sudden stress-drop at the onset of the process followed by a nearly flat stress region and a nearly sudden increase of the stress at the end of the process. Macroscopically, this region has a localized deformation character (figure

7c), i.e., during a constant movement of the cross-heads, the deformation amplitude can be different within different parts of the specimen. Further more, during unloading within this region, a new stress-barrier is created against further deformation, suggesting that lattice defects have been generated or rearranged during deformation. Microscopically, martensite reorientation of <011> type II twins following formation of (100) compound twins and the detwinning of martensite <011> type II twins have been observed. Rearrangements of lattice dislocations have also been found along the junction plane areas, which may be responsible for the stress-barrier observed during re-loading (Figure 8). TEM observations further show that, within this region, not all martensite twins are detwinned.

By further deformation into the region III beyond the stress-plateau, further reorientation and detwinning of martensite twins, which are less favorable to the applied force, can be expected. In this region, deformation of martensite twins is accompanied by a further increase in the applied stress. In addition, a high density of dislocations has been generated. A schematic representation of the deformation details of martensite twins from region I to region III is shown in figure 10b. Plastic deformation will be induced in the martensite plates having orientations unfavorable to the applied stress. It is unavoidable that the TEM observations may represent only the local microstructure features. Abstraction of these localized observations, although being a large number to some extent, into a generalized pattern having major representative features is not always an easy task. Figure 10a summarizes the experimental observations that are believed to represent the major features of the microstructural changes during tension, while figure 10b tries to abstract these changes into a more generalized pattern. Extensive computational research is needed for a thorough description of the detwinning process.

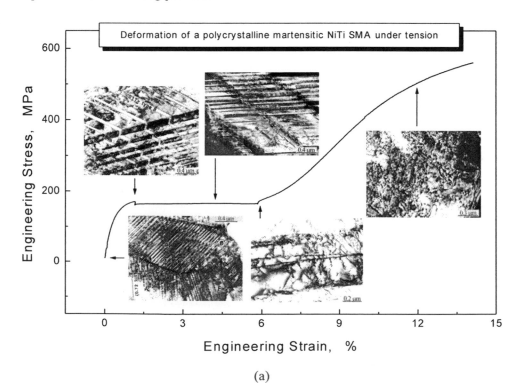

(a)

Figure 10. Continued on next page.

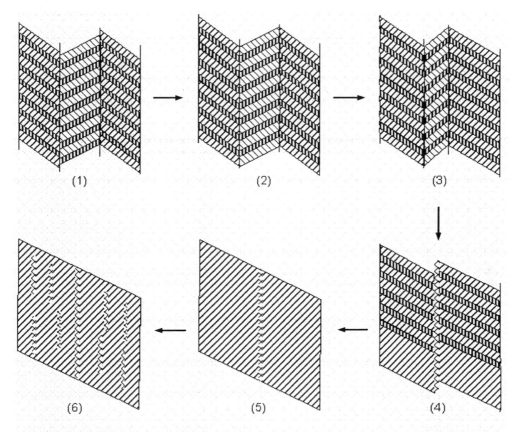

(1) Undeformed state.
(2) Inelastic accommodation of twins.
(3) Martensite reorientation.
(4) Further martensite reorientation and detwinning.
(5) Further detwinning process.
(6) Generation of lattice dislocations through further deformation.

(b)

Figure 10. Schematic representation of the re-configuration of martensite twins under tension in polycrystalline NiTi SMAs.

The detwinning stress is also a function of testing temperature. As shown in Figure 11, when tested below M_s, the detwinning stress increases linearly with lowering the temperature [Liu 2002a]. Experimental results show that, at testing temperatures between points A and B in Figure 11, a stress-plateau occurred during tensile deformation. From temperature C above, no stress-plateau exists in the stress-strain curve. The yield stress increases with increasing temperature from C to D. The plateau stress has the lowest value at temperature B which might be the highest testing temperature before the reverse transformation starts. It needs to be mentioned that, in this NiTi sheet, no superelasticity is observed within the testing temperature range.

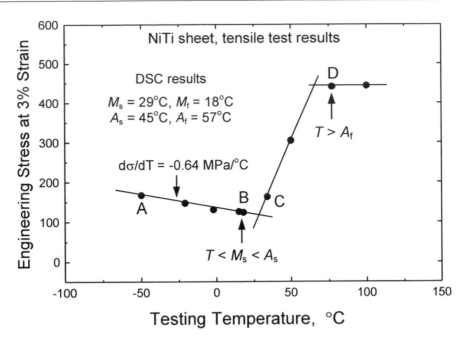

Figure 11. Stress of a NiTi sheet at 3% strain as a function of testing temperature. From A to B, a stress-plateau due to martensite detwinning is observed. The detwinning stress increases with decreasing temperature. Above C, no stress plateau exists while the yield stress increases with increasing testing temperature.

Detwinning Anisotropy between Tension and Compression

For polycrystalline SMAs, the stress-strain curves are different for tension and compression, exhibiting an "asymmetry" as shown typically in Figure 12. In contrast to the case of tension, under compression the SMA is quickly strain hardened and no flat stress-plateau is observed. This result clearly suggests that, when application of SMAs is concerned, attention should be paid to the loading mode in their usage for both design and performance prediction. The mechanism of the mechanical asymmetry is further studied using TEM to reveal the microstructures developed under deformation [Liu et al. 1998b].

Comparing to what has been observed in the tensioned samples, the martensite microstructure in specimens compressed to 4% strain is different. The martensite variants are still self-accommodated with mainly <011> type II twins, while a high density of lattice defects, mainly dislocations, has been generated both inside the martensite twin bands and inside the variant accommodation areas as typically shown in Figures 13a. A large amount of dislocations are also present inside the $(11\bar{1})$ type I twin bands (Figs 13b). It was noted that the martensite accommodation is similar to that in the undeformed samples. No neighboring martensite plates are found to be twin related to each other after compression as typically shown in Figure 13a. Thus, the junction planes between martensite plates were not mobile and, as a result, no migration of these junction planes has been observed. This observation explains the difference in the stress-strain curves between tension and compression (Figure 12) and the asymmetry of stress-strain curves during cyclic deformation (Figure 14) [Xie et *al.* 1998; Liu *et al.* 1999d]. The different deformation mechanisms between tension and compression further lead to the difference in the shape recovery.

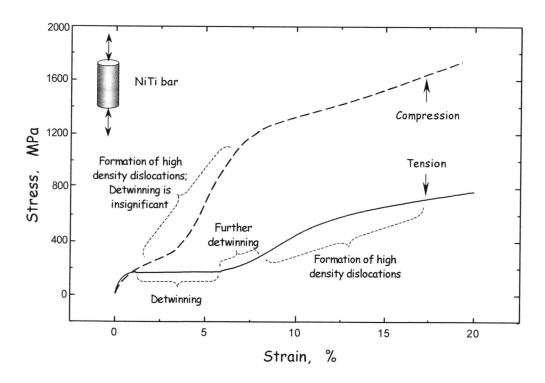

Figure 12. Anisotropy of martensite deformation process between tension and compression. Microstructure study shows that tension deformation leads to martensite detwinning while compression deformation results in formation of a high density of dislocations and no significant detwinning process took place.

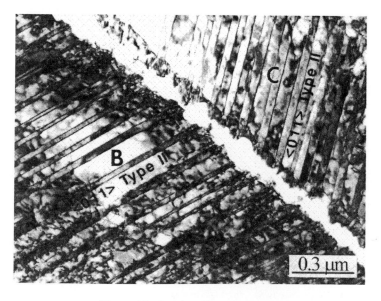

Figure 13. Continued on next page

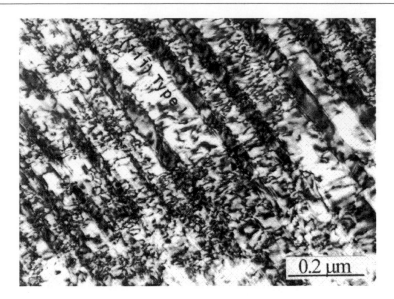

Figure 13. BF micrograph (a) of NiTi martensite in specimens compressed to 4% strain showing a large amount of lattice defects inside twin bands and plastic deformation in variant accommodation area.

Results obtained by the present authors (Xie *et al.* 1998; Liu *et al.* 1999d) have shown that, under tension-compression cyclic deformation within ±4% strain (Figure 14a), $(11\bar{1})$ type I martensite twins in stead of <011> type II martensite twins have been frequently observed. In addition, various lattice defects have been generated both in the junction plane areas of martensite plates and within the martensite twins. Results from the partial cyclic tests (Figure 14b) also show that in the polycrystalline NiTi SMA, the martensite deformation process on both sides of the strain involves different mechanism. During cycling within the positive strain, the stress-strain curves are symmetric and the stress levels during tension and compression are nearly the same. This shows that, within the positive side of the strain, the deformation mechanism is similar between tension and compression, which may also suggest that, in this case, the microstructural variation is nearly reversible during alternative tension and compression loads. This can be understood from the TEM observations (Figure 10). The microstructural change within the positive side of the strain involves mainly the reversible movement of the mobile interfaces between adjacent martensite plates. However, when the material is cycled on the negative side of the strain, the stress-strain curves are asymmetric and the stress levels differ significantly between tension and compression. Hence, the deformation under compression is more difficult than under tension. TEM observations on specimens monotonically compressed to 4% show that two neighboring martensite plates containing <011> type II twins were not twin related to each other when compressed, and no evidence of interfacial migration have been observed. However, martensite deformation through generation and rearrangement of lattice defects especially dislocations were clearly observed. Thus, under cyclic deformation on the negative side of the strain, it is unlikely that a reversible movement of the junction planes of martensite plates has been involved.

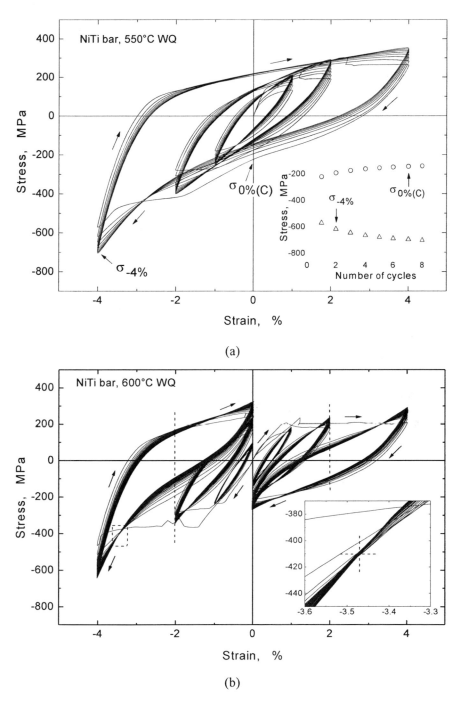

Figure 14. Stress-strain curves of a NiTi SMA under tension-compression cyclic deformations. (a) Full loop cyclic tests, strain rate is 1.6×10^{-2} s^{-1}. Cyclic hardening and softening effects are highlighted in the figure. (b) Partial loop cyclic tests, strain rate is 8×10^{-3} s^{-1}. One specimen was subsequently cycled between 0 and 1%, 0 and 2%, 0 and 4% strain (each for 50 cycles). Another specimen was subsequently cycled between 0 and -1%, 0 and -2%, 0 and -4% strain (each for 50 cycles). A crossing point in the stress-strain curve is highlighted in the enlarged framed area.

The textures of the NiTi bar before and after various deformation have been measured. The axial direction of the bar was set to be parallel to the normal direction (ND) in the pole figure goniometer. The (001) pole figure of the NiTi bar before mechanical test is shown in Figure 15a. It can be seen that in the as-annealed condition, the NiTi bar consists of [001] fiber texture. As shown in the standard [001] stereographic projection (Figure 15b), the normal direction of (001) plane deviates from the [001] direction for about 7° angle due to the monoclinic crystal structure. Comparing Figures 15a and 15b, the [001] fiber texture has about an 20° angle away from the ND. After tension deformation to 4% the [001] fiber texture has about an 10° angle away from the ND (Figure 16a), meaning that the [001] fiber texture has rotated towards to the ND. By further deformation to about 6% under tension, the [001] fiber texture is well aligned to the ND of the NiTi bar (Figure 16b) and the texture is enhanced as a result of martensite detwinning/reorientation. However, if the NiTi bar is compressed to 4%, the [001] texture became less significant as shown in Figure 16c. This shows that under compression some of the crystals have rotated to directions away from the fiber texture direction.

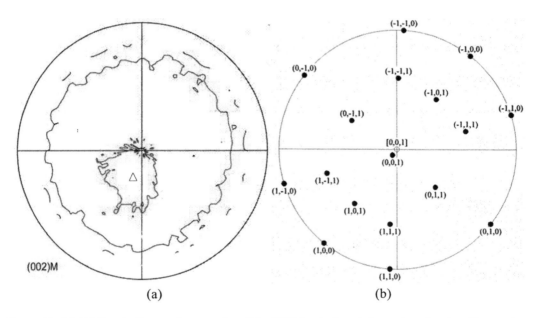

(a)   (b)

Figure 15. (a) (001) pole figure of martensite of the NiTi bar before deformation showing fiber texture. (b) Standard [001] stereographic projection of B19' NiTi martensite.

As has been reported, when deformed in the martensitic state, a two-way memory effect can be developed in the SMAs (Liu *et al.* 1999c). Due to the different deformation mechanisms between tension and compression, the resulted two-way memory effect is found also different [Liu 2001]. Figure 16 highlights the results obtained from two pre-deformed NiTi samples, one was under tension to 18.5 % strain (Sample A) and the other was under compression to 20% strain (Sample B). After unloading to a stress free state, both samples were then heated to above 200°C and their shape recovery processes were recorded. Upon heating, Sample A contracts while Sample B elongates. As shown in Figure 17, for both samples, pre-deformation leads to a shift of the austenite transformation temperatures to higher temperature range, i.e., the martensite phase is stabilized. For both samples, a two-way memory effect was developed, however, differs in magnitude in both the shape recovery and the two-way memory strain. In addition, the magnitude of martensite stabilization and the reverse transformation features are also different. During the 1st heating step, Sample A recovers 4% while Sample B recovers about 3.5%. For Sample A, the reverse transformation begins at about 125°C which is about 50°C higher than that of undeformed condition, and the (M_s-M_f) is about 18°C. For Sample B, the reverse transformation begins at about 110°C, however, the (M_s-M_f) is about 65°C. In addition, the two-way memory strain developed in Sample A is 2.7%, while in Sample B it is only 0.8%. It is also noted that the transformation hysteresis is much narrower for Sample A than for Sample B.

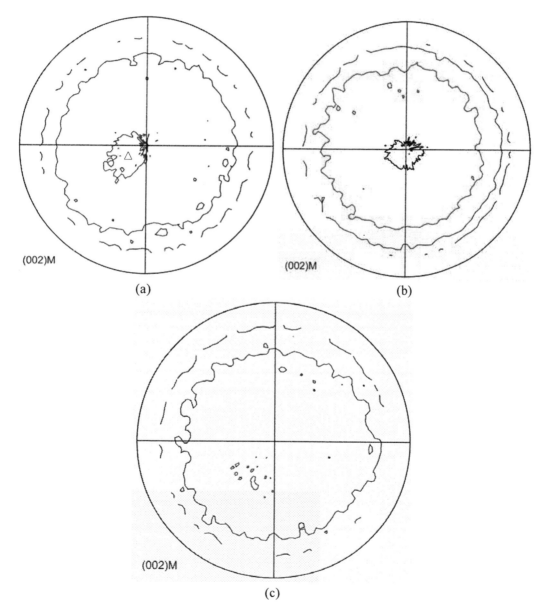

Figure 16. Martensite textures of NiTi bar after tensile deformed to (a) 4%, (b) 6%, and compressed to 4% (c).

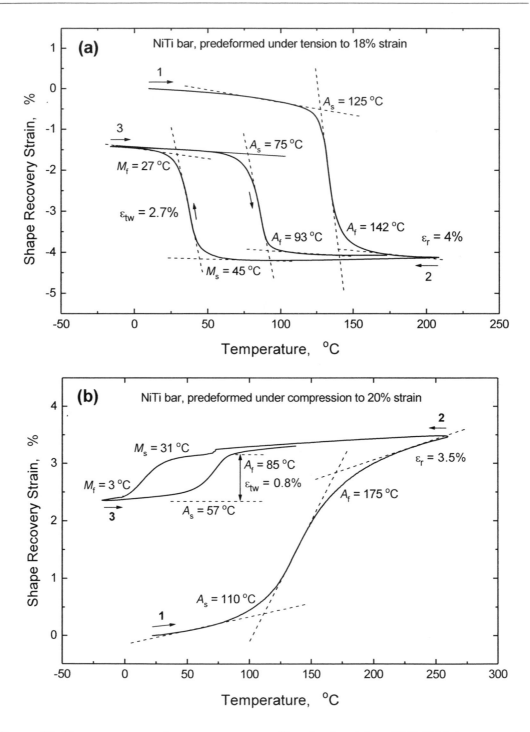

Figure. 17. Shape recovery and two-way memory effect developed in a NiTi SMA by martensite deformation under (a) tension to 18.5% strain and (b) compression to 20% strain. Anisotropy in both shape recovery and two-way memory strain is found between tension and compression. The transformation temperatures are roughly estimated by a slope line method in the graph and can be different from the values determined by other methods, e.g. DSC. In the undeformed condition the transformation temperatures are $M_s =59°C$, $M_f = 39°C$, $A_s = 74°C$ and $A_f = 93°C$, respectively.

In order to understand the effect of texture on the tension-compression anisotropy of NiTi, a NiTi ingot of less textured was also studied and compared to the results of a highly textured NiTi bar. Typical shape recovery behavior of both NiTi bar and ingot under tension and compression are shown in Figure 18. Together shown are the corresponding stress-strain curves. The stress-strain curves are asymmetric between tension and compression for both materials and the shape recovery magnitude is also anisotropic with respect to the loading mode irrespective to the magnitude of texture existing in the materials. Tension deformation results in a stress-plateau and a higher recovery strain. While compression deformation leads to a quick strain hardening and a less shape recovery. It can be concluded that deformation mode not only affects the detwinning process and the resulted mechanical behavior of the SMAs, but also strongly affects the subsequent shape recovery behavior and transformation characteristics of SMAs. Detwinning under tension corresponds to a better shape recovery and higher two-way memory strain, while a heavy dislocation generation due to compression corresponds to a lower shape recovery and lower two-way memory strain.

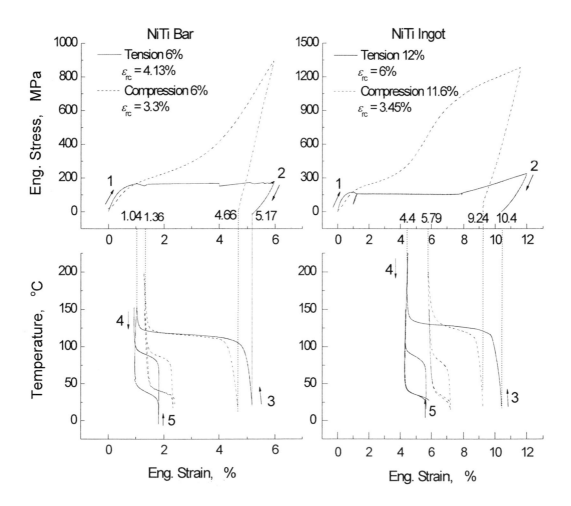

Figure 18. Shape recovery anisotropy between tension and compression for a NiTi bar and a NiTi ingot.

Detwinning Anisotropy due to Textures

In most cases, SMAs in application are polycrystalline materials prepared by either rolling or drawing or other heavy deformations. Thus, textures exist in most of the SMAs. In general, possess of textures is a wanted microstructural state if the material is used properly. However, it can also lead to unwanted poor performance if the textured materials are used improperly. From an application point of view, prediction of the stress-strain relation and shape recovery strain under given texture(s), loading direction and deformation amplitude is of primary importance. Understanding the rule of textures in the anisotropy of various properties of SMAs can help to optimize the properties for desired applications through optimization of the material processing route.

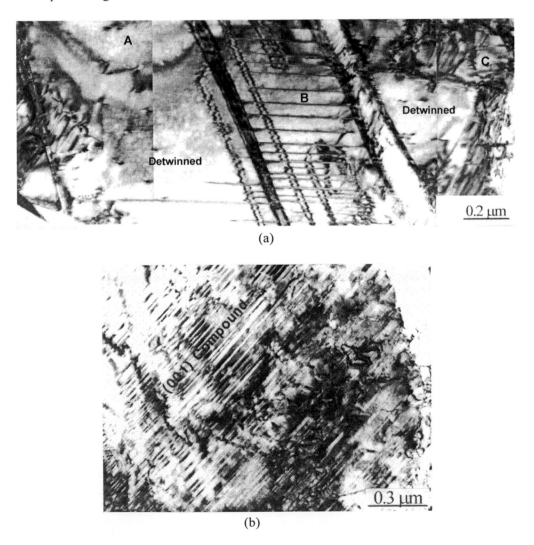

(a)

(b)

Figure 19. Response of (a) <011> type II twin and (b) (001) compound twins in a martensitic NiTi rolled sheet under tension to 6% strain along rolling direction. Type II twin was significantly detwinned while detwinning of compound twin was insignificant. A large amount of lattice defects were formed in the (001) compound twin bands.

Various results [Melton 1990; Mulder *et al.* 1993; Inoue *et al.* 1996; Zhao 1997; Liu *et al.* 1999e] have shown that, textured SMA rolled sheet exhibits a considerable difference in plateau-strain and shape recovery strain when tested along different directions. Microstructure study has also shown that, when deformed along either rolling or transverse direction, the responses of different types of twins are different [Liu *et al.* 1999e]. Particularly, when under tension along rolling direction (RD), it is found that the <011> type II twin has detwinned while no significant detwinning of the (001) compound twin was observed. On the other hand, when under tension along the transverse direction (TD), the situation was reversed. Meanwhile, dislocations are generated. Typical TEM observations are shown in Figures 19 and 20.

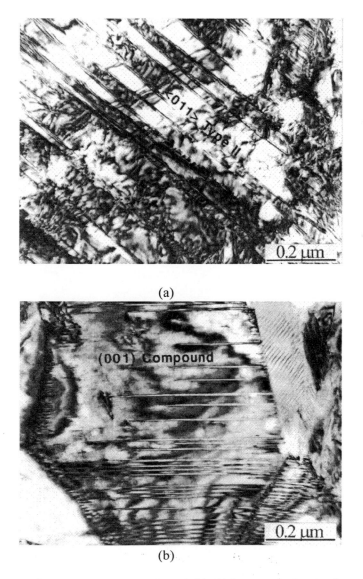

(a)

(b)

Figure 20. Response of (a) <011> type II twin and (b) (001) compound twins in a martensitic NiTi rolled sheet under tension to 6% strain along transverse direction. Detwinning of (001) compound twins was found. Detwinning of <011> type II twin was insignificant with formation of dislocations.

Figure 21 summarizes the mechanical behavior and corresponding microstructure changes as a function of loading direction. Tension along rolling direction leads to a stress-plateau and martensite detwinning. While tension along transverse direction results in dislocation generation and strain hardening. The stress-plateau occurring during deformation along RD is related to the detwinning of mainly the <011> type II twins. The detwinning of (001) compound twin is favorable to the loading along TD. In general, when tested along TD the shape recovery strain is the lowest if compared to those measured along other directions. For a rolled sheet, the direction where the highest shape recovery strain is obtained corresponds to the direction where the longest plateau-strain is obtained during martensite deformation. While the test direction where a poor shape recovery is obtained corresponds to a martensite stress-strain curve with no stress-plateau.

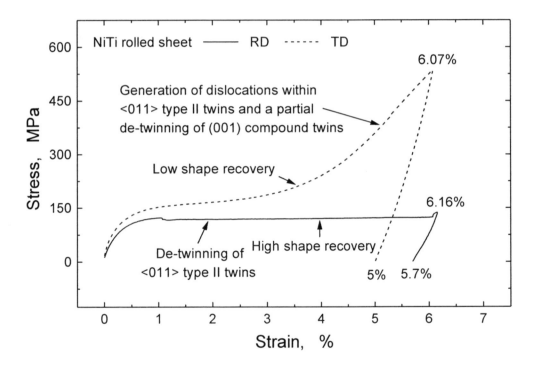

Figure 21. Anisotropy of martensite deformation process in a textured rolled NiTi SMA sheet. Tension along rolling direction (RD) leads to detwinning of <011> type II twins. While tension along transverse direction (TD) leads to formation of dislocations and partial detwinning of (001) compound twins.

In order to understand such detwinning and shape recovery anisotropy, the texture of the NiTi sheet was measured. Two martensite texture components existed in the material were found to be more pronounced, they are $(010)[001]_M$ and $(001)[010]_M$. Further analysis (Table 2) shows that for these two texture components, the shear direction of <011> type II twins has angles of respectively 41.7° and 48.3° to the RD, while it has respectively 90.2° and 95° angles to the TD. For the (001) compound twins, the SD has an angle of 6.56° and 0° to the TD, while it has 96.8° and 90° angle to the RD for the $(010)[001]_M$ and $(001)[010]_M$ texture components, respectively. Such relation can be visualized in Figure 22.

Table 2. Orientation relation between shear direction (SD) of three types of martensite twins, rolling (RD) and transverse (TD) directions of four types of textures of NiTi cold rolled sheet.

Textures and ND directions →	(010)[001]$_M$ ND [010]		(001)[010]$_M$ ND [105]		(111)[$\bar{2}$11]$_M$ ND [522]		(11$\bar{1}$)[2$\bar{1}$1]$_M$ ND [63$\bar{2}$]	
SD of twins ↓	RD [001]	TD [14 0 1]	RD [010]	TD [100]	RD [$\bar{2}$11]	TD [1$\bar{6}$5]	RD [2$\bar{1}$1]	TD [$\bar{1}$ 27 26]
<011> type II SD [011]	41.7°*	90.2°	48.3°	95°	40.7°	89°	88.6°	1.4°
(001) compound SD [100]	96.8°	6.56°	90°†	0°	135.7°	89.7°	49.7°	96°
(11$\bar{1}$) type I SD [112]	29.4°	73.9°	66.2°	79.9°	58.1°	71.6°	65.3°	24.7°

* Textures correspond to TEM observations.

† Textures not correspond to TEM observations.

Thus, deformation along RD is easier to shear the <011> type II twins and, in contrast, the <011> type II twin is difficult to detwin if the force is applied along TD. However, for the (001) compound twins, the situation is opposite, deformation along TD will easily shear the (001) compound twins. This relation is considered to be responsible for the differences in the microstructural changes during deformation along both directions and is also responsible for the difference in the respective stress-strain curves. This should also be responsible for the shape recovery anisotropy due to texture.

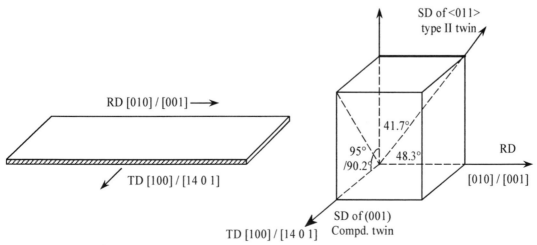

Figure 22. Schematic relationship among SD of twins, RD and TD for two martensite texture components, (001)[010]$_M$ and (010)[001]$_M$. SD of <011> type II twins is parallel to [011]$_M$ which has 48.3°/41.7° angle to RD and 95°/90.2° angle to TD for (001)[010]$_M$/(010)[001]$_M$ textures. SD of (001) compound twins is nearly parallel to the TD and perpendicular to the RD for both (001)[010]$_M$ and (010)[001]$_M$ textures.

Prediction of the Detwinning Anisotropy

To predict the shape recovery strain as a function of loading direction for textured SMAs, efforts have been made from an approach of lattice correspondence between martensite and austenite by taking into account the transformation strain as a function of lattice orientation [Inoue *et al.* 1996; Bhattacharya & Kohn 1996; Shu & Bhattacharya 1998; Lu & Weng 1998; Thamburaja *et al.* 2005].

In a recent research, Zheng and Liu [2002] have tried to predict the martensite detwinning process by taking into consideration of texture distribution, twinning shear and loading direction. Based on a proposed model of detwinning, a crystallographic analysis was performed on the resolved shear stress along the shear direction of <011> type II and (001) compound twins for two texture components, namely, (010)[001]$_M$ and (001)[010]$_M$. Further, the deformation kinetics of both types of twins for loading along different directions was examined by assuming several Orientation Distribution Functions (ODFs) of the texture. Finally, the simulation results of the stress-strain curves under deformation along both RD and TD are compared to experimental results in Figure 21.

In order to describe the relation between texture orientation, twinning shear and loading direction, a Cartersian coordinate system was introduced as shown in Figure 23. Where axes x_1, x_2 and x_3 coincide with the RD, TD and the normal direction (ND) of the rolled sheet, respectively.

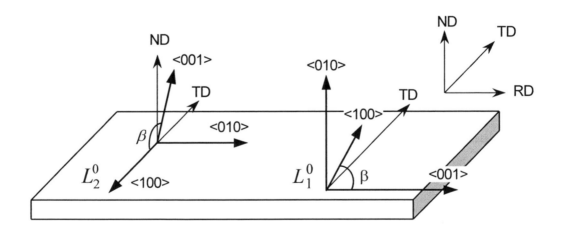

Figure 23. Schematic representation of two types of lattices, L_1^0 and L_2^0 lattices, in accordance with two textures, (010)[001]$_M$ and (001)[010]$_M$ textures, respectively.

In conjunction with the (010)[001]$_M$ and (001)[010]$_M$ textures, two lattices of L_1^0 and L_2^0 are assigned so that (001) and (010) directions are coincidence with RD and ND for lattice L_1^0, and with ND and RD for lattice L_2^0, respectively.

$$L_1^0 \rightarrow \quad \langle 100 \rangle = a \begin{bmatrix} \cos \beta \\ \sin \beta \\ 0 \end{bmatrix}, \quad \langle 010 \rangle = b \begin{bmatrix} 0 \\ 0 \\ 1 \end{bmatrix}, \quad \langle 001 \rangle = c \begin{bmatrix} 1 \\ 0 \\ 0 \end{bmatrix}$$

$$L_2^0 \rightarrow \quad \langle 100 \rangle = a \begin{bmatrix} 0 \\ -1 \\ 0 \end{bmatrix}, \quad \langle 010 \rangle = b \begin{bmatrix} 1 \\ 0 \\ 0 \end{bmatrix}, \quad \langle 001 \rangle = c \begin{bmatrix} 0 \\ -\cos \beta \\ \sin \beta \end{bmatrix}$$

A schematic representation of the atomic arrangement of lattice twins is shown in Figure 24a. As schematically illustrated, detwinning proceeds through shearing of the twinned lattice relative to matrix (see Figure 24b). The lattice shear is taken place opposite to the shear direction of the lattice twin. Consider that the detwinning process is proceeded in the opposite direction of twinning by a lattice shear of the same magnitude of the twinning shear strain, γ, as shown in Figure 24c. It is obvious that, during the detwinning process, ideally, only a half of the lattices can shear from one position (having mirror plane symmetry) to the other position (loss of mirror plane symmetry). Thus, only those lattices having an angle less than 90 degree inclined to the externally applied shear stress are able to shear (*shearable*), otherwise, they are unable to be sheared (*unshearable*) as also illustrated in Figure 24c. Clearly, only these two possibilities exist under shear stress and the detwinning process is due to the lattice shear of the shearable portion of the twined lattices. Twinning plane K_1, shear strain γ and shear direction η_1 are intrinsically related to the lattice structure as well as the twinning type.

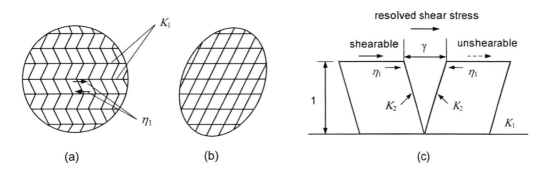

(a) (b) (c)

Figure 24. Schematic illustration of the detwinning process. (a) Lattice twins formed thermally as a result of phase transformation and (b) detwinning through shear along shear direction of twins. (c) Under given loading mode, only half of the paired lattice (twinned lattice) can be (need to be) sheared in order to detwin.

Details of the analysis of the detwinning process can be found from [Zheng & Liu 2002]. The analysis result agrees that the anisotropy of detwinning process in textured SMAs is responsible for the anisotropy of their mechanical behavior. The anisotropy of detwinning process is due to a combination of texture distribution and twinning type. In general, the stronger is the texture the stronger is the anisotropy of the detwinning process. The resolved shear stress along the shear direction of the thermally formed twins is the driving force for their detwinning. As soon as the resolved shear stress in some grain(s) of the twinned lattice reaches the critical value, the detwinning process will begin. The barrier stress of the detwinning process is dependent on the twinning type and is a function of its twinning shear strain.

The results further suggest that two types of detwinning processes exist, *domino detwinning* and *assisted detwinning*. These two types of detwinning mechanisms define the macroscopically observed mechanical behavior of the material. *Domino detwinning* is characterized by a self-propagating manner of detwinning under constant external load, i.e., when the resolved shear stress for detwinning reaches a critical value for some favorably oriented martensite twins, detwinning of these twins will take place. Along with this initial detwinning, neighboring martensite twins of less favorably oriented are triggered to detwin due to increase in the localized internal stress. This detwinning process will continue until a limit of the orientation is reached where the increase in the internal stress will be unable to further induce the detwinning of the remaining twins. In this case, an increase in the external load is required for a further detwinning process to take place and this type of detwinning is termed as *assisted detwinning*.

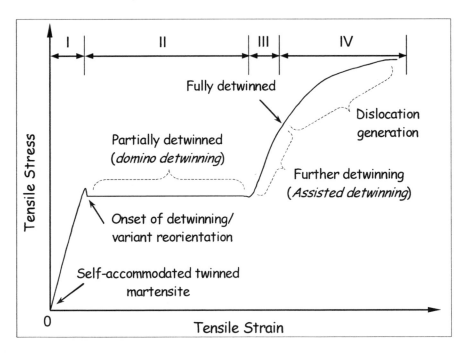

Figure 25. Current understanding of the martensite deformation process in shape memory alloys. Thermally formed martensite in SMAs consists of mainly self-accommodated lattice twins. Under tension, the lattice twins are detwinned leading to macroscopic deformation up to 8% strain. Further deformation is realized through dislocation generations.

Domino detwinning is responsible for the stress-plateau, while the *assisted detwinning* takes place at higher stresses beyond the stress-plateau region. Along with the *assisted detwinning* process, dislocation generation will take place as soon as a critical resolved shear stress is reached. In highly textured SMAs, the *domino detwinning* is strong along certain directions and longer plateau-strains can be expected. This prediction agrees well with the experimental observations (Figure 21). Since a higher plateau-strain corresponds to a higher shape recovery strain, *domino detwinning* is thus expected to promote shape recovery. The results in Figure 12 and Figure 21 can thus be explained. Tension promotes *domino detwinning*, while compression results in *assisted detwinning*; tension along RD promotes *domino detwinning* while tension along TD leads to *assisted detwinning* in the materials presently studied. Figure 25 summarizes the deformation procedures of martensitic SMA under tension.

Dynamic Deformation of SMA

The rate dependence of the detwinning mechanism is of significant interest for both application and for fundamental understanding the shape memory phenomenon. The unique combination of novel properties provides SMAs with the potential to be used as a primary actuating mechanism for devices with dimensions in the micro-to-millimeter range (especially devices requiring large forces over relatively large displacements). Three factors control the actuation speed namely, detwinning speed, heating rate and cooling rate. For actuation mechanisms using thin-film SMAs, in which heating and cooling rates can be sufficiently high with the help of novel design, the speed limit of the detwinning process is thus of primary importance. So far, it has been difficult to determine experimentally the speed limit of the detwinning process.

The authors and collaborators have studied the rate dependence of the martensitic deformation under both dynamic compression and dynamic tension [Liu *et al.* 1999f; 2002b,c]. Dynamic compression of a NiTi bar under strain rate up to 3000 s^{-1} shows no significant difference in deformation mechanism if compared to specimens deformed under quasi-static compression mode. However, since the deformation mechanisms in SMA bar are different between tension and compression along axial direction, the results on high rate compression can not provide relevant information on martensite detwinning mechanism. As shown in previous sections, different from under tension, under compression the deformation mechanism is mainly associated with dislocation generation rather than a detwinning process. Thus, an experimental research on the deformation under dynamic tension was conducted. Standard servohydraulic testing and a tension Kolsky bar are used to subject the SMA to tensile deformations at high strain rates [Liu et al. 2002b]. The highest strain rate achieved was 300 s^{-1}, which was also the highest strain rate ever used to study SMAs under tension.

Dog-bone shaped specimens were prepared from a NiTi bar having initial diameter of 13mm. The diameter of the middle portion of the specimens was 4mm. Several samples were deformed under dynamic tension at strain rates up to 300 s^{-1} and strain amplitudes up to 15%. The results of both the quasi-static and dynamic tests are shown in Figure 26a. Several dynamic tests have been repeated and the stress-strain curves obtained are essentially identical. Figure 26b highlights the initial stages of the deformation. In the stress-strain curves at a strain rate of 300 s^{-1}, a stress-plateau (Region II) that provides up to 4% strain is visible. Within the stress-plateau region the stress level differs moderately for strain rates of 300 s^{-1} and 150 s^{-1}, and it is higher than that observed in a quasi-static test (Figure 26c) [Liu *et al.* 2002c]. The difference between the high-rate and quasi-static data in Region II is not very significant if considering the large difference in the strain rates, and the same can be said of Region III. However, when the deformation strain reaches Region IV, the difference in stress levels between the dynamic and quasi-static curves increases more obviously with increasing strain amplitude as can be seen in Figure 26a. It is known that in the initial deformation stage

(Region II), the deformation mechanism involves mainly a detwinning process. Beyond about 10% strain, in the present material, dislocation mechanisms are most likely to operate. Correspondingly, the difference in stress levels between dynamic and quasi-static tests increases, which is a general deformation feature of metals associated with dislocation mechanisms [Nicholas 1980; Stouffer & Dame 1996]. For example for low-carbon steels at about 4% strain, the stress level under a strain rate of 300 s^{-1} is nearly twice as high as that during quasi-static deformations [Rajendran & Bless 1985].

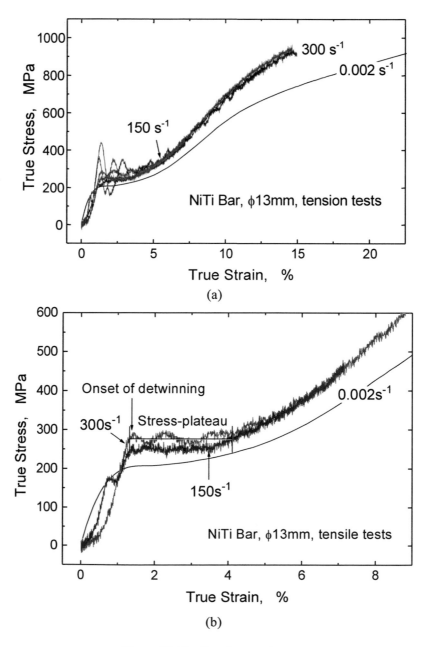

Figure 26. Continued on next page.

(a)

(b)

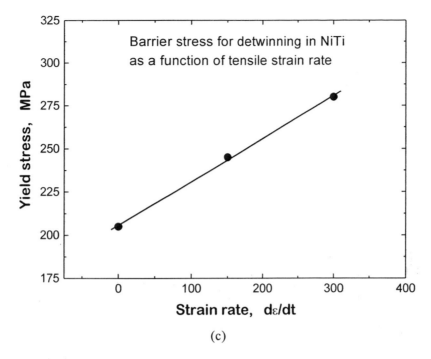

(c)

Figure 26. Comparison of the stress-strain curves of martensitic NiTi SMAs under tension at strain rates between 0.002 s^{-1} and 300 s^{-1}. (b) Enlarged portion of the initial deformation stages in (a). (c) Yield stress of martensite in NiTi as a function of strain rate.

The results shown in Figure 26 indicate that (i) the stress-plateau still exists even at a strain rate as high as 300s^{-1} and (ii) the deformation process that takes place in Region II is not as sensitive to strain rate as dislocation process taking place at higher strain amplitude (Region IV). Note that during dynamic deformation the specimen's temperature may increase significantly because insufficient time is available for heat conduction.

For specimens deformed to both 7% and 15% strains, a shape recovery process is observed during the first heating process after the deformation (Figure 27). A shape recovery strain of about 3.6 % is observed for both specimens. During the subsequent cooling and 2nd thermal cycles, a two-way memory effect (TWME) has been recorded for both specimens, being 0.9% strain for 7% deformed sample and 2.5% for samples deformed to 15% strain. These results are identical to that observed in a highly textured NiTi SMA under quasi-static deformations [Liu et al. 1999c].

It has been suggested that the TWME is related to the formation of certain dislocation networks that guide the growth of the martensite variants to preferential orientations. Detwinning is a necessary condition for having shape memory but not sufficient for obtaining a TWME. An internal stress field due to the presence of dislocation networks is needed to generate a two-way memory effect by directing the growth of martensite variants. However, excessive deformation will decrease the TWME [Liu et al. 1999c]. A weak detwinning process and formation of high density of dislocations will result in poor shape recovery and poor TWME. A maximum shape recovery and TWME were obtained when the material was deformed to about 12% under tension.

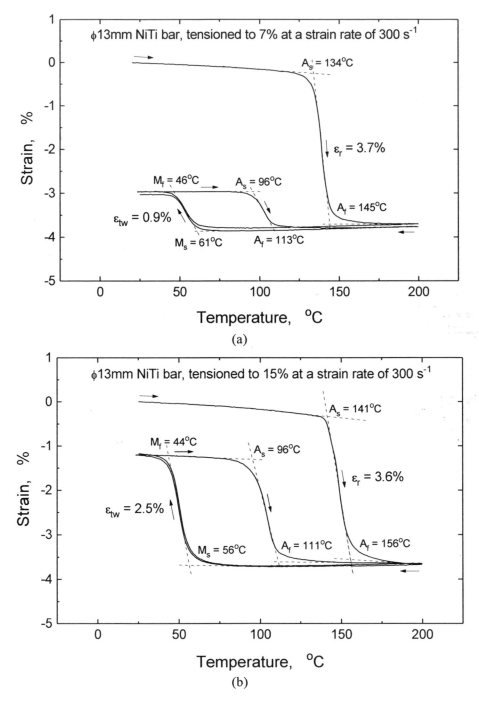

Figure 27. Shape recovery of the dynamically deformed samples during subsequent heating as determined by TMA. Upon the following thermal cycles, a two-way memory effect has occurred in (a) sample A deformed to 7% strain and, (b) sample B deformed to 15% strain.

As shown in Figure 26b, the flat plateau strain is only about 3.5% in the present material. Thus, a very high shape recovery strain and TWME strain should not be expected from the present material. The TMA results show that the characteristics of shape recovery and TWME in dynamically deformed samples are very similar to that of quasi-statically deformed samples. Thus, one can expect that the deformation details remain nearly the same for both high and low rate deformation. Detwinning has taken place during dynamic deformation, and the dislocation formation was insignificant when deformed to 7% strain.

As shown by a flat stress-plateau during dynamic deformation (Figure 26b), a "*domino detwinning*" process may have been involved. The initial detwinning process is able to progress at a rather high rate under constant load without significant involvement of dislocation processes.

References

Bhattacharya, K. and Kohn, R. V. (1996). Symmetry, texture and the recoverable strain of shape-memory polycrystals. *Acta Mater.*, **44**, 529-542.

Buehler, W. J. and Cross, W. B. (1969). 55-nitinol, unique wire alloy with a memory, presented at a technical session of the Non-Ferrous Division Meeting of The Wire Association, April 3-4, 1968, at Huntsville, *Ala*, USA. Published in June 1969.

Buehler, W.J. and Wiley, R.C. (1961). Report Noltr 61-75 (AD 266607) U.S. *Naval Ordnance Laboratory.*

Burkart, M.W. and Read, T.A. (1953). Diffusionless phase change in the Indium-Thallium system. Trans AIME J. *METALS J. Metals,* **197**, 1516-1524.

Chang, L.C. and Read, T.A. (1951). Plastic deformation and diffusionless phase changes in metals – the gold-Cadmium beta phase. J. Metals, *Trans. AIME*, **191** (47), 47-52.

Chen, C.W. (1957). Some characteristics of the martensite transformation, *Trans AIME J. Metals*, **209**, 1202-1203.

Christian, J.W. (1965). The Theory of Transformations in Metals and Alloys, *Oxford, Pergamon Press.*

Christian, J.W. (2002). The Theory of Transformations in Metals and Alloys (3rd edition), Oxford, *Pergamon Press.*

Hornbogen, E and Wassermann (1956). UBER DEN EINFLUSS VON SPANNUNGEN UND DAS AUFTRETEN VON UMWANDLUNGSPLASTIZITAT BEI DER BETA-1-BETA-"-UMWANDLUNG DES MESSINGS. *G. Z. Metallkunde*, **47**(6), 427-433.

Inoue, H., Miwa, N. and Inakazu, N. (1996). Texture and shape memory strain in TiNi alloy sheets. *Acta Mater.*, **44**, 4825-4834.

Knowles, K. M. and Smith, D. A. (1981). The crystallography of the martensitic transformation in equiatomic nickel-titanium. *Acta metall.*, **29**(1), 101-110.

Knowles, K. M. (1982). A high-resolution electron microscope study of nickel-titanium martensite. *Phil. Mag. A*, **45**(3), 357-370.

Liu, Y., Liu, Y. and Van Humbeeck, J. (1998a). Lüders-like deformation associated with martensite reorientation in NiTi. *Scripta mater.*, **38**, 1047-1055.

Liu, Y., Xie, Z. L., Van Humbeeck, J. and Delaey, L. (1998b). Asymmetry of stress-strain curves under tension and compression for NiTi SMAs. *Acta Materialia*, **46**, 4325-4338.

Liu, Y., Xie, Z. L., Van Humbeeck, J. and Delaey, L. (1999a). Some results on the detwinning process in NiTi shape memory alloys. *Scripta Materialia*, **41**, 1273-1281.

Liu, Y., Xie, Z. L., Van Humbeeck, J. and Delaey, L. (1999b). Deformation of shape memory alloys via twinned domain re-configurations. *Mater. Sci. Eng. A*, **273**-275, 679-684.

Liu, Y. N., Liu, Y. and Van Humbeeck, J. (1999c). Two-way memory effect developed by martensite deformation in NiTi. *Acta Materialia*, **47**, 199-209.

Liu, Y., Xie, Z. L. and Van Humbeeck, J. (1999d). Cyclic deformation of NiTi shape memory alloys. *Mater. Sci. Eng. A*, **273**-275, 673-678.

Liu, Y., Xie, Z. L., Van Humbeeck, J. and Delaey, L. (1999e). Effect of texture orientation on the martensite deformation of NiTi shape memory alloys. *Acta Materialia*, **47**, 645-660.

Liu, Y., Li, Y. L., Ramesh, K.T. and Van Humbeeck, J. (1999f). High rate deformation of martensitic NiTi shape memory alloy. *Scripta Mater.*, **41**, 89-95.

Liu, Y., Xie, Z. L., Van Humbeeck, J., Delaey, L. and Liu, Y. N. (2000). On the deformation of twinned domain in NiTi shape memory alloys. *Phil. Mag. A*, **80**, 1935-1953.

Liu, Y. (2001). Detwinning process and its anisotropy in shape memory alloys. In Smart Materials, Wilson, A.R. and Asanuma, H., Editors, *Proceedings of SPIE*, **4234**, 82-93.

Liu Y. (2002a). On the Detwinning Mechanism in Shape Memory Alloys. In Proceedings of IUTAM Symposium on Mechanics of Martensitic Phase Transformation in Solids, Ed. Q-P Sun, *Kluwer Academic Publishers*, Dordrecht, ISBN 1-4020-0741-8, pp. 37-44.

Liu, Y., Li, Y. L. and Ramesh, K. T. (2002b). Rate dependence of deformation mechanisms in shape memory alloy. *Phil. Mag. A*, **82**, 2461-2473.

Liu, Y., Li, Y. L., Xie, Z. L. and Ramesh, K. T. (2002c). Dynamic deformation of shape memory alloy: evidence of domino detwinning? *Phil. Mag. Let.*, **82** 511-517.

Liu, Y. and Xie, Z. L. (2003). Twinning and detwinning of <011> type II twin in shape memory alloy. *Acta mater.*, **51**(18), 5529-5543.

Liu, Y. and Xie, Z.L. (2006). The rational nature of type II twin in NiTi shape memory alloy, invited paper for a special issue of Journal of Intelligent Material Systems and Structures, accepted.

Lu, Z. K. and Weng, G. J. (1998). A self-consistent model for the stress–strain behavior of shape-memory alloy polycrystals. *Acta Mater.*, **46**, 5423-5433.

Melton, K. N. (1990). Ni-Ti Based Shape Memory Alloys. In: Duerig, T.W., Melton, K. N., Stöckel, D. and Wayman, C. M., editors. Engineering Aspects of Shape Memory Alloys, p. 21. London, Butterworth-Heinemann.

Miyazaki, S., Otsuka, K. and Wayman, C. M. (1989a). Shape memory mechanism associated with the martensitic transformation in Ti-Ni alloys. I. Self-accommodation. *Acta metall.*, **37**, 1873-1884.

Miyazaki, S., Otsuka, K. and Wayman, C. M. (1989b). Shape memory mechanism associated with the martensitic transformation in Ti-Ni alloys. II. Variant coalescence and shape recovery. *Acta metall.*, **37**, 1885-1890.

Mohamed, H. A. and Washburn, J. (1977). Deformation behaviour and shape memory effect of near equi-atomic NiTi alloy. *J. Mater. Sci.*, **12**, 469-480.

Mulder, J. H., Thoma, P. E. and Beyer, J. (1993). Anisotropy of the shape memory effect in tension of cold-rolled 50.8 Ti 49.2 Ni (at.%) sheet. *Z. Metallkd.*, **84**(7), 501-508.

Nicholas, T. (1980). Dynamic tensile testing of structural materials using a split Hopkinson bar apparatus, AFWAL-TR-80-4053, Materials Laboratory, Wright-Patterson AFB, OH.

Nishida, M., Yamauchi, K., Itai, I., Ohgi, H. and Chiba, A. (1995a). High resolution electron microscopy studies of twin boundary structures in B19' martensite in the Ti-Ni shape memory alloy. *Acta metall. mater.*, **43**, 1229-1234.

Nishida, M., Ohgi, H., Itai, I., Chiba, A. and Yamauchi, K. (1995b). Electron microscopy studies of twin morphologies in B19′ martensite in the Ti-Ni shape memory alloy. *Acta metall. mater.*, **43**, 1219-1227.

Nishida, M., Hara, T., Chiba, A., Hiraga, K. (1998). Transformation electron microscopy of twins in Ti-Ni ad Ti-Pd shape memory alloys. In: Inoue, K., Mukherjee, K., Otsuka, K. and Chen, H., editors. Displacive Phase Transformations and Their Applications in Materials Engineering. *The Minerals, Metals & Materials Society*, pp. 257-266.

Onda, T., Bando, Y., Ohba, T. and Otsuka, K. (1992). Electron microscopy study of twins in martensite in a Ti-50.0 at%Ni alloy. *Mater. Trans. JIM*, **33**, 354-359.

Otsuka, K. and Ren, X. (2005). Physical metallurgy of Ti-Ni-based shape memory alloys. *Progress in Materials Science*, **50**(5), 511-678.

Owen, W. S. (1975). Shape memory effect and applications: an overview. In: Perkins J, editor. Shape Memory Effects in Alloys, The Metallurgical Society of AIME, New York, *Plenum Press*, 1975, 305-325.

Ölander, A. (1932a). An electrochemical investigation of solid Cadmium-Gold alloy. *J. Am. Chem. Soc.*, **54**, 3819-3833.

Ölander A., (1932b). The crystal structure of AuCd. *Zeit. Für Kristol.*, **83**, 145.

Rajendran, A.M. and Bless, S.J. (1985). High strain rate material behavior. AFWAL-TR-85-4009, Materials Laboratory, Wright-Patterson AFB, OH.

Shu, Y. C. and Bhattacharya, K. (1998). The influence of texture on the shape-memory effect in polycrystals. *Acta Mater.*, **46**, 5457-5473.

Stouffer, D.C. and Dame, L.T. editors, (1996). *Inelastic Deformation of Metals*. New York, John Wiley & Sons.

Suoninen, E. J. (1954). An investigation of the martensitic transformation in metastable beta brass. MIT thesis.

Thamburaja, P., Pan, H. and Chau, F. S. (2005). Martensitic reorientation and shape-memory effect in initially textured polycrystalline Ti–Ni sheet. *Acta Materialia*, **53**, 3821-3831.

Wayman, C. M. (1975). Deformation, mechanisms and other characteristics of shape memory alloys. In: Perkins J., editor. Shape Memory Effects in Alloys. The Metallurgical Society of AIME, New York, *Plenum Press*; 1975, p. 1.

Wayman, C. M. and Duerig, T. W. (1990). An introduction to martensite and shape memory. In: Duerig, T.W., Melton, K. N., Stöckel, D. and Wayman, C. M., editors. *Engineering Aspects of Shape Memory Alloys*. London, Butterworth-Heinemann, p. 3.

Xie, Z. L., Liu, Y. and Van Humbeeck, J. (1998). Microstructure of NiTi shape memory alloy due to cyclic deformation. *Acta Materialia*, **46**, 1989-2000.

Xie, Z. L. and Liu, Y. (2004). HRTEM study of <011> type II twin in NiTi shape memory alloy. *Phil. Mag.*, **84**, 3497-3507.

Zhao, L. (1997). Ph.D. thesis, University of Twente, Enschede, *The Netherlands*.

Zheng, Q. S. and Liu, Y. (2002). Prediction of the detwinning anisotropy in textured NiTi shape memory alloy. *Phil. Mag. A*, **82**, 665-683.

In: Progress in Smart Materials and Structures
Editor: Peter L. Reece, pp. 67-113

ISBN: 1-60021-106-2
© 2007 Nova Science Publishers, Inc.

Chapter 4

SMART IONIC POLYMER-METAL COMPOSITES: DESIGN AND THEIR APPLICATIONS

Sangki Lee[a], Kwang J. Kim[a]*,
Hoon Cheol Park[b] and Il-Seok Park[a]
[a]University of Nevada, Reno, NV, USA
[b]Konkuk University, Seoul, South Korea

Abstract

Currently, IPMCs (Ionic Polymer-Metal Composites)--as biomimetic actuators and sensors--are being rigorously studied because of their enormous potential for engineering applications in medical, electrical, mechanical, and aerospace engineering. IPMCs have been considered as suitable, soft actuators/sensors to develop biologically-inspired smart structures and artificial-muscle systems since they can create large deformations under low input voltages and generate sensor signals under proper mechanical deformations. This article describes general aspects and recent progress in the application of IPMCs and also deals with several design examples in detail. We introduce fundamentals, manufacturing techniques, actuation principles, and electrical properties of typical IPMCs. Further, the equivalent bimorph beam model is introduced, and several design examples--such as flapping wing, IPMC diaphragm, ZNMF (Zero-Net-Mass-Flux) pump, valveless micropump, and muscle-like linear actuators--are presented.

Introduction

Some of the reasons why electroactive polymers (EAPs) are becoming an increasingly studied actuator technology is due to their large electrically induced strains (bending or extensional), low density, ease of processing, and mechanical flexibility that offer advantages over traditional electroactive materials [1]. Generally, EAPs have the ability to induce strains that are two-to-three orders of magnitude greater than the movements possible with rigid and fragile electroactive ceramics. EAP materials have higher response speeds, lower densities

---

*. E mail address: kwangkim@unr.edu. (Corresponding author, Associate Professor and Director of AMPL)

and improved resilience when compared to shape memory alloys. They have shown great promise towards the field of actuators with respect to soft actuation and biomimetics prompting further research.

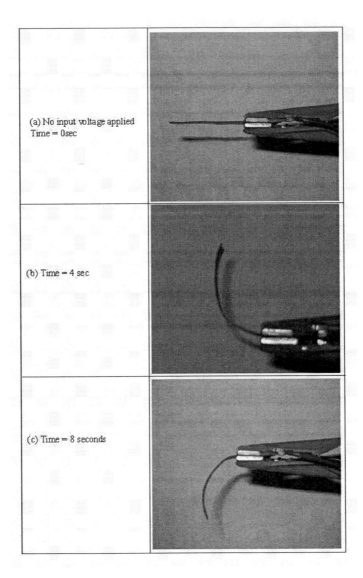

Figure 1. Typical IPMC placed under an applied AC voltage step function input of 3.0 volts at a frequency of 0.25Hz. (*top*) no voltage applied, (*middle*) positive polarity, and (*bottom*) negative polarity (from ref. [7]).

Ionic Polymer-Metal Composites (IPMCs) are a unique polymer transducer that when subjected to an imposed bending stress exhibits a measurable charge across the chemically and/or physically placed effective electrodes of the electro-active polymer. IPMCs are also known as bending actuators capable of large bending motion when subjected to a low applied electric voltage across the metalized or conductive surface (see Figure 1). The voltage found across the IPMC under an imposed bending stress is one to two orders of magnitude smaller than the voltage required to replicate the bending motion input into the system. This leads into the observation that the material is quite attractive by showing inclination for possible transduction as well as actuation [2-6].

There are various characteristics associated with the uniqueness of IPMCs. The transducer material has a necessity to be hydrated by a polar solvent (typically water or ionic liquids) and if the level of hydration is altered, the subsequent ionic conductivity and transductive qualities are altered as well. In Figure 2, the effect of water contents are presented in terms of cyclic voltagram at varying water content. Also, surface effects related to the platinized surface (effective electrodes) of the IPMC due to reactions caused by reduction/oxidation can result in mass transfer at the membrane surface in transduction measurement. Although this is not the dominating factor in transduction/actuation of the material, it has a finite contribution that alters performance of the material on many levels.

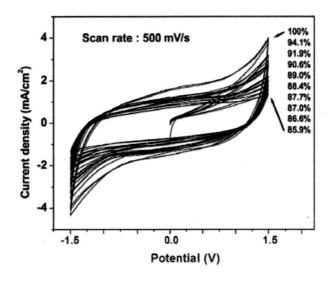

Figure 2. Cyclic voltammetry for IPMC with varying degrees of hydration (from ref. [8]).

IPMC Fabrication

The basis for the fabrication of an IPMC typically begins with an ion exchange membrane (IEM), which provides the bulk of the material in polymer form. This is due to the unique nature of the fixed ionomeric perfluorinated polymer backbone of the IEM that is permeable to cations but impermeable to anions (in case of cation exchange materials). Different membranes can be used, the most common being Nafion™ from DuPont, USA or Flemion from Asahi, Japan. It is called ion exchange membrane because the protons, in the case of Nafion, H^+ can be exchanged with other cations.

If H^+ ions are exchanged with metal ions that are then reduced by appropriate reducing agents, a sandwich structure is produced with the top and bottom being effective electrodes composed of the metal ion that was introduced into the IEM. The two surfaces being metallic are highly conductive and can be varied and manipulated depending on the chemical process. A micrograph showing the cross section of a typical IPMC sample is shown in Figure 3. This illustrates the sandwich like appearance of the material where the metal composite comprises the outside layers and the polymer matrix is the center.

Actuation Principle

Upon application of an electric field upon the membrane such as shown in Figure 1, there is an observed fast bending motion towards one of the electrodes. This can be explained by examining the overall solvent (=water in general) flux within the complex network of the polymer. Generally, when an electric field is applied across the membrane, there is a fast bending motion towards one of the electrodes followed by a slow relaxation towards the other electrode in most cases. It is commonly believed that this bending motion is the result of an electro-osmotic pressure induced by the migration of hydrated cations and free water from these cluster regions through the nanochannels towards the electrode side (in the case of a positive polarity, the cathode side). This results in a pressure differential causing a bending stress that translates into bending towards the electrode (in the case of a positive polarity, the anode side) as is illustrated in Figure 4 where shows the bending motion of an IPMC placed in cantilevered configuration placed under a positive electric field.

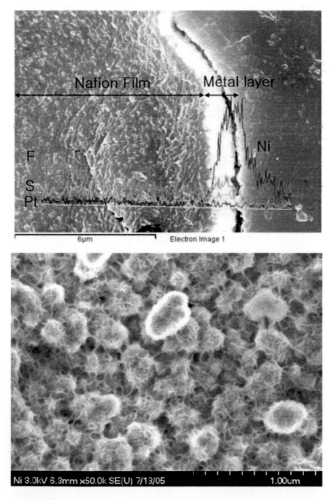

Figure 3. (*top*) An EDS analysis on a cross sectional view of the nickel doped IPMC with scanning elements of Ni, S, F, and Pt. (*bottom*) A micrograph of the nickel doped crystals on an ion exchange membrane. Nickel particles, forming as effective electrodes, are connected by needle shape branches.

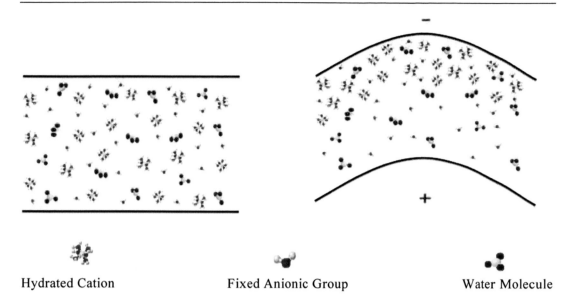

Hydrated Cation   Fixed Anionic Group   Water Molecule

Figure 4. The cation migration results in a transfer of charge across the membrane and a subsequent current is developed. The IPMC experiences bending stress (*left* to *right*).

Electrode Compositing

There are various chemical reduction processes that can be used to produce the desired deposition of metal particles (or conductive medium) into the surface of the membrane. The reason why noble metals are commonly used is the acidic and corrosive environment in the IEM (Nafion™). Other metals (or conducting mediums) that can be used include palladium, silver, gold, carbon, graphite, and nanotubes. The most popular method is by chemical platinization of a Nafion™ IEM [9].

For an IEM to function as an IPMC, the surfaces of Nafion™ needs to be metallized. There are several ways for depositing electrodes of the membrane surface. The electrodes may be: (a) mechanically pressed onto the IEM, (b) electrochemically deposited, or (c) chemically deposited. An anionic metal ion in a solution in contact with one face of the IEM is reduced by a reducing agent which diffuses through the membrane from a solution in contact with the opposite face. Liu et al. [10] used a two-step, impregnation-reduction procedure in which a cationic metal ion is first impregnated into the Nafion and is subsequently reduced *in situ* in a following operation. Metal ions are dispersed throughout the hydrophilic regions of the polymer, and are subsequently reduced to the corresponding metal state. Liu also reported that for an ideal electrode, it should have the following characteristics, (a) good interparticle contact for low electronic resistance, (b) porous structure so that mass transfer through the deposit will not be a limiting factor, (c) large electrode-IEM contact area to provide electrochemically active surface area, and (d) metal-film deposition predominantly within the Nafion in a thin layer adjacent to the membrane surface. The average thickness of a Nafion membrane is approximately 0.20 mm. The deposition penetration depth of the metal particles into the ionomer is typically on the order of 10 μm so there will be no contact between the two electrodes. It can be seen that the importance of the electroding process determines the effective capacitance, resistance and current density of the IPMC, and the resulting responsive behavior of the material. An SEM micrograph of a typical IPMC that has been platinized through a chemical impregnation/reduction process as described above was previously shown in Figure 3. As would be anticipated by the chemical impregnation/reduction process, the surface appears nonuniform. It can clearly be seen that the density of the particles is much greater at the surface of the membrane. This is due to the reducing agent's inability to penetrate deep into the membrane. The density decreases at the dispersion distance is increased.

Electrical Properties of Typical IPMCs

Figures 5 and 6 show the typical impedance behavior for a Pt electroded IPMC and an equivalent circuit model for a single layer IPMC, respectively. By observing the impedance characteristics of Figure 5, it is clear to see the capacitive/resistive behavior of the IPMC. This allows for the material to be modeled using an RC circuit. The basic unit cell model is also shown within Figure 6. The RC circuit connected in parallel on the top of the unit cell represents the surface capacitance and resistance of the particular IPMC and the RC circuit on the bottom is that of the bottom surface electrode. There is an additional resistance placed in the center of each individual unit cell. This resistance represents the resistance that the solvent flux sees during actuation due to migration of hydrated cations and free water through very narrow pores and channels. Resistors were placed between the unit cells to emulate the contact resistance incurred by placing consecutive IPMCs upon each other as well as the increased resistance observed as the length of the IPMC specimen is transversed.

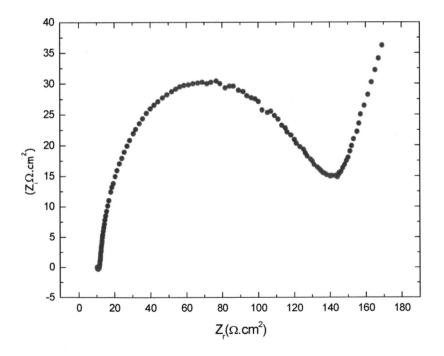

Figure 5. Electrochemical impedance behavior of IPMC (from ref. [11]).

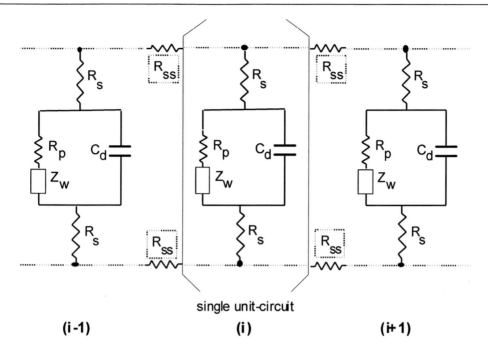

single unit-circuit

(i-1)   (i)   (i+1)

Figure 6. Equivalent circuit schematic representing IPMCs (from ref. [6]).

The ion exchanging capability of the base material (ion-exchange polymer, such as Nafion[TM]) is what allows for the resulting force of an IPMC when an electric field is applied. This is due to the unique ionic nature of the fixed perfluorinated polymer backbone, which is permeable to cations but impermeable to anions (in case of cation exchange materials). There is a general consensus that, when an electric field is applied, hydrated cations migrate towards the cathode, resulting in an electro-phoretic pressure that leads the IPMC to bend to the anode side. This behavior has been studied and documented numerously. The equivalent circuit model can be extended to handle multi-layer IPMCs for combined senor-actuator applications. In an effort to investigate the behavior exhibited in the results shown in Figure 7 in the form of a large voltage drop across the middle IPMC of the sandwich structure, a cyclic voltammogram analysis of the system was performed for a single IPMC, a double layered IPMC and a three layered IPMC. A solution form of Nafion (for casting) was used to adhere the layers of IPMC together as illustrated. The Nafion was applied in an effort to lower the large resistance observed. The cyclic voltammetry for the various IPMC multi-layered systems is also shown and the resulting deviation from the single layer IPMC in current density. It can be seen that the difference in current density increases as the number of layers of IPMCs increases. This is most likely due to the increased number of layers preventing charge transfer between electrodes hence increasing the resistance at the interlayer surfaces and a reduction in the platinum oxidation phenomena that occurs in a single layered IPMC.

Figure 8 shows the output potential and resulting displacement of a single Ni doped IPMC. It shows chronopotentiometric signals responding to a square pulse input of a 2 second duration at ±200 mA. The output potential range of about ±3.7V was observed. Seemingly, a capacitive response can be seen.

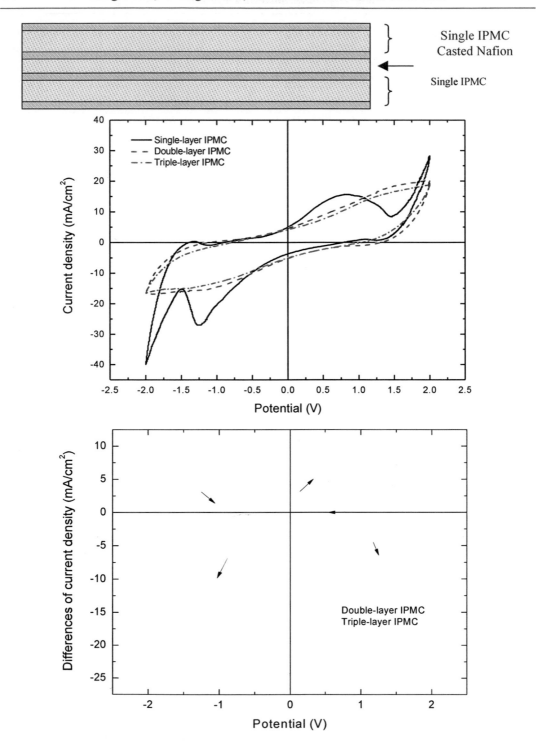

Figure 7. (top) Schematic diagram of a double layered IPMC used in experimental investigation. The middle Nafion layer was formed from its solution form. (middle) Potential-current profiles of various configurations of multi-layered IPMCs. (bottom) The deviation in current density for the double layer and triple layer IPMC configurations. The difference in current density (mA/cm^2) was calculated by the current density of single layer IPMC minus that of the double of triple layer. (from ref. [11]).

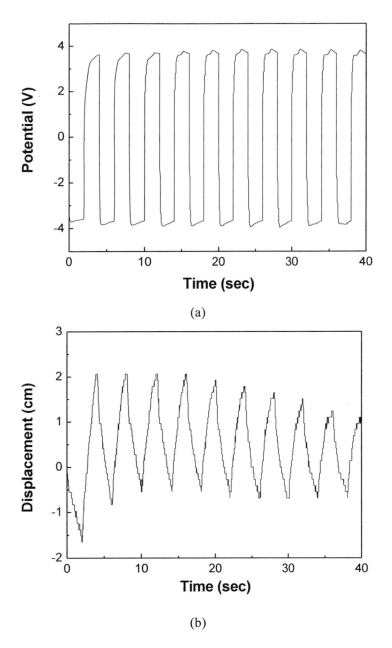

(a)

(b)

Figure 8. Output potential (a) and tip-displacement (b) of a Ni doped IPMC responding to a square wave input of -200/+200 mA in a cantilever configuration (1 x 5 cm IPMC sample).

Equivalent Bimorph Beam Model and Design Examples

Equivalent Bimorph Beam Model for IPMC Actuators

Models for IPMCs are usually categorized by *physical* models, *black box* models, and *gray box* models [4,12]. Physical models, based on physics and chemistry, were proposed by Shahinpoor et al. [13] and Nemat-Nasser et al. [14, 15]. They may predict behaviors of

IPMCs quite accurately, but the models require many adjusted physical quantities that are non-trivial in experiments. On the other hand, black box models, presented by Kanno et al. [16] and Xiao et al. [17], are convenient and are able to estimate curvatures and actuation displacements of IPMCs; however, they are only applicable to specific shapes and boundary conditions from which the models are empirically extracted. Alternatively proposed by Kanno [18] et al. and DeGennes et al. [19], gray box models, can more easily predict the behaviors of IPMCs with general shapes and boundary conditions and are based on several physical laws combined with limited physical properties [20] determined experimentally or analytically.

With statically analyzed IPMC actuators, an equivalent bimorph beam model is herein presented as a gray box model. The equivalent bimorph beam model is combined with important physical properties of IPMC's: Young's modulus and electro-mechanical coupling coefficient--determined from the bimorph beam equations and measured force-displacement data of a cantilevered IPMC actuator. The finite element analysis (FEA), along with the estimated physical properties of IPMC, is used to accurately predict the force-displacement relationship of an IPMC actuator-the data can be used to effectively design many engineering devices of interest.

One estimating method used to determine the equivalent Young's modulus of IPMCs and their bending stiffness is through the bimorph beam model and force-displacement relationships of an IPMC actuator. It should be noted that the elastic modulus of IPMC contributes to its bending stiffness and is quite difficult to be experimentally determined due to the complexity of the morphology of IPMC under the electric field [6].

Measurement of the Force-Displacement Relationship

The electro-mechanical properties considered are determined by the force-displacement relationship of an IPMC actuator. The method used to measure these properties is graphically depicted in Figure 9. An IPMC actuator is cantilevered at one end, and the other end is constrained, as shown in Figure 9 (a). The reaction force (or actuation force) at the right end of the actuator is generated with application of an electrical field. We measure the reaction force with a small force transducer. After the right-end constraint is moved up with amount of the displacement s, the same test is conducted. In this way, the actuation force corresponding to the end displacement s can be measured, as illustrated in Figure 9 (b). Finally, without the constraint, the free end displacement can be determined. Following this procedure, the force-displacement relationship was obtained as shown in Figure 10. Figure 10 shows the measured force-displacement relationship for an IPMC actuator for two- and three-volt inputs across the IPMC. Region **A** and **B** in Figure 10 include the maximum actuation forces and the maximum displacements, respectively. The specimen tested was a Nafion[TM]-based IPMC in Li^+ form and plated with platinum. The length of the IPMC actuator was 20 mm, with a width of 5 mm, and a thickness of 0.3 mm.

The curve fitted relationships are as follows:

$$F(s) = -0.4712s + 1.3766 \quad \text{for 2 V}$$
$$F(s) = -0.5150s + 1.7708 \quad \text{for 3 V}$$

(1)

where s stands for the tip displacement, and F is the measured reaction force (or actuation force) at the right-end constraint. Thus, the force F for $s = 0$ (region **A** in Figure 10) is the reaction force for the case shown in Figure 9(a), and the measured displacement s when $F = 0$ (region **B** in Figure 10) stands for the tip displacement without the right-end constraint.

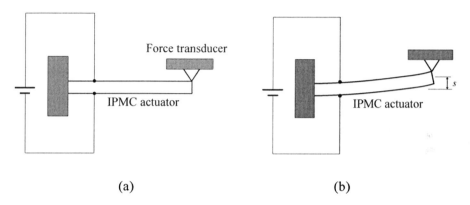

(a)   (b)

Figure 9. Test set-up for the force-displacement relationship (from ref. [21]).

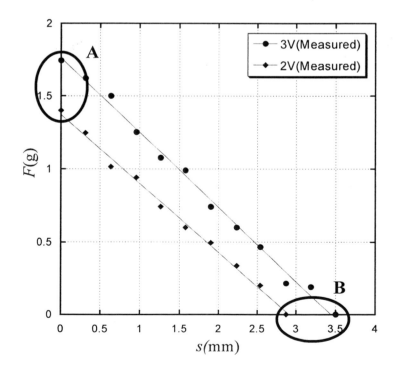

Figure 10. Force-displacement relationship of an IPMC actuator (from ref. [21]).

Equivalent Bimorph Beam Modeling

In the equivalent bimorph beam model, shown in Figure 11, it can be assumed that an IPMC has two virtual layers with the same thickness. Under an imposed electric field across the material, the upper layer and the lower layer of an IPMC expand or contract, opposing each other, to produce IPMC's bending motion.

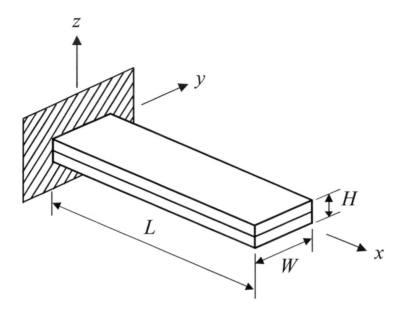

Figure 11. Equivalent bimorph beam for an IPMC actuator (from ref. [21]).

Generally, for a cantilevered bimorph beam with a sandwiched elastic layer, the tip displacement s and the blocking force F_{bl} can be written as follows [22]:

$$s = \frac{3L^2}{2H} \frac{(1+b)(2b+1)}{ab^3 + 3b^2 + 3b + 1} d_{31} E_3 \qquad (2)$$

$$F_{bl} = \frac{3WH^2 E}{8L} \frac{2b+1}{(b+1)^2} d_{31} E_3 \qquad (3)$$

where a and b are the Young's modulus ratio and the thickness ratio, respectively, of the sandwiched elastic layer and the outer layers, E_3 the electric field, and d_{31} the electromechanical coupling coefficient in which the subscripts 1 and 3 stand for the x-direction and the z-direction, respectively. In our assumption for an IPMC actuator, there is no sandwiched elastic layer; therefore, the ratios are zeros ($a = b = 0$). From Equation (2), a relationship between the input voltage V and the induced tip displacement s of an IPMC based on an equivalent bimorph beam model can be written as follows:

$$s = \frac{3L^2 d_{31}}{2H^2} V . \qquad (4)$$

By using Equation (4), the equivalent electro-mechanical coupling coefficient d_{31} is expressed as follows:

$$d_{31} = \frac{2sH^2}{3L^2V}(= d_{32})$$

(5)

where the subscript 2 stands for the y-direction. Substituting the measured tip displacement for $F = 0$ (region **B** in Figure 10) into the s in Equation (5), the electro-mechanical coupling coefficient can be obtained for a given input voltage. Meanwhile, in the equivalent bimorph beam model, the Young's modulus E contributing to the bending stiffness of an IPMC is determined by using the blocking force Equation (3). Disregarding the sandwiched elastic layer, a relationship between the input voltage V and the blocking force F_{bl} of a cantilevered IPMC actuator can be written as follows:

$$F_{bl} = \frac{3WHEd_{31}}{8L}V .$$

(6)

In order to determine the equivalent Young's modulus, Equation (6) is rewritten as follows:

$$E = \frac{8LF_{bl}}{3WHd_{31}V} ,$$

(7)

where F_{bl} is the measured blocking force when $s = 0$ (region **A** in Figure 10), and d_{31} the equivalent electro-mechanical coupling coefficient calculated by Equation (5).

The equivalent Young's modulus, computed by Equation (7), may be interpreted as the real elastic modulus that contributes to the bending stiffness of an IPMC since it has been driven from the measured force-displacement data of the IPMC actuator. Equivalent properties of an IPMC for the finite element analysis were calculated from Equations (5) and (7). The equivalent electro-mechanical coupling coefficients for 2- and 3-volt inputs were 2.153×10^{-7} and 1.750×10^{-7} m/V, respectively. It should be noted that since the thickness of the IPMC strip was 0.3 mm, the electric fields, which are directly related to the electro-mechanical coupling coefficients (2.153×10^{-7} and 1.750×10^{-7} m/V), are 6.7 and 10 V/mm for 2- and 3-volt inputs, respectively. The equivalent Young's moduli for 2- and 3-volt inputs were 1.133 and 1.158 GPa, respectively. (This IPMC was heavily loaded with platinum [~6 vol % Pt]. The Pt loading technique was uniquely designed to enhance the humidity control of IPMC [23]). The estimated equivalent electro-mechanical coefficients are different for the each of the input voltages. The coefficients demonstrate the nonlinear behavior characteristic of an IPMC as opposed to the applied electric potential (or electric field). The difference in the equivalent Young's moduli for the different input voltages seems to be caused by the different water contents [15] that may occur during the tests.

Verification of the Equivalent Bimorph Beam Model

The force-displacement relationship of an IPMC actuator shown in Figure 10 was numerically reproduced to validate the equivalent bimorph beam model. For the numerical analysis, a commercial finite element analysis program, MSC/NASTRAN [24], was used. To calculate the actuation force (or reaction force) at the tip of an IPMC, contact elements were used (see Figure 12), and a nonlinear static analysis was performed. Since MSC/NASTRAN doesn't support the electro-mechanical coupling analysis, a thermal analogy technique [25, 26] was utilized to implement the electro-mechanical coupling effect into the finite element model. In the thermal analogy technique, the electro-mechanical coupling coefficient d_{31} is converted into the thermal expansion coefficient α_1 by using Equation (8):

$$\alpha_1 = \frac{d_{31}}{t}\frac{\Delta V}{\Delta T}, \tag{8}$$

where t is the thickness of an layer across which an electric potential is applied, ΔV the electric potential, and ΔT the temperature difference. Since an IPMC can be assumed to have two virtual layers in the equivalent bimorph beam model, as shown in Figure 11, the thickness t is defined as $H/2$. The electric potential applied across a virtual layer is half of the total voltage that is applied across the total thickness of an IPMC. More details and verifications for the thermal analogy technique can be found in elsewhere [25, 26].

A finite element model for an IPMC (with the Poisson's ratio assumed to be zero) is shown in Figure 12(a), and its deformed shape with tip displacement, s =2.54 mm, for a 3 volt input, is presented in Figure 12(b). The numerically reproduced force-displacement relationship of an IPMC is plotted in Figure 13 and is compared with the measured data. For each input voltage, the estimated actuation forces of an IPMC by the finite element analysis agree with the measured data for various tip displacements. When $F=0$, relative errors between the calculated values and the measured data for 2 and 3 V are 2.8 % and 3.7 %, respectively.

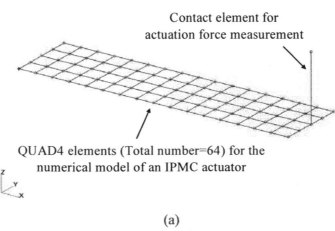

(a)

Figure 12. Continued on next page.

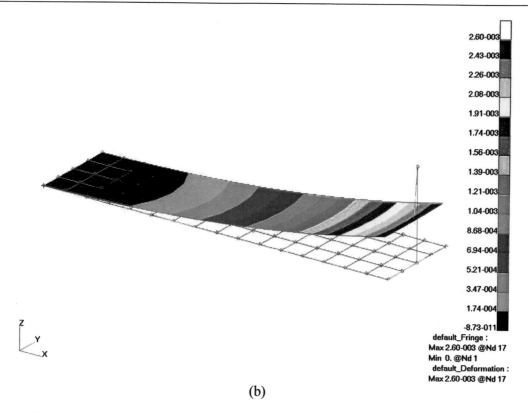

(b)

Figure 12. (a) A finite element model of an IPMC actuator and (b) its deformed shape with tip displacement, S =2.54 mm for 3 V input (from ref. [21]).

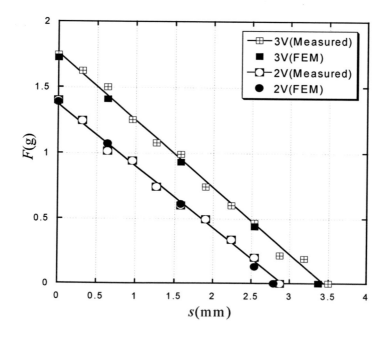

Figure 13. Reproduction of force-displacement relationship of an IPMC actuator (from ref. [21]).

Flapping Wing

Pneumatic and motor-driven actuators have been widely adopted and used in aerospace applications as well as other industrial robot systems. However, these actuators are not plausible for use in small or micro-scale flying and locomotive vehicles due to their large payload and system-complexity. Furthermore, they are not suitable for mimicking bird or insect wings' flapping motion. Using micro-robots to create flapping flight is attractive due to their maneuverability that could not be obtained by conventional, fixed or rotary wing aircraft. An electroactive polymer, IPMC is a good candidate for the flapping motor because it is lightweight and can create large deformation under low electric voltage input. Birds/insects wings can generate lift and thrust at the same time during flapping motion, since the wing can be actively flapped and twisted during the flapping motion [27]. To mimic the motion, the artificial flapping mechanism should also be able to simultaneously create flap and twist. Also, the width of a bird wing tip is pointed compared with the remaining parts of the wing. This reduces drag during the up-/down-strokes of the wing and strength of the tip vortex. Thus, the actuation mechanism and shape are both important for successfully mimicking a bird wing. The IPMC can generate this particular motion if it has a specially designed plan form.

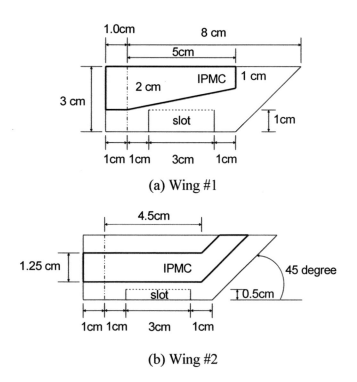

(a) Wing #1

(b) Wing #2

Figure 14. Flapping wing and patterns of the IPMC actuators (from ref. [28]).

Since the flapping wing must create a twisting motion as well as a bending up and down motion for thrust generation [27], the IPMC actuators have non-symmetric shapes, as shown in the two wings in Figure 14. The wing shapes and dimension of the wings are also shown in Figure 14. Note that the areas of the IPMC actuators in the two wings are kept the same for fair comparison in the actuation displacement analysis. The wing itself is made of a thin plastic film.

The numerical deformation analysis has been conducted to determine the shape of the IPMC actuator such that the designed wing can produce maximum bending and twisting motion at the same time. Deformation of the wing has been estimated by using the equivalent bimorph beam model and MSC/ NASTRAN with the thermal analogy. For the finite element modeling, QUAD4 elements were used for both Wing #1 and #2 as shown in Figure 15(a) and Figure 16(a). Material properties and thicknesses for the calculations are shown in Table 1.

Table 1. Material properties and thicknesses.

	Young's Modulus (GPa)	Poisson's Ratio	$d_{31} = d_{32}$ (m/V)	t (mm)
IPMC	1.158*	0.487	1.750×10^{-7}	0.3
Plastic film	0.1	0.3	N/A	0.1

* Pt (~6%) heavy IPMC

(a) Finite element model

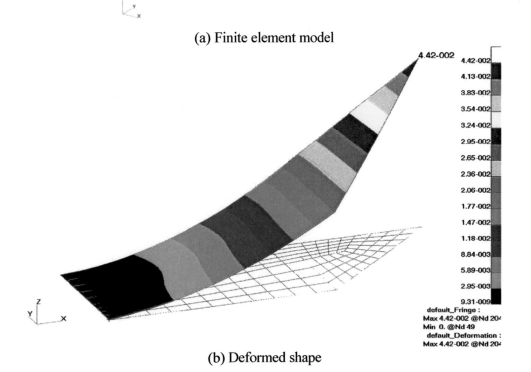

(b) Deformed shape

Figure 15. Flapping simulation for flapping wing #1 (Total number of elements=164) (from ref. [28]).

The flap-up displacement and twisting angle at the tip under 3 V (i.e. $E_3 = 10$ V/mm) are calculated as 4.42 cm and 3.4 degrees, for Wing #1, and 4.68 cm and 9.1 degrees, for Wing #2, respectively. The deformed shapes are shown in Figure 15(b) and Figure 16(b) for Wing #1 and #2, respectively. Wing #2 is the better design for the flapping wing in terms of a twisting angle. It should be noted that our analysis is based on the linear elasticity and thus may not accurately predict the actuation displacement. However, the present approach provides a simple but effective design tool to determine the shape of the IPMC actuator for a specific purpose.

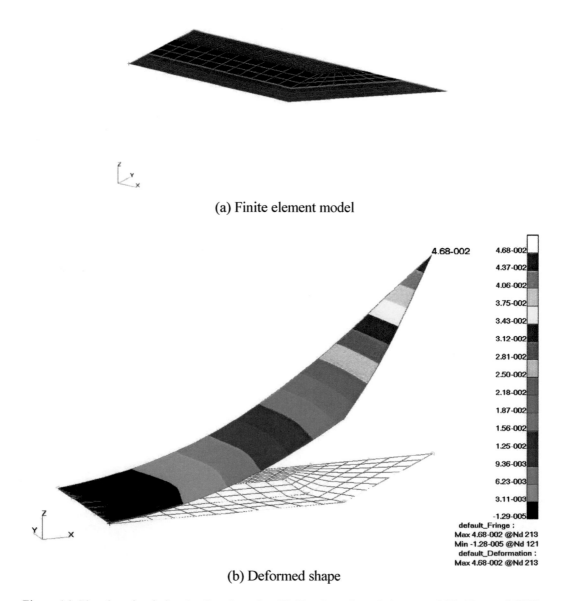

(a) Finite element model

(b) Deformed shape

Figure 16. Flapping simulation for flapping wing #2 (Total number of elements=202) (from ref. [28]).

Pump Devices

Effectively designed IPMC diaphragms can be used as motors for zero-net-mass-flux (ZNMF) pumps and micropumps. After investigating several characteristics of optimal IPMC diaphragms, a ZNMF pump and a micropump were designed and analyzed.

IPMC Diaphragm

Deformations of circle-shaped IPMC diaphragms were estimated for the circle-shaped and ring-shaped electrodes, respectively. Through parametric studies, an electrode shape was chosen for the optimal diaphragm, which generates maximum stoke volume. In order to show the effectiveness of the circle-shaped diaphragm, its stroke volume was compared to that of square-shaped diaphragm maintaining the same actuator area. In addition, both the normal mode analysis and pressure effect on the selected IPMC diaphragm are introduced.

Circle-shaped Electrode vs. Ring-shaped Electrode

Parametric studies on two kinds of electrodes for circle-shaped diaphragms with a radius of 10 mm were conducted with the material properties and thicknesses (see Table 2). The material properties E and d_{31} of IPMC were determined through the equivalent bimorph beam model. The elastic modulus of Nafion[TM] and Poisson's ratios were obtained from literature [8,21].

Table 2. Material properties and thicknesses for an IPMC diaphragm.

	Elastic Modulus (GPa)	Poisson's Ratio	$d_{31} = d_{32}$ (m/V)	t (mm)
IPMC	1.158	0.487	1.750×10^{-7}	0.2
Nafion[TM]	0.05	0.487	N/A	0.2

Figure 17 shows shapes of the two electrodes in a 1/4 finite element model of diaphragms. The left model is (a) the circle-shaped electrode and the right one is (b) the ring-shaped electrode. The total number of elements (Quad4 [24]) used for each model was 400.

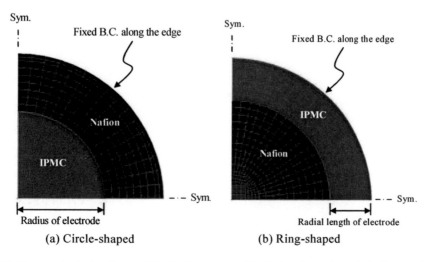

Figure 17. Shapes of electrodes for IPMC diaphragms (1/4 finite element model) (from ref. [29])

Symmetry boundary condition was applied to the vertical and horizontal lines, and fixed boundary condition to the outside edge. As shown in Figure 17, each IPMC diaphragm consists of an IPMC part and a Nafion^{TM} part. Therefore, when a voltage is applied on the IPMC part, the vertical interface between the IPMC and the Nafion^{TM} can rotate easier to produce a large bending deformation since Nafion^{TM} has a lower elastic modulus than the IPMC.

Under an applied 2 V input (i.e. $E_3 = 10$ V/mm) and fixed boundary conditions along the outside edge, the center displacements of the diaphragms were calculated with variations of the electrode length in the radial direction. The calculated results are provided in Figure 18. For the diaphragm with the circle-shaped electrode in which the radius of electrode was 8.5 mm, the maximum center displacement was 0.966 mm. The maximum center was only 0.686 mm for the diaphragm with the ring-shaped electrode in which the radial length of electrode was 5.5 mm. Such a comparison shows the circle-shaped electrode is more efficient than the ring-shaped electrode in terms of deformation. The parametric studies suggest that there is an optimal radius and radial length to each electrode to create the maximum deflections. For the two optimal electrode cases (radius, 8.5 mm, for the circle-shaped electrode; radial length, 5.5 mm for the ring-shaped electrode), stroke volumes were calculated from the deformed shapes as shown in Figure 19. Note that the diaphragm with the circle-shaped electrode is bent upward while the diaphragm with the ring-shaped electrode is bent downward for the same electrical input of 2 V. Considering 2 V AC input, calculated stroke volumes (Definition of the stroke volume is described in Figure 24) for the circle-shaped and the ring-shaped electrode cases were 261 and 104 μl, respectively.

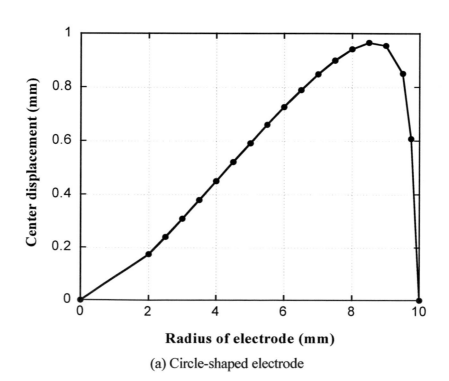

(a) Circle-shaped electrode

Figure 18. Continued on next page.

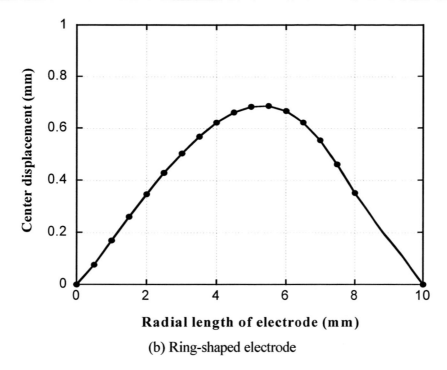

(b) Ring-shaped electrode

Figure 18. Center displacements of IPMC diaphragms for each electrode case (from ref. [29]).

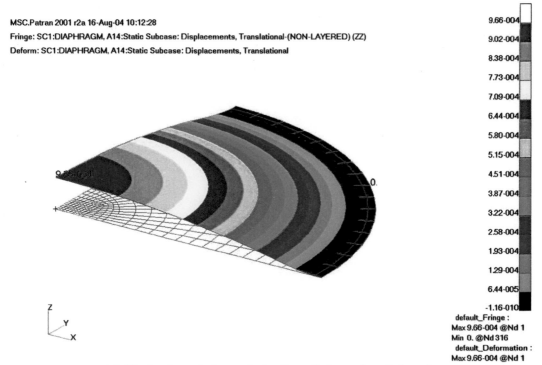

(a) Circle-shaped electrode (radius of electrode = 8.5 mm)

Figure 19. continued on next page.

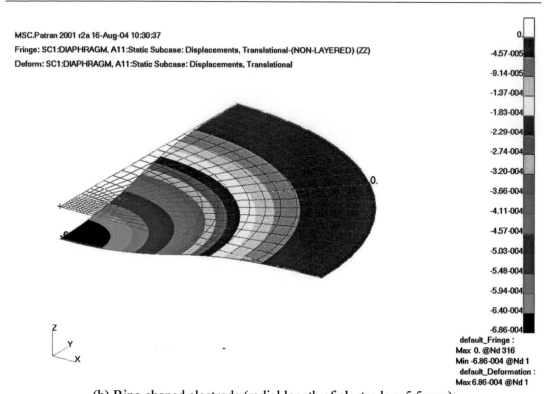

MSC.Patran 2001 r2a 16-Aug-04 10:30:37
Fringe: SC1:DIAPHRAGM, A11:Static Subcase: Displacements, Translational-(NON-LAYERED) (ZZ)
Deform: SC1:DIAPHRAGM, A11:Static Subcase: Displacements, Translational

0.
-4.57-005
-9.14-005
-1.37-004
-1.83-004
-2.29-004
-2.74-004
-3.20-004
-3.66-004
-4.11-004
-4.57-004
-5.03-004
-5.48-004
-5.94-004
-6.40-004
-6.86-004
default_Fringe :
Max 0. @Nd 316
Min -6.86-004 @Nd 1
default_Deformation :
Max 6.86-004 @Nd 1

(b) Ring-shaped electrode (radial length of electrode = 5.5 mm)

Figure 19. Deformed shapes of IPMC diaphragms (from ref. [29]).

Circle-shaped Diaphragm vs. Square-shaped Diaphragm
The shape-effect of the diaphragm on the stoke volume was investigated. A square-shaped diaphragm was modeled and analyzed to calculate its center displacement and stroke volume, and the results of the square-shaped diaphragm were compared with those of circle-shaped

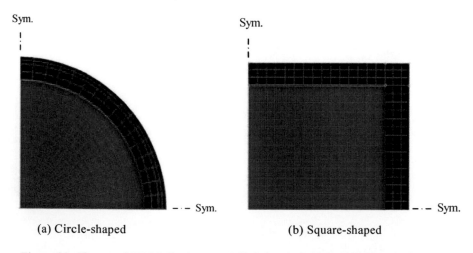

(a) Circle-shaped   (b) Square-shaped

Figure 20. Shapes of IPMC diaphragms (1/4 finite element model) (from ref. [29]).

diaphragm. The areas of IPMC and Nafion™ were maintained for the circle-shaped diaphragm--the optimal case (i.e. the radius of diaphragm: 10 mm and the radius of electrode: 8.5 mm). Figure 20 shows shapes of the two diaphragms (1/4 finite element model). For the finite element modeling, 400 and 324 elements (Quad4 [24]) were used for the circle-shaped diaphragm and the square-shaped diaphragm, respectively.

Under a 2 V input, for the square-shaped diaphragm, the calculated center displacement and stoke volume were 0.760 mm and 196 μl, respectively. Note that the calculated values for the circle-shaped diaphragm are 0.966 mm and 261 μl, respectively. From the results it is evident that the use of the circle-shaped diaphragm is advantageous over the square-shaped one in order to generate larger stroke volumes.

Normal Mode Analysis
The normal mode analysis was performed for the optimal circle-shaped diaphragm with the circle-shaped electrode (radius of electrode: 8.5 mm). For the calculation, the density of Nafion™ was 2.078×10^3 kg/m³ [14] and that of IPMC was assumed to be 2.5×10^3 kg/m³. Figure 21 shows the first and second mode shapes of the diaphragm. The computed first (i.e. fundamental) and the second natural frequencies are 430 and 1,659 Hz, respectively. If we consider the driving frequency range of the IPMC diaphragm as less then 40 Hz, the calculated fundamental frequency is much higher than the driving frequency range. Therefore, the resonance will not affect the stroke volume in that driving frequency range.

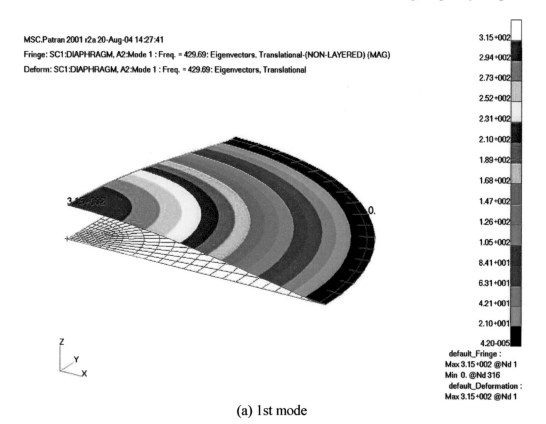

(a) 1st mode

Figure 21. Continued on next page.

The results imply that we can linearly control the flow rates of an IPMC-driven pump within the driving frequency (~40 Hz) of interest since the flow rate of a pump linearly increases with increasing driving frequency in a low-driving frequency range [30].

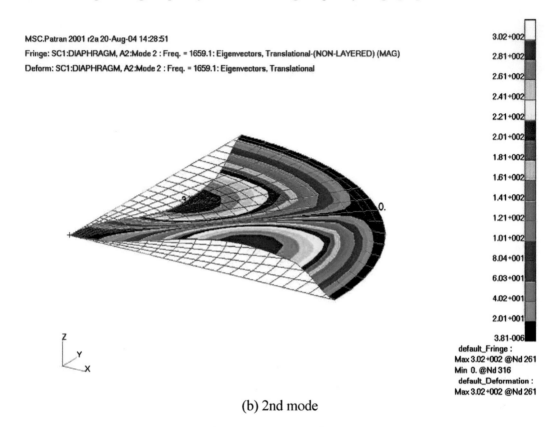

(b) 2nd mode

Figure 21. Normal mode analysis results for an IPMC diaphragm (radius of electrode = 8.5 mm) (from ref. [29]).

Pressure Effect on the Stroke Volume

The external pressure effect on the circle-shaped diaphragm having a circle-shaped electrode was investigated. The external pressure could be considered as the chamber pressure of a pump. In order to numerically calculate the stroke volume under external pressure, uniform pressures were applied to the FE model for the optimal IPMC diaphragm as shown in Figure 22.

Figure 23 shows the estimated half stroke volumes of the optimal circle-shaped diaphragm under applied pressures and a 2 V input. In Figure 23, the "opposite direction" indicates the half stroke volume when the diaphragm's bending and the pressure are in the opposing directions. The "same direction" indicates the half stroke volume when the diaphragm's bending and the pressure are in the same direction. According to the result, in the case of "opposite direction," the IPMC diaphragm could generate a half-stroke volume under the pressure until around 2,300 Pa.

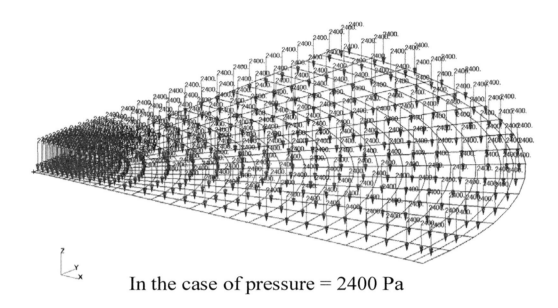

In the case of pressure = 2400 Pa

Figure 22. Diaphragm under uniform pressure (1/4 FEA model) (from ref. [29]).

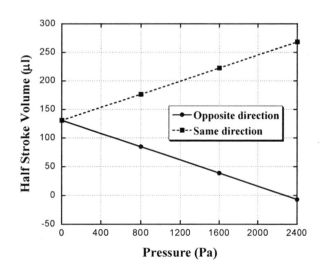

Figure 23. Half stroke volumes of the circle-shaped IPMC diaphragm (from ref. [29]).

ZNMF Pump for Flow Control

Currently, zero-net-mass-flux (ZNMF) pumps, also called synthetic jet pumps, are being widely studied for flow control devices [31]. A number of numerical and experimental works [32-36] show that oscillating excitations close to the separation points on airfoils can increase the lift-to-drag ratio of the wings because of the added flow momentum by the ZNMF pump delays separations toward the trailing edges. Seifert et al. [32, 33] conducted a number of experimental tests on oscillatory excitation and suggested effective values for the non-dimensional frequency and the momentum coefficient increase the aerodynamic performance of airfoils. By using numerical simulations, Shatz et al. [34] showed that flow separation could be delayed by periodic excitation through a slot near the leading edge. Gilarranz et al. [35] developed a DC motor-driven synthetic jet actuator for a specific wing and demonstrated its performance in a wind tunnel. Mallinson et al. [36] performed numerical studies to investigate a synthetic jet interacting with a laminar hypersonic boundary layer and reported that the areas of separated region were enormously reduced by the synthetic jet.

For the development of a ZNMF pump, piezoceramic actuators are commonly considered for actuating diaphragms [31, 37, 38] because piezoceramic material is reliable and easily acquired from manufacturers. However, since a piezoceramic actuator produces a small deformation under a high electric field, it is inadequate to generate enough stroke volumes for the boundary layer control of a wing. An ionic polymer-metal composite (IPMC) is a new emerging material used for actuating diaphragms in small size synthetic pumps, as IPMC actuators can generate a large bending deformation under a low input voltage (~2 V) and are easier to manufacture in a small size and operate in air [39].

An IPMC-driven ZNMF Pump with a Slot

Figure 24 shows a schematic of ZNMF pump with a slot for the flow control of a wing. Based on the previous results for the optimal circle-shaped IPMC diaphragm (i.e. radius of diaphragm: 10 mm and radius of electrode: 8.5 mm), a prototype IPMC-driven ZNMF pump with a slot was drawn in Figure 25. The inner and outer radii of the pump are 10 and 12 mm, respectively. The height is 5 mm. The area of slot is 1.5×10^{-6} m^2 (width: 0.5 mm, and length: 3 mm) and the outer radius of diaphragm is 12 mm. The effectiveness of the designed ZNMF pump for flow control of small air vehicles is evaluated in the following subsection.

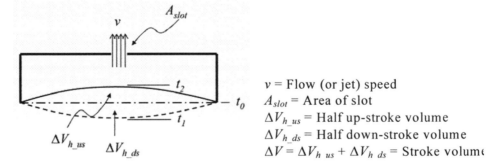

v = Flow (or jet) speed
A_{slot} = Area of slot
ΔV_{h_us} = Half up-stroke volume
ΔV_{h_ds} = Half down-stroke volume
$\Delta V = \Delta V_{h\ us} + \Delta V_{h\ ds}$ = Stroke volume

Figure 24. Schematic of an IPMC-driven ZNMF pump (from ref. [29]).

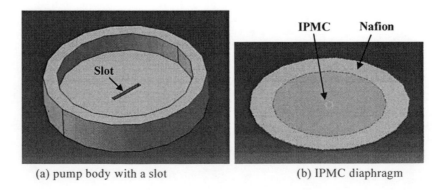

(a) pump body with a slot   (b) IPMC diaphragm

Figure 25. IPMC-driven pump with a slot (from ref. [29]).

Feasibility Study of the Designed ZNMF Pump

In order to verify the feasibility of the designed IPMC-driven ZNMF pump as a flow control device, the non-dimensional actuation frequency and the momentum coefficient [33] were calculated by considering the flow speed through the slot and the flight speed of a micro air vehicle (MAV).

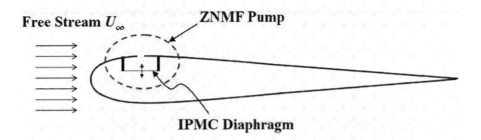

Figure 26. Schematic of an airfoil with an IPMC-driven ZNMF pump (from ref. [29]).

A wing body-type MAV, "SPOT" [40], was considered as a platform for the IPMC-driven ZNMF pump. The flight speed and chord length of "SPOT" are around 8~12 m/s and approximately 15 cm, respectively. The MAV flies in the air with a Reynolds number of 10^5 [40]. To prevent the wing from stalling at a high angle of attack, the ZNMF pump is usually mounted in the leading edge area [32-35], as shown in Figure 26.

The non-dimensional actuation frequency F^+ is calculated as follows [33]:

$$F^+ = f \frac{x_{te}}{U_\infty},\qquad(9)$$

where f is the driving frequency, x_{te} the distance from the actuator pump to the trailing edge, and U_∞ the free stream velocity. By substituting the flight speed and the distance from the actuator pump to the trailing edge of the considered MAV into U_∞ and x_{te}, in Equation (9), the non-dimensional frequency F^+ was estimated to be approximately 0.45, where the driving frequency f of the IPMC-driven ZNMF pump was set at 30 Hz and the distance x_{te} was assumed to be 0.15 m. According to Gilarranz et al. [35], F^+ should be on the order of ~1 to delay airflow separation on a wing. Also, Schatz et al. [34] suggested that around 0.5 under a low Reynolds number (1.6×10^5) is good for F^+ to increase the lift-to-drag ratio of a wing. The estimated value of $F^+ = 0.45$, from the present work, is an attractive value for the flow control of the considered MAV's wing. Also, the momentum coefficient C_μ is calculated by using Equation (10) [33]:

$$C_\mu = \frac{w(1 + T_\infty / T_j)v^2_{RMS}}{cU^2_\infty},\qquad(10)$$

where w is the slot width, T_∞ the free stream temperature, T_j the jet temperature at the slot, v_{RMS} the root mean square of jet velocity through the slot, c the airfoil chord, and U_∞ the free stream velocity.

If we assume that the free steam temperature and the jet temperature are the same, Equation (10) is rewritten as follows:

$$C_\mu = \frac{2w}{c}\left(\frac{v_{RMS}}{U_\infty}\right)^2. \tag{11}$$

Based on Figure 24 and the half stroke volumes of the optimal diaphragm under external pressure as shown in Figure 23, the average jet speed $v_{average}$, which is related to v_{RMS}, can be calculated. Since the chamber pressure of the ZNMF pump affects deformed shapes of the diaphragm, the pressure was estimated by using the Bernoulli's equation. According to the Bernoulli's equation, the amount of the pressure drop in the pump chamber could be considered as the dynamic pressure P_d, as follows:

$$P_d = \frac{1}{2}\rho_\infty U_\infty^2, \tag{12}$$

where U_∞ is the free stream velocity, and ρ_∞ the air density. By using Equation (12) and the flight speed of the MAV, the calculated pressure drop P_d in the pump chamber was 60.25 Pa, where the air density of 1.205 kg/m^3 was considered at sea level condition. From Figure 23, the stroke volume at 60.25 Pa was calculated to be:

$$\Delta V = \Delta V_{h\_us} + \Delta V_{h\_ds} = 134 + 127 = 261(\mu l).$$

Based on the assumptions of incompressible and inviscid flow, the average jet speed through the slot can be calculated by using the continuity equation:

$$A_{slot}v_{average} = Av, \tag{13}$$

where A_{slot} is the area of slot, $v_{average}$ the average flow speed at the slot during the half cycle time; $t = t_2 - t_1 = 1/(2f)$. Since Av can be considered as the stroke volume (ΔV) per unit time, as was demonstrated in Equation (13), the average jet speed for a given driving frequency f is calculated as follows:

$$v_{average} = \frac{\Delta V}{A_{slot}t} = \frac{2\Delta Vf}{A_{slot}}. \tag{14}$$

By using the calculated stroke volume, 261 μl (= 2.61×10^{-7} m^3) for the optimal circle-shaped diaphragm and the area of slot, 1.5×10^{-6} m^2, the calculated average jet speed $v_{average}$ was 10.44 m/s from Equation (14) under the driving frequency, 30 Hz. From the relationship between the root mean square value and the average value for a sinusoidal wave, $v_{RMS} = (\pi v_{average})/(2\sqrt{2})$, we can obtain the root mean square of jet speed v_{RMS} to be 11.6 m/s. Finally, substituting the calculated jet speed v_{RMS} into Equation (11), the momentum coefficient C_{μ} was calculated to be 0.009. Gilarranz et al. [35] recommended that the momentum coefficient should be at least 0.002 for any substantial effects on the flow. The calculated value 0.009 appears to be reasonable for the flow control of the considered MAV's wing. If the dimensions of the slot are altered, the value of C_{μ} can be adjusted to maximize the effect of ZNMF pump. To accurately understand the effect of the designed ZNMF pump on the MAV's wing, it should be evaluated using the computational fluid dynamics or experiments.

Valveless Micropump
Micropumps are attractive devices since they can be used for dispensing therapeutic agents, cooling microelectronic systems, developing micro total analysis systems (μ TAS), and propelling microspacecraft, etc. [41-43]. For such a variety of applications, many types of micropumps have been developed, but generally, they fall into two categories: mechanical micropumps (i.e. piezoelectric, electrostatic, thermopneumatic, magnetic, etc.) and non-mechanical micropumps (i.e. electroosmotic, electrophoretic, electrohydrodynamic, magnetohydrodynamic, etc.). In view of inlet/outlet mechanisms, categories of micropumps are also divided into with-valve micropumps and without-valve (or valve-less) micropumps [41, 42, 44]. Valve-less micropumps, using nozzle/diffuser elements, are easily created in small sizes and can avoid the wear and fatigue of moving parts.

In order to generate stroke volumes of mechanical-type micropumps, diaphragms are widely used [41, 42]. Piezoelectrically-actuated diaphragms [45, 46] usually produce high actuation forces and fast mechanical responses, but they need high input voltages. The diaphragms produce relatively small stroke volumes. Thermopuematically-actuated diaphragms [47] need low input voltages, generate high pump rates, and can be very compact, but high power consumption and long thermal time constants are the main disadvantages. Electrostatically-actuated diaphragms [48] have the merits of fast response time, micro-electro-mechanical system (MEMS) compatibility, and low power consumption, but small actuator stroke, degradation of performance, and high input voltage are the main deterrents to using this diaphragm. Electromagnetically-actuated diaphragms [49] have a fast response time, but they are not well compatible with MEMS and require high power consumption.

The ionic polymer-metal composite (IPMC) is a new, promising material used for actuating diaphragms in micropumps [50, 51]. It is anticipated that the micropump manufacturing process with IPMC is convenient and the manufacturing cost of the IPMC micropump is competitive to the other competing technologies described above.

Figure 27 shows a schematic of the IPMC-driven micropump with conical nozzle/diffuser elements. The flow rate of the micropump can be estimated by considering conical nozzle/diffuser elements at very low Reynolds numbers (~ 50). We consider the optimal IPMC diaphragm (i.e. radius of diaphragm: 10 mm, radius of electrode: 8.5 mm) as the actuating diaphragm of the micropump for pumping water and its driving frequency as 0.1 Hz.

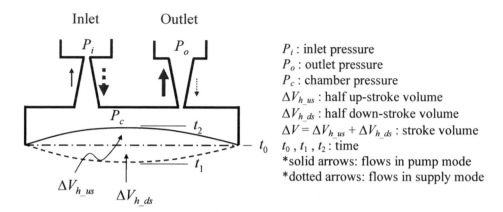

P_i : inlet pressure
P_o : outlet pressure
P_c : chamber pressure
ΔV_{h_us} : half up-stroke volume
ΔV_{h_ds} : half down-stroke volume
$\Delta V = \Delta V_{h\_us} + \Delta V_{h\_ds}$: stroke volume
t_0 , t_1 , t_2 : time
*solid arrows: flows in pump mode
*dotted arrows: flows in supply mode

Figure 27. A schematic of IPMC-driven micropump with nozzle/diffuser elements (from ref. [51]).

Flow Resistance Coefficients for Conical Nozzle/Diffuser Elements

Figure 28 shows the conical nozzle/diffuser elements where D is the diameter, v the flow speed, α the conical angle, L the length, Re the Reynolds number, and μ the kinematic viscosity. The subscript 0 and 1 indicate the small diameter part and the large diameter part, respectively. The subscript n and d stand for the nozzle and the diffuser, respectively. As shown in Figure 28, the same element can be considered as a nozzle or a diffuser in accordance with the flow direction.

Under low Reynolds numbers ($1 < Re < 50$) and conical angles ($\alpha < 40°$), the flow resistance coefficient of the diffuser can be written as follows [52, 53]:

$$\xi_d = \frac{A_d}{Re_d} \quad (A_d = \frac{20(D_1^2 / D_0^2)^{0.33}}{(\tan \alpha)^{0.75}}).$$ (15)

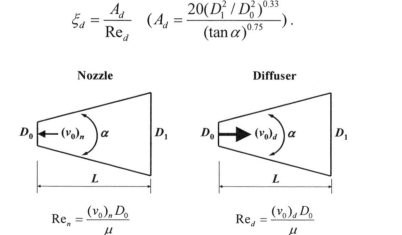

Figure 28. Conical nozzle and diffuser elements (from ref. [51]).

For the nozzle under low Reynolds numbers (1<Re<50) and conical angles (5° <α < 40°), the flow resistance coefficient can be expressed as follows [52, 53]:

$$\xi_n = \frac{A_n}{Re_n} \quad (A_n = \frac{19}{(D_0^2 / D_1^2)^{0.5}(\tan\alpha)^{0.75}}). \tag{16}$$

By using Equation (15) and (16), the ratio of flow resistance coefficients η is written for the conical nozzle/diffuser elements as follows:

$$\eta = \frac{\xi_n}{\xi_d} = \frac{A_n}{A_d} \frac{Re_d}{Re_n}. \tag{17}$$

Meanwhile, the flow resistance coefficients are related to the pressure drops across the diffuser and the nozzle as follows: [54]:

$$\Delta P_d = \frac{\rho(v_0)_d^2}{2} \xi_d, \tag{18}$$

$$\Delta P_n = \frac{\rho(v_0)_n^2}{2} \xi_n, \tag{19}$$

where ΔP_d and ΔP_n are the pressure drops across the diffuser and the nozzle, respectively, and ρ the density of fluid. Plugging Equations (15) and (16) into Equations (18) and (19), the pressure drops are rewritten at low Reynolds numbers, as follows:

$$\Delta P_d = \frac{\rho(v_0)_d^2}{2} \frac{A_d}{Re_d}, \tag{20}$$

$$\Delta P_n = \frac{\rho(v_0)_n^2}{2} \frac{A_n}{Re_n}. \tag{21}$$

If the inlet and outlet pressure P_i and P_o are can both be neglected compared to the chamber pressure P_c (see Figure 27 for the pressure notations), the pressure drops are to be $\Delta P_d = \Delta P_n = P_c$ [54], and from Equations (20) and (21), following equation is derived:

$$\frac{A_n}{A_d} = \frac{(v_0)_d^2}{(v_0)_n^2} \frac{Re_n}{Re_d}. \tag{22}$$

Since the ratio of Reynolds numbers for the nozzle and diffuser is $\mathrm{Re}_n / \mathrm{Re}_d = (v_0)_n / (v_0)_d$ (see the equations in Figure 28), Equation (22) can be rewritten as follows:

$$\frac{A_n}{A_d} = \frac{\mathrm{Re}_d}{\mathrm{Re}_n}, \tag{23a}$$

or

$$\frac{A_n}{A_d} = \frac{(v_0)_d}{(v_0)_n}. \tag{23b}$$

From Equations (17) and (23), the ratio of flow resistance coefficients is rewritten as follows:

$$\eta = \frac{\xi_n}{\xi_d} = \left(\frac{A_n}{A_d}\right)^2, \tag{24a}$$

or

$$\eta = \frac{\xi_n}{\xi_d} = \left(\frac{(v_0)_d}{(v_0)_n}\right)^2. \tag{24b}$$

According to Equations (15), (16), and (24a), the ratio η is determined by only the geometry of nozzle/diffuser elements at low Reynolds numbers. In addition, Equation (24b) can be directly obtained from Equations (18) and (19).

Figure 29(a) and (b) show the calculated ratio of flow resistance coefficient η, with respect to the diameter D_0, the conical angle α, and the length L of conical nozzle/diffuser elements. The coefficient ratio η decreases as the diameter D_0 increases; on the other hand, it increases as the conical angle α and the length L of nozzle/diffuser elements increase. If we consider only the efficiency of nozzle/diffuser elements, the smaller diameter D_0 with the larger conical angle α and the larger length L are better for the flow of low Reynolds numbers. Note that the diameter D_1 is 8.55 mm in the case of $D_0 = 2$ mm, $\alpha = 40°$, and $L = 9$ mm in Figure 29(b).

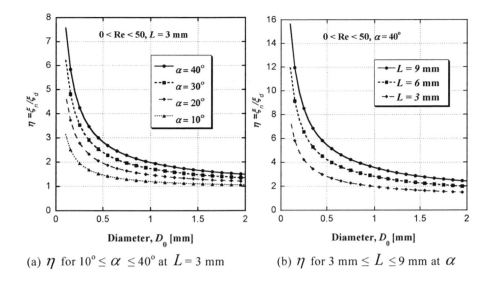

(a) η for $10° \leq \alpha \leq 40°$ at $L = 3$ mm   (b) η for 3 mm $\leq L \leq 9$ mm at α

Figure 29. Ratio of flow resistance coefficients of nozzle to diffuser (from ref. [51]).

Mean Output Flow Rate of the Micropump

If we consider the flow speeds through the nozzle/diffuser elements as averaged values, the stroke volume (see Figure 27 for the definition of stroke volume) during pump or supply modes is related to the flow speed, as follows:

$$\Delta V = (\Delta V_{out})_{outlet} + (\Delta V_{out})_{inlet} = \left\{ F_0 (v_0)_d \frac{T}{2} \right\}_{oulet} + \left\{ F_0 (v_0)_n \frac{T}{2} \right\}_{inlet} \quad (25a)$$

in pump mode
or

$$\Delta V = (\Delta V_{in})_{outlet} + (\Delta V_{in})_{inlet} = \left\{ F_0 (v_0)_n \frac{T}{2} \right\}_{oulet} + \left\{ F_0 (v_0)_d \frac{T}{2} \right\}_{inlet} \quad (25b)$$

in supply mode,

where $(\Delta V_{out})_{outlet}$ and $(\Delta V_{out})_{inlet}$ are the output volumes through the outlet and inlet during pump mode, respectively, $(\Delta V_{in})_{outlet}$ and $(\Delta V_{in})_{inlet}$ the input volumes through the outlet and inlet during supply mode, respectively. F_0 is the area of nozzle/diffuser at the small diameter part ($\pi D_0^2 / 4$), and T the period.

For both pump and supply modes, we can rewrite Equations (25a) and (25b) as follows:

$$\Delta V = F_0 \left\{ (v_0)_d + (v_0)_n \right\} \frac{T}{2} . \quad (26)$$

Note that $(v_0)_d$ of the outlet in pump mode is equal to $(v_0)_d$ of the inlet in supply mode, and $(v_0)_n$ of the inlet in pump mode is equal to $(v_0)_n$ of the outlet in supply mode.

The net output volume ΔV_{net} through outlet part during one period T is defined as follows:

$$\Delta V_{net} = (\Delta V_{out})_{outlet} - (\Delta V_{in})_{outlet} = F_0\{(v_0)_d - (v_0)_n\}\frac{T}{2} \qquad (27)$$

Defining the mean output flow rate Q as $\Delta V_{net}/T$, we can rewrite Equation (27) as follows:

$$
\begin{aligned}
Q &= F_0\{(v_0)_d + (v_0)_n\}\frac{1}{2}\left\{\frac{(v_0)_d - (v_0)_n}{(v_0)_d + (v_0)_n}\right\} \\
&= \frac{1}{T}F_0\{(v_0)_d + (v_0)_n\}\frac{T}{2}\left\{\frac{(v_0)_d/(v_0)_n - 1}{(v_0)_d/(v_0)_n + 1}\right\}
\end{aligned}
\qquad (28)
$$

By applying Equations (24b) and (26) into Equation (28), the mean output flow rate Q during one period T can be predicted as follow [53,54]:

$$Q = \frac{2\Delta V_h}{T}\frac{\sqrt{\eta} - 1}{\sqrt{\eta} + 1}, \qquad (29)$$

where ΔV_h is the half stroke volume ($\Delta V/2$).

Since Equations (15) and (16) are only valid under low Reynolds numbers ($1 < \mathrm{Re} < 50$), we should know the Reynolds numbers at the nozzle/diffuser elements for a valid prediction of the mean output flow rate. The flow speed through nozzle is always less than that through the diffuser at low Reynolds numbers. Therefore, only the flow speed through diffuser is calculated to predict Reynolds number in this study.

From Equations (23b) and (26), the average flow speed through diffuser is calculated as follows:

$$(v_0)_d = \frac{2\Delta V}{F_o T(A_d/A_n + 1)}. \qquad (30)$$

By using the flow speed of Equation (30), the Reynolds number of flow through diffuser is calculated as follows:

$$\mathrm{Re}_d = \frac{2\Delta V D_0}{\mu F_o T(A_d/A_n + 1)}. \qquad (31)$$

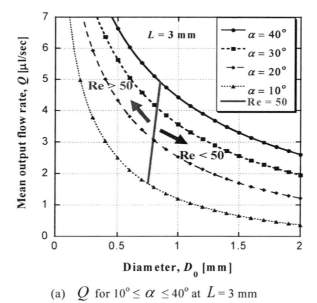

(a) Q for $10° \leq \alpha \leq 40°$ at $L = 3$ mm

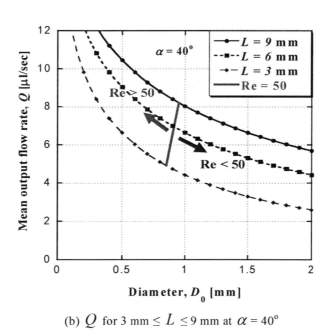

(b) Q for 3 mm $\leq L \leq 9$ mm at $\alpha = 40°$

Figure 30. Estimated mean output flow rate of IPMC actuator-driven micropump at low Reynolds numbers (from ref. [51]).

Figure 30(a) and (b) show the estimated mean output flow rate of the IPMC actuator-driven micropump. For the calculation of the flow rate, we chose the driving frequency of IPMC diaphragm, $f = 0.1$ Hz and used the half-stroke volume, $\Delta V_h = 130.6$ μl, of the optimal IPMC diaphragm. The kinematics viscosity μ used was 1.0×10^{-6} m^2/s for water at 20 $°$C. Since the flow rate estimation is only valid for the range of low Reynolds numbers (1 < Re < 50), the valid estimation limit of Re = 50 was marked in each graph. As shown in

Figure 30(a) and (b), the mean output flow rate Q increases as the diameter D_0 decreases, and it increases as the conical angle α and the length L of nozzle/diffuser elements increase. In Figure 30(b), if we consider $\alpha = 40°$ and $L = 9$ mm for the nozzle/diffuser elements, the mean output flow rate could be estimated as 8.2 μl /sec, where the diameter and Reynolds number are 0.95 mm and 50, respectively.

There are numerous design parameters for the IPMC actuator-driven micropump, including geometry, input voltage, driving frequency for the IPMC diaphragm, and also equations for the nozzle/diffuser elements. All of the design parameters should be adjusted and optimized for the development of an IPMC actuator-driven micropump for each specific application.

Muscle-Like Linear Actuator

Amongst the many types of EAP (Electro-Active Polymer) actuators [55], IPMC is considered as a plausible actuator suitable for use in biological and/or biomimetic applications [3,4]. However, the actuation principle of IPMCs is based on an electromechanically induced "bending" motion while, in nature, biological muscles utilize chemo-mechanical strain and stress-creating "linear" actuation. Herein, we attempt to investigate a strategy to create a linear motion out of bending based IPMC.

Several researchers introduced muscle-like linear actuators composed of IPMCs, which utilize bending motions of IPMCs to produce linear motions [3, 56-59]. Shahinpoor et al. [3] fabricated a platform-type linear actuator driven by eight IPMC legs; the team also created a bi-strip type linear actuator made with two IPMC strips. Jung et al. [56] manufactured AMuLA (Artificial Muscle-like Linear Actuator), which mimics the actin-myosin structure of biological muscles, and experimentally evaluated the actuation characteristics of it. Kamamichi et al. [57, 58] fabricated an elementary unit for an IPMC linear actuator and applied it to a waking robot. Recently, Malone et al [59] presented freeform fabricated IPMC actuators and suggested a concept of three-dimensional IPMC geometry which converts bending motion to linear motion.

Even though the above works were dedicated to developing linear actuators made with IPMCs, the behavior analyses and optimizations for the linear actuators were not conducted based on mathematical models for IPMC actuators. In this work, design, modeling, and optimization of a muscle-like linear actuator composed of IPMC and Nafion^{TM} are introduced.

Deformation Characteristic of an IPMC Strip

A basic deformation characteristic of an IPMC actuator strip is investigated. For estimating the deformation, the finite element method was conveniently utilized, based on the equivalent bimorph beam model for IPMC actuators. In order to validate the finite element model for an IPMC actuator strip, numerically-calculated bending displacement was compared with an analytical beam solution.

Finite Element Analysis of an IPMC Strip
An IPMC strip (LxWxH; 10x2x0.2 mm) was modeled and analyzed to investigate its deformation characteristic. Supported ends were considered for the boundary condition as shown in Figure 31(a) and the deformed shape under 2V input (i.e. $E_3 = 10$ V/mm) is shown in Figure 31(b). For the finite element model of the IPMC strip, 320 elements (Quad4 [24]) were used. The Poisson's ratio was assumed to be zero for the comparison with the analytical beam solution.

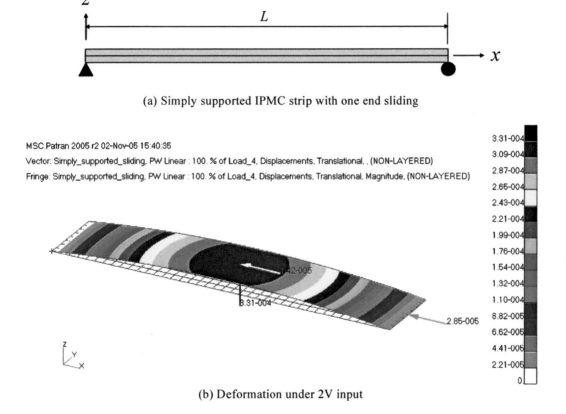

(a) Simply supported IPMC strip with one end sliding

(b) Deformation under 2V input

Figure 31. Basic deformation characteristic of an IPMC strip (10x2x0.2 mm) (from ref. [60]).

Since the deformed shape shown in Figure 31(b) was obtained from a geometrically, nonlinear analysis [24], the lateral displacement u at the right end, and the bending displacement w at the center, were obtained at the same time. According to the result, the calculated bending displacement, 0.331 mm, was much larger than the lateral displacement, 0.029mm. It should be noted that in case of linear analysis the bending displacement at the center was 0.333mm. Hence, if we efficiently utilize the bending displacement of an IPMC strip, we can construct a linear actuator with a larger actuation displacement or free strain.

Analytical Solution
In order to validate the finite element model in the previous section, bending displacement at the center of an IPMC strip was analytically derived, based on the equivalent bimorph beam assumption for IPMC actuators. Figure 32(a) shows a cross section of an IPMC strip and

Figure 32(b) shows the electro-mechanically induced stress σ. The electro-mechanically induced bending moment M_0 for an IPMC actuator can be expressed as follows:

$$M_0 = \int_{-H/2}^{H/2} z\sigma dA = \int_{-H/2}^{H/2} zE\alpha W dz = \frac{E\alpha W H^2}{4}, \qquad (32)$$

where E is the Young's modulus, and α the electro-mechanically induced strain ($d_{31}V/H$).

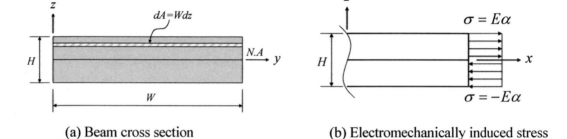

(a) Beam cross section   (b) Electromechanically induced stress

Figure 32. Beam cross section and electromechanically induced stress of an IPMC strip.

A linear relation between the second derivative of the bending displacement w and the bending moment M at each section along the beam is rewritten as follows:

$$M = EI\frac{d^2w}{dx^2}, \qquad (33)$$

where I is the moment of inertia ($WH^3/12$) of beam cross-sectional area. Since M is M_0 and is a constant value along the IPMC strip, the bending displacement w becomes:

$$w(x) = \frac{M_0}{2EI}x^2 + c_1x + c_2. \qquad (34)$$

where c_1 and c_2 are constants of integration arising from the two integrations.

For the supported boundary condition as shown in Figure 31(a), we assume:

$$w(0) = 0 \text{ and } w(L) = 0. \qquad (35)$$

Making use of Equations (34) and (35), we find:

$$c_1 = -\frac{M_0L}{2EI} \text{ and } c_2 = 0. \qquad (36)$$

Substituting Equation (36) into Equation (37), the bending deflection is rewritten as follows:

$$w(x) = \frac{M_0}{2EI}(x^2 - Lx).$$ (37)

Considering the electro-mechanically induced moment in Equation (32), the bending displacement at the center ($x = L/2$) of an IPMC strip is obtained as follows:

$$w(L/2) = \frac{M_0 L^2}{8EI} = \frac{3L^2 d_{31}}{8H^2}V$$ (38)

By using Equation (38), the analytical bending displacement 0.328 mm is calculated, and it agrees well with the value 0.331 mm from the previous finite element analysis.

Design of a Muscle-Like Linear Actuator
A design for muscle-like linear actuators made with IPMC is presented here, and its actuation characteristics such as free strain (or actuation displacement) and blocked stress (or blocked force) are numerically investigated.

Design Concept
Based on the preliminary analysis results of an IPMC strip, a muscle-like linear actuator was conceptually constructed, shown in Figure 33. The linear actuator is composed of rectangular-shaped elementary units along the axial direction, and each elementary unit expands or contracts under an applied voltage. By using the accumulated bending displacements from all elementary units, a large actuation displacement can be achieved in the axial direction. In addition, a bundle of parallelized linear actuators can be made for generating a large actuation force.

Finite Element Modeling of the Elementary Unit
Dimensions of an elementary unit are shown in Figure 34, and its finite element model and boundary conditions are shown in Figure 35. The total number of Quad4 [24] elements was 736. As shown in Figure 35, the elementary unit consists of IPMC and Nafion[TM]. Because of the low modulus of Nafion[TM], the interfaces of IPMC and Nafion[TM] can rotate easily to produce large bending deformation.

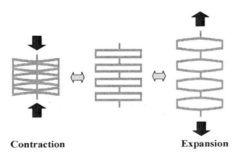

Contraction   Expansion

Figure 33. Schematic of a muscle-like linear actuator with a rectangular elementary unit(from ref. [60]).

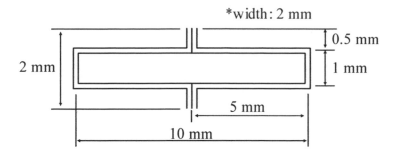

Figure 34. Dimensions of an elementary unit (from ref. [60]).

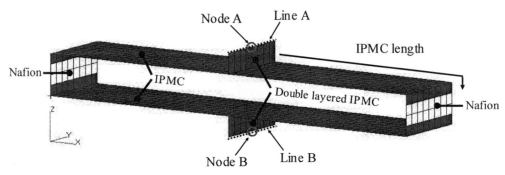

<Boundary Conditions>
- Node A: Displacement u=0, v=0, w=0 in x, y, z–directions
- Line A: u=0 & Multipoint Constraint (keeping up with the displacement w of node A)
- Node B: u=0 & free or fixed in z-direction
- Line B: u=0 & Multipoint Constraint (keeping up with the displacement w of node B)

Figure 35. Finite element model for the elementary unit (from ref. [60]).

Actuation Characteristics of the Elementary Unit

In this section, axial stiffness, free strain, and blocked stress for the elementary unit are discussed. For the calculations, the IPMC length was increased from 0 to 5.5 mm (see Figure 35 for the definition of IPMC length). When the IPMC length is 5.5 mm, the whole elementary unit consists of IPMC only.

Figure 36 shows a deformed shape of the elementary unit under an axial force of 0.01 N. The calculated axial displacement (in z-direction) at Node B and the corresponding axial stiffness are shown in Figure 37. By using the Hook's law in Equation (39), the axial stiffness, K, of an elementary unit can be estimated:

$$\sigma = E'\varepsilon \Rightarrow F = \frac{E'A}{H'}w \Rightarrow K = \frac{F}{w}(= \frac{E'A}{H'}), \tag{39}$$

where σ is the equivalent axial stress, E' the equivalent modulus of an elementary unit in axial direction, A the equivalent cross-sectional area, H' the axial length of an elementary unit, F the applied force, and w the axial displacement at Node B (see Figure 35 for the location of Node B). For the present elementary unit, A and H' were 10×2 mm² and 2 mm, respectively. As shown in Figure 37, the axial stiffness slowly increased as the IPMC length increased until about 5mm, but the increase of the IPMC lengths between 4 to 5 mm did not affect the axial stiffness as much. The axial stiffness rapidly increased the IPMC lengths of 5 ~ 5.5 mm; the IPMC covers each side wall of the elementary unit. Due to this, the low modulus of Nafion[TM] on each side wall drastically reduces the axial stiffness of the elementary unit.

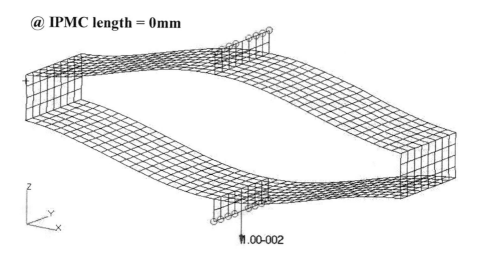

Figure 36. A deformed shape under 0.01N of axial force (from ref. [60]).

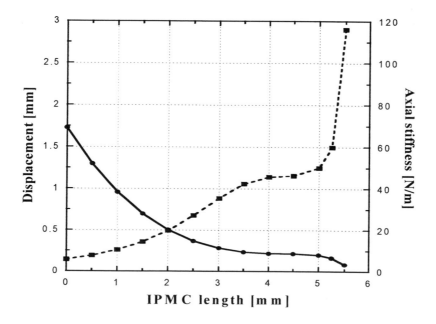

Figure 37. Displacement under 0.01 N and axial stiffness (from ref. [60]).

A deformed shape of the elementary unit under a 2 V input is shown in Figure 38, and the calculated actuation displacement, w at Node B and the corresponding free strain, $\varepsilon_{free}\, (= w / H')$, are shown in Figure 39. When the IPMC length was 4.5 mm, the elementary unit produced the maximum free strain of 25%. If we consider an alternating current of ± 2 V input, a free strain of 50% may be achievable from the present elementary unit. Since real muscles [61] can strain up 30~70%, the estimated free strain, 50%, of the present elementary unit is a positive value to be a substitution for real muscles. In addition, we could increase the maximum free strain of the elementary unit by reducing the axial length H' because the free strain is inversely proportional to the axial length.

@ IPMC length = 4.5mm

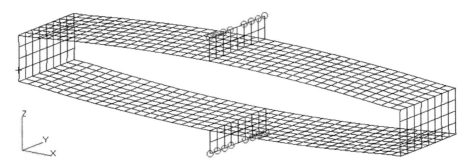

Figure 38. A deformed shape under 2 V input (from ref. [60]).

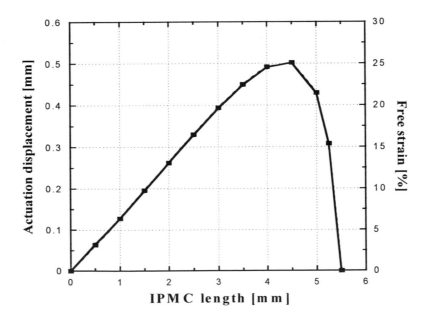

Figure 39. Actuation displacement and free strain (from ref. [60]).

@ IPMC length = 4.5mm

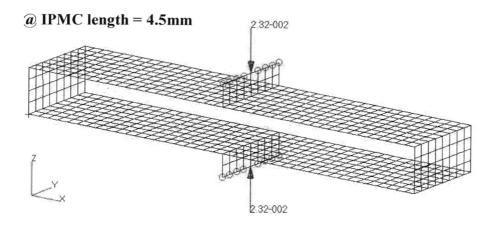

Figure 40. A blocked force under 2V input (from ref. [60]).

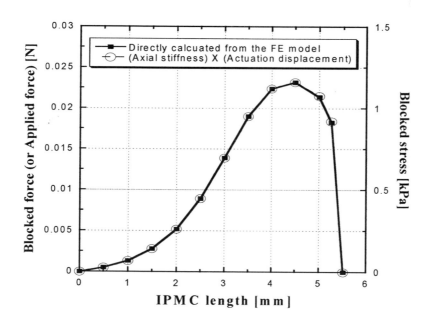

Figure 41. Blocked force and blocked stress (from ref. [60]).

In order to estimate blocked forces of the elementary unit, Node B was fixed in the z-direction. Figure 40 shows a deformed shape with a calculated blocked force under a 2 V input. The calculated blocked force, F_{bl}, and the corresponding blocked stress, σ_{bl} ($= F_{bl} / A$), were plotted in Figure 41. At the IPMC length of 4.5mm, the maximum blocked stress of 1.16 kPa was achieved, as shown in Figure 41. (It should be noted that for the maximum free strain of the elementary unit, the IPMC length was also 4.5 mm; therefore, the IPMC length of 4.5 mm is a unique optimal length for the present elementary unit.) In comparison with the stresses 0.1~0.4 MPa of real muscles [61], the maximum stress, 1.16 kPa, of the elementary unit was very small. Meanwhile, in accordance with the force-displacement relation [the 2nd equation in Equation (39)], we can reversely calculate the applied force needed to produce the same amount of axial displacement (from Figures 37 and 39). As shown in Figure 41, the calculated applied forces are the same as blocked forces which were directly calculated from the finite element model. The blocked force can be considered as the applied force to an electrically un-activated elementary unit.

Conclusion

In this book chapter, the identified potential design methodologies and fundamental properties of ionic polymeric-metal composites (IPMCs) as smart biomimetic sensors, actuators and artificial muscles were presented. This paper presented a summary of recent progresses and their engineering applications.

Acknowledgements

This work was partially supported by U.S. Office of Naval Research. Additional financial support from Korea Research Foundation and Artificial Muscle Research Center is appreciated. The authors thank the laboratory work done by graduate research students, D. Kim and J. Paquette of Active Materials and Processing Laboratory at the University of Nevada, Reno.

References

[1]   Kim, K. J. & Tadokoro, S. (ed.). (in print, 2006*). Electroactive Polymers for Robotic Applications.* London, Springer.

[2]   Kim, K. J. & Shahinpoor, M. (2002). *Applications of Polyelectrolytes in Ionic Polymeric Sensors, Actuators, and Artificial Muscles.* Review Chapter in Handbook of Polyelectrolytes, edited by S. Tripathy, J Kumar, and H. S. Nalwa, American Scientific Press (ASP), Vol. 3, Chapter 1, pp. 1-22.

[3]   Shahinpoor, M. & Kim, K. J. (2005). Ionic Polymer-Metal Composite-IV: Industrial and Medical Applications (Review Paper). *Smart Materials and Structures*, Vol. 14, pp. 197-214.

[4]   Shahinpoor, M. & Kim, K. J., (2004). Ionic Polymer-Metal Composites: III. Modeling and Simulation as Biomimetic Sensors, Actuators, Transducers, and Artificial Muscles (Review Paper). *Smart Materials and Structures*, Vol. 13, pp. 1362-1388.

[5]   Kim, K. J. & Shahinpoor, M. (2003). Ionic Polymer-Metal Composites - II. Manufacturing Techniques (Review Paper). *Smart Materials and Structures*, Vol. 12, No. 1, pp. 65-79.

[6]   Shahinpoor, M. & Kim, K. J. (2001). Ionic Polymer-Metal Composites – I. Fundamentals (Review Paper). *Smart Materials and Structures*, Vol. 10, pp. 819-833.

[7]   Paquette, J. & Kim, K. J. (2004). Ionomeric Electro-Active Polymer Artificial Muscle for Naval Applications. *IEEE Journal of Oceanic Engineering* (JOE), Vol. 29, No. 3, pp. 729-737.

[8]   Nam, J. D. & Lee, J. H. & Lee, J. H. & Choe, H. K. & Kim, J. & Tak, Y. S. (2005). Water Uptake and Migration Effects of Electroactive IPMC (Ion-exchange Polymer Metal Composite) Actuator. *Sensors and Actuators A: Physical*, Vol. 118, pp. 98-106.

[9]   Kim, K. J. & Shahinpoor, M. (2002). Development of Three Dimensional Ionic Polymer-Metal Composites as Artificial Muscles. *Polymer*, Vol. 43(3), pp. 797-802.

[10]   Liu, R. & Wei-Hwa, H. & Fedkiw, P. S. (1992). In Situ Electrode Formation on a Nafion Membrane by Chemical Platinization. *Journal of Electrochemical Society*, Vol. 139, No. 1, pp. 15-23.

[11]   Paquette, J. W. & Kim, K. J. & Kim, D. & Yim, W. (2005). The Behavior of Ionic Polymer-Metal Composites in a Multi-Layer Configuration. *Smart Materials and Structures*, Vol. 14, pp. 881-888.

[12]   Newbury, K. (2002). *Characterization, modeling, and control of Ionic polymer transducers*. Dissertation, Virginia Polytechnic Institute and State University.

[13]   Shahinpoor, M. (1995). Micro-electro-mechanics of Ionic polymer gels as electrically controllable artificial muscles. *Journal of Intelligent Material* Systems *and Structures*, Vol. 6, pp. 307-317.

[14]   Nemat-Nasser, S. & Li, J. Y. (2000). Electromechanical response of ionic polymer-metal composites. *Journal of Applied Physics*, Vol. 87, pp. 3321-3331.

[15]   Nemat-Nasser, S. (2002). Micromechanics of actuation of ionic polymer-metal composites. *Journal of Applied Physics*, Vol. 92, pp. 2899-2915.

[16]   Kanno, R. & Kurata, A. & Hattori, M. & Tadokoro, S. & Takamori, T. & Oguro, K. (1994). Characteristics and modeling of ICPF actuators. *Proceedings of the Japan-USA Symposium on Flexible Automation*, Vol. 2, pp. 691-698.

[17]   Xiao, Y. & Bhattacharya, K. (2001). Modeling electromechanical properties of ionic polymers. *Proceedings of the SPIE*, Vol. 4329, pp. 292-300.

[18]   Kanno, R. & Tadokoro, S. & Takamori, T. & Hattori, M. (1996). Linear approximate dynamic model of ICPF actuator. *Proceedings of the IEEE International Conference on Robotics and Automation*, pp. 219-225.

[19]   DeGennes, P. & Okumura, K. & Shahinpoor, M. & Kim, K. (2000), Mechanoelectric effects in ionic gels. *Europhysics Letters*, Vol. 40, pp. 513-518.

[20]   Farinholt, K. & Leo, D. (2004). Modeling of electromechanical charge sensing in ionic polymer transducers. *Mechanics of Materials*, Vol. 36, pp. 421-433.

[21]   Lee, S. & Park, H. C. & Kim, K. J. (2005). Equivalent modeling for ionic polymer-metal composite actuators based on beam theories. *Smart Materials and* Structures, Vol. 14, pp.1363-1368.

[22]   Wang, Q. & Zhang, Q. & Xu, B. & Liu, R. & Cross, E. (1999). Nonlinear piezoelectric behavior of ceramic bending mode actuators under strong electric fields. *Journal of Applied Physics*, Vol. 86, pp. 3352-3360.

[23]   Shahinpoor, M. & Kim, K. J. (May 2, 2002). Solid-state polymeric sensors, transducers, and actuators. *US Patent Application* 20020050454.

[24]   MSC.Software Corp. (2005). MSC/NastranTM quick reference guide: Volume I & II.

[25]   Taleghani, B. K. & Campbell, J. F. (1999). Non-linear Finite Element Modeling of THUNDER Piezoelectric Actuators. *NASA/TM*-1999-209322.

[26]   Lim, S. M. & Lee, S. & Park, H. C. & Yoon, K. J. & Goo, N. S. (2005). Design and demonstration of a biomimetic wing section using a lightweight piezo-composite actuator (LIPCA). *Smart Materials and Structures*, Vol. 14, pp. 496-503.

[27] Alexander, D. E. (2003). Nature's Flyers, Chapter 4. London, *The Johns Hopkins University Press*.

[28] Park, H. C. & Lee, S. & Kim, K. J. (2005). Equivalent modeling for shape design of IPMC (Ionic Polymer-Metal Composite) as flapping actuator. *Key Engineering Materials*, Vol. 297-300, pp. 616-621.

[29] Lee, S. & Kim, K. J. and Park, H. C. (2006, in print). Modeling of an IPMC actuator-driven zero-net-mass-flux pump for flow control. *Journal of Intelligent Material Systems and Structures*.

[30] Fan, G. & Song, G. & Hussain, F. (2005). Simulation of a piezoelectrically actuated Valve-less micropump. *Smart Materials and Structures*, Vol. 14, pp. 400-405.

[31] Kral, L. D. (2000). Active flow control technology. *ASME Fluids Engineering Technical Brief*.

[32] Seifert, A. & Pack, L. (1999). Oscillatory control of separation at high Reynolds numbers. *AIAA Journal*, Vol. 37, pp. 1062-1071.

[33] Seifert, A. & Pack, L. (1999). Oscillatory excitation of unsteady compressible flows over airfoils at flight Reynolds numbers. *AIAA Paper* 99-0925.

[34] Schatz, M. & Thiele, F. (2001). Numerical study of high-lift flow with separation control by periodic excitation. *AIAA Paper* 2001-0296.

[35] Gilarranz, J. L. & Rediniotis, O. K. (2001). Compact, high-power synthetic jet actuators for flow separation control. *AIAA Paper* 2001-0737.

[36] Mallinson, S. G. & Reizes, J. A. & Hillier, R. (2001). The interaction between a compressible synthetic jet and a laminar hypersonic boundary layer. Flow, *Turbulence and Combustion*, Vol. 66, pp. 1-21.

[37] Lockerby, D. (2001). Numerical simulation of boundary-layer control using MEMS actuation. Ph.D. *Thesis*, The University of Warwick, UK.

[38] Lee, C. & Hong, G. & Ha, Q. P. & Mallinson, S. G. (2003). A piezoelectrically actuated micro synthetic jet for active flow control. *Sensors and Actuators A*, Vol. 108, pp. 168-174.

[39] Bennett, M. & Leo, D. L. (2004). Ionic liquids as novel solvents for ionic polymer transducers. Proceedings of SPIE-*Smart Structures and Materials*, Vol. 5385, pp. 210-220.

[40] Hwang, H. C. & Chung, D. K. & Yoon, K. J. & Park, H. C. & Lee, Y. J. & Kang, T. S. (2002). Design and flight test of a fixed wing MAV. *AIAA Paper* 2002-3413.

[41] Laser, D. J. & Santiago, J. G. (2004). A review of micropumps. *Journal of Micromechanics and Microengineering*, Vol. 14, R35-R14.

[42] Woias, P. (2005). Micropumps-past, progress and future prospects. *Sensors and Actuators B*, Vol.105, pp.28-38.

[43] Thielicke, E. & Obermeier, E. 2000. *Microactuators and their technologies. Mechatronics*, Vol. 10, pp. 431-445.

[44] Olsson, A. (1998). Valve-less diffuser micropumps. Ph.D. Thesis, *Royal Institute of Technology*, Stockholm.

[45] Spencer, W. J. & Corbett, W. T. & Dominguez, L. R. & Shafer, B. D. (1978). An electronically controlled piezoelectric insulin pump and valves. *IEEE Transactions on Sonics and Ultrasonics*, SU-25, pp.153-156.

[46] thinXXS Microtechnology, (2004). http://www.thinxxs.com/products/index_products2. html, Germany.

[47]   van de Pol, F. C. M. & van Lintel, H. T. G. & Elwenspoek, M. & Fluitam, J. H. J. (1990). A thermopeumatic micropump based on micro-engineering techniques. *Sensors and Actuators*, A21-A23, pp.198-202.

[48]   Zengerle, R. & Ulrich, J. & Kluge, S. & Richter, M. & Richter, A. (1995). A bidirectional silicon micropump. *Sensors and Actuators* A, Vol. 50, pp. 81-86.

[49]   Böhm, S. & Olthuis, W. & Bergveld, P. (1999). A plastic micropump constructed with conventional techniques and materials. *Sensors and Actuators* A, Vol. 77, pp. 223-228.

[50]   Pak, J. J. & Kim, J. & Oh, S. W. & Son, J. H. & Cho, S. H. & Lee, S. K. & Park, J. Y. & Kim, B. (2004). Fabrication of ionic-polymer-metal-composite (IPMC) micropump using a commercial Nafion. *Proceedings of SPIE-Smart Structures and Materials*, Vol. 5385, pp. 272-280.

[51]   Lee S. & Kim, K. J. (in review, 2006). Design of IPMC actuator-driven valve-less micropump and its flow rate estimation at low Reynolds numbers. *Smart Materials and Structures.*

[52]   Idelchik, I. E. (1986). *Handbook of hydraulic resistance* (Second edition). Hemisphere Publishing Corporation.

[53]   Jiang, X. N. & Zhou, Z. Y. & Huang, X. Y. & Li, Y. & Yang, Y. & Liu, C. Y. (1998). Micronozzle/diffuser and its application in micro Valve-less pumps. *Sensors and Actuators A*, Vol. 70, pp. 81-87.

[54]   Stemme, E. & Stemme, G. (1993). A Valve-less diffuser/nozzle-based fluid pump. *Sensors and Actuators A*, Vol. 39, pp. 159-167.

[55]   Bar-Cohen, Y. (2001). Electroactive polymer(EAP) *Actuators as artificial muscles -* Reality, Potential, and Challenges. USA: *SPIE Press*

[56]   Jung, K. & Ryew, S. & Jeon, J. W. & Kim, H. & Nam, J. D. & Choi, H. (2001). Experimental investigations on behavior of IPMC polymer actuator and artificial muscle-like linear actuator. *Proceedings of SPIE-Smart Structures and Materials*, Vol. 4329, pp. 449-457.

[57]   Kamamichi, N. & Kaneda, Y. & Yamakita, M. & Asaka, K. & Luo, Z. W. (2003). Biped walking of passive dynamic walker with *IPMC* linear actuator. *SICE Annual Conference in Fukui,* Fukui University, Japan, August 4-6.

[58]   Kaneda, Y. & Kamamichi, N. & Yamakita, M. & Asaka, K. & Luo, Z. W. (2003). Control of linear artificial muscle actuator using *IPMC*. SICE Annual Conference in Fukui, Fukui University, Japan, August 4-6.

[59]   Malone, E. & Lipson, H. Freeform fabrication of Ionomeric polymer-metal composite actuators. *Proceedings of the 16th Solid Freeform Fabrication Symposium*, Austin TX, Aug. 2005.

[60]   Lee, S. & Kim, K. J. (in review, 2006). Muscle-like linear actuator by using ionic polymer-metal composite and its actuation characteristics. *Smart Materials and Structures.*

[61]   Huber, J. E. & Fleck, N. A. & Ashby, M. F. (1997). The selection of mechanical actuators based on performance indices. *Proc. R. Soc. Lond. A*, Vol. 453, pp. 2185-2205.

In: Progress in Smart Materials and Structures
Editor: Peter L. Reece, pp. 115-149

ISBN: 1-60021-106-2
© 2007 Nova Science Publishers, Inc.

Chapter 5

SPECTRAL FINITE ELEMENTS FOR ACTIVE WAVE CONTROL IN SMART COMPOSITE STRUCTURES

S. Gopalakrishnan

Department of Aerospace Engineering, Indian Institute of Science,
Bangalore, India

Abstract

Smart structures require actuation and this actuation is provided by a control system. Hence, control of structure is an important element in the study of smart structures. A control system is driven by a set of control algorithms called the control law. Traditionally, control law is designed based on the reduced order finite element model. By this approach, one cannot handle multi-modal phenomenon such as wave propagation, wherein the transients may cause steep increase in the responses causing the failure of structures. In this chapter, a new method of designing control law for laminated composite structure is addressed using the novel Active Spectral Finite Element for controlling responses caused by high frequency impact type loading. The advantage of the spectral element approach is that, it gives a very small system sizes due to its ability to represent the inertia distribution exactly and hence all modes are contained in this small system size. As a result, one need not resort to modal order reduction that is normally associated with any finite element approach for control design. In this work, the design of a PID controller for a smart laminated composite beam embedded/surface mounted using Piezo-Fiber Composite (PFC) actuators, is addressed. The efficiency of the proposed method is demonstrated on a few problems.

Introduction

The concept of broadband control in flexible structure has evolved in recent time. Tremendous technological success in the field of smart structures and Micro Electro-Mechanical Systems (MEMS) has laid the path towards implementation of such concepts. Especially structures made off multi-functional composites have provided a wide range of platforms for precision sensing, distributed actuation and control related applications. Most of the mathematical frameworks behind control system normally use frequency domain characteristics of the system. Since the basic foundation of Spectral Finite Element (SFEM) is in the frequency domain, these aspects will be fully exploited in the present chapter while developing finite element model with integrated control algorithm for control of distributed flexible structure.

Broadband control becomes important for many flexible structures with stringent vibration and noise level specifications and subjected to broadband loading. In most of the aerospace structures such as helicopters, launch vehicles, satellites and spacecrafts, component-level vibration contribute in the system-level noise spectrum which is broad banded. This is unlike control requirement of steady-state vibration in machineries, where passive devices can perform efficiently. Almost all the broad band control system requires active devices to augment the band-limited performance of passive devices.

Some of the examples where high amplitudes of structural vibration and noise are generated are the gearbox transmitted noise in helicopters, launch load induced noise in launch vehicle fairing etc. In these systems, a very high-powered fatigue-tolerant active control system is an absolute necessity. The concept of transmission path treatment by providing active struts for gearbox mounts has been found suitable for Active Vibration/Noise Control in most helicopters. A *10 – 20 dB* reduction in helicopter cabin noise spectrum within frequency band *10 Hz – 5 kHz* has been targeted [1]. Launch vehicles impart high levels of vibration to spacecraft during launch. The vibration environment is defined over several frequency bands (1) Transient vibration *80 Hz* (2) Random vibration *20 – 2000 Hz* and (3) pyro-technique shock *100 – 10000 Hz*. Loads from transient vibration define spacecraft design of primary structures such as spacecraft bus, solar panels and antenna support, instruments mounts etc. Loads from random vibration define the design for spacecraft light structures such as antenna and solar panels, and the shock loads define the design of electronic components and instruments. The spacecraft must survive the combination all vibration environment. This involves broadband control requirement and needs to be cost-effective for short launch duration [2]. On the other hand, instrument jitter during air and space-borne measurements, micro-gravity isolation in satellites and spacecrafts etc. fall in the other category, where low-powered, light-weight but high precision active control system is required throughout the design life [3].

This chapter is organized as follows. First the available methodologies for vibration and wave control is discussed. Next, active material system (piezoelectric and PFC) integrated with composite beam structure is then presented. Different active composite material models are developed. Spectral Finite Element Model (SFEM) for laminated composite structures and the Active Spectral Finite Element Model (ASFEM) are then developed for broadband control of vibration and waves in skeletal structures with generalized sensor/actuator configuration. Numerical simulation on a optimal broadband control of composite beam network with Piezoelectric Fiber Composite (PFC) actuators and point sensors for non-collocated feedback is carried out.

General Techniques for Vibration and Wave Control

Design of smart structural systems based on control of first few resonant modes individually are the most common in practice. For many vibration control applications, this will serve as the main control objective, since the modal energy is distributed over first few resonant modes only. The basic steps behind development of such active control system models can be described as follows.

1. Assume appropriate kinematics and constitutive model. For actuators or load cells mounted on the host structure, appropriate lumping of the control force and actuator inertia can be considered. For surface-bonded or embedded layered sensors/actuators, the same kinematics as the host structure with additional constraints (e.g. shear-lag to model active/passive constrained layers, discontinuous function to represent interfacial slip while handling inclusions, air-gap etc.) are to be used.

2. One should adopt application specific control scheme. For known harmonic disturbance, control force can be applied in open-loop having optimal phase difference with the mechanical disturbance. Actuator force can be directly specified to add on the equivalent mechanical force vector. For unknown dynamic loading (as required in most of the stable controller design), closed-loop control scheme are to be adopted. Initial configuration of the error sensors, whose placements and numbers are to be fixed based on optimal control performance (observability and controllability), can be used for feedback or feed forward control. These error measurements are considered as input to the controller that is under design. The controller output vector is to be used as input electrical signal to the actuators. For off-line optimal control design based on conventional optimization technique, the above steps are to be repeated at every iteration while extremizing the cost function(s). For off-line optimal control design based on soft-computing tools (e.g. genetic algorithm), the solution space can be explored directly.

3. Once all the system parameters (stiffness, mass, damping, electro-mechanical properties of sensors and actuators, sensor locations, actuator locations, actuator input etc.) for a particular configurations are available, one can develop the global model for the passive structure and senor/actuator segment using analytical, finite element or boundary element techniques. Under certain cases of electro-mechanical coupling, the system matrices can be decoupled into passive and active components. For fully coupled electromechanical case, analytical solution can be obtained for only few electro-mechanical boundary conditions and hence the detail finite element model is the best candidate. However, for mounted actuator or load cells, this is not a problem while lumping the effect of actuator stiffness, inertia and force on the respective system matrices.

4. Next, one should adopt suitable methods of system solution in temporal or modal space. When the discretized system size is large, appropriate reduced order modeling technique can be used. Dynamic Condensation, Proper Orthogonal Decomposition (POD) or System Equivalent Reduction Expansion Process (SEREP), among many reduced order modeling techniques, are found useful. Based on the formalism of the control cost function construction, a state-space model (first order representation) is often used instead of direct second order representation. This is particularly suitable for conventional designs based on quadratic regulator approach, where the state-space plant matrix, the input/output matrix along with required weighting matrices are introduced. Peak response specifications are generally found to be linear matrix functions of the design variables, which allow them to be incorporated within the design framework without increasing the complexity of the optimization [4].While

designing optimal control system, the control cost function is minimized including special control system features (e.g. gain scheduling, feedback delay etc.). When modal analysis is adopted, the modified dynamic stiffness matrix (including the contribution of sensor, controller and actuator parameters) is to be optimized so that the prescribed modes are controlled. In this approach, the control efficiency is quantified in terms of reduction in the modal amplitude level in the frequency response.

5. Once the range of control system parameters and the sensor/actuator collocation pattern is obtained, sensitivity and stability studies can be carried out. Sensitivity studies are important to identify the most effective solution space of design parameters. This also helps in visualizing the deviation in the desired response due to control uncertainty and measurement noise. With the narrowed-down solution space of design parameters thus obtained above, locus of the roots of the characteristic system, i.e. poles (resonances) and zeros (anti-resonances) of the system transfer function for varying design parameters are studied. The range of design parameters that produces the root locus on the right-half phase-plane are unstable and are avoided in the final design. A secondary objective is often placed for control of transient disturbances, which is to minimize the transient response time of the controller.

6. For real-time automatic control system, the off-line design discussed above is augmented by adaptive filter that tune the control gains in presence of measurement errors and uncertainty [5]. Also, there are certain drawbacks of the finite dimensional design to control a distributed parameter system, such as control spillover. This is the result of insufficient modes considered in the MIMO state-space model. Although adaptive filters can augment the performance of off-line design based on finite number of states, better modeling techniques for distributed parameter system are often advantageous. Many such recent techniques for structural vibration and waves based on exact solution to the wave equations can be referred, and is the main aspect of study in the present chapter. In this method, a finite element model is created in the frequency domain, which is called the spectral finite element model. Since the dynamic stiffness matrix of this matrix is created with the exact solution to the governing wave equation, all the system modes are exactly represented over a very small system size. Hence, one need not resort to modal order reduction to the design of controllers.

Most commonly used system for structural vibration control consists of active/passive Tuned Mass Damper (TMD) devices [6 & 7]. Several types of TMD devices are available in industry. Many of them consist of combination of lumped masses, springs and viscoelastic rod/beam/solid block type structures. Shape Memory Alloys (SMA) has also been used as TMD for building structural control [8]. The initial TMD devices used the combined effect of spring, mass and passive damper to reduce vibrations. A single such device will be capable to control single mode or few closely-spaced modes at most. On the other hand, in TMD devices with constrained viscoelastic layers, modes within a broader frequency band can be controlled. Multiple TMD devices can be used for multi-mode control. A general treatment of finite element computational model based on complex modes for large structure having multiple TMD devices can be found in [9].

In active devices, typically based on PZT layers/stacks, a much wider design space can be effectively used. Anderson and his co-workers [3] developed a packaged flexible PZT strip with integrated electronics for active vibration control. *5-10%* damping of single mode within the frequency range of *10-200 Hz* was reported while using this device. Although conventional active/passive TMD devices find their applications mostly in civil engineering structural controls, the active damping systems with integrated electronics are now being used extensively in aerospace applications, ranging from instrument jitter control to launch vehicle vibration isolation to spacecraft component-level and system-level stabilization [3].

Commonly, both the active/passive devices as subsystem or active layers bonded or embedded in composite for distributed actuation requires accurate modeling and analysis. Simpler structure with distributed actuator layers can be modeled using semi-numerical approaches, such as Rayleigh-Ritz method, assumed modes method, Galerkin's method and collocation method [10]. However, for complex configurations, finite element method is used. For narrowband applications involving the control of few lower modes, the solution of finite element system in modal coordinate is computationally efficient compared to direct time integration [9]. However, in case of displacement, velocity and acceleration feedback, non-symmetric system matrices are generated, which may cause loss of orthogonality. Due to this reason, special care is required while using the natural modes. Also, the number of steps in matrix computation increases. Beside this, one would require fine finite element mesh to minimize any unwanted discretization error in capturing the coupled electro-mechanical or magneto-mechanical field. As the number of modes under consideration increases, especially in broadband applications, the computational cost becomes significantly higher.

In contrast to the modal approach for broadband control, the wave approach is based on solution of finite set of traveling waves. Instead of the active/passive control systems for structural vibration (standing waves) as discussed above, wave-absorbing controllers can be used for controlling disturbance propagation in flexible structures. In References [11-13], a transfer function based model and point control forces was developed to cancel the traveling waves in skeletal structures. The basic objective here is to completely cancel the unwanted component(s) of the original traveling waves.

There are some transfer function based models, which are used for active wave control. These models are semi-analytical in nature and hence applicable to specific sensor/actuator collocation and structural boundary. The implementation of generic closed-loop control model for this system is very difficult. For broadband control application, first the fundamental limits on the performance of available active control systems are found at low and high frequencies. The low frequency limit is caused by near field of the secondary source corrupting the output of error sensor [14]. In all the feed forward wave cancellation-based active control systems available in literature [15-17], this happens because the near field information is not inherent to the controller input. Therefore, restriction regarding relative placements of sensors and actuators must be made while adopting such control system. Also, the dispersive wave speed (group speed) increases with increasing frequency and hence arrival of near field noise is difficult to identify in complex interconnections of structural components. Experiment on metallic beam shows that such high frequency limit for error signal delay increases for reduced flexural rigidity [14]. Also at component level, this effect becomes manifold, which must be captured accurately from the integrated mechanics. As a complementary strategy, one can consider adaptive tuning of the filter to compensate the cumulative error in phase and amplitude of actuator input signal [14 & 18]. However, such adaptation may not be always fast enough against high frequency transient loading. In another direction, feedback control of wave transmission has been studied in References [19-21]. In these studies, stabilization of the close-loop plant was carried out. This can be viewed as augmentation of the feed forward wave cancellation of incoherent noise on error sensors by minimizing certain cost functions. Similar wave absorbing techniques using transfer function-based methods were proposed, which considered exact solution to the wave equation in frequency domain [12-13]. Essentially, in these studies, and also in our present approach in this chapter, the main objective of the feedback control effort is local or low-authority control (LAC) of transmitted waves.

However, effect of distributed actuator dynamics related to such application has been studied in very few literatures [22]. Although analytical in nature, and restricted to continuous cantilever beams, similar reported studies have shown the possibility of using fewer distributed actuators with strain sensors for control of multiple waves. While mounting packaged TMD devices or integrated active layers, the original poles and zeros of the system transfer function (even under open-loop condition) can experience significant sliding due to added component-level dynamics [23].

Although significant development in robustness of the control system has been reported in literature, the effect of finite actuator dynamics in the control of distributed parameter systems still remains a core problem area from stability point of view. Such stability problems encountered in distributed space structures were discussed in Reference [24]. Noyer and Hanagud [25] proposed a Laplace domain model for optimal control of beam structure including actuator dynamics. In the present study, we prefer on the visualization of the closed-loop responses (hence locations of poles and zeros along the frequency axis) directly, since the Active Spectral Finite Element Model (ASFEM) proposed in this chapter is computationally more efficient compared to equivalent state-space or Laplace domain models.

In recent works by author and his co-workers [26-29], frequency domain models based on SFEM have been employed for open-loop feed forward and closed-loop PID feedback control. Exact actuator dynamics and electro-mechanical boundary scattering have been considered in these models. This enables one to analyze multiply connected beams with arbitrary geometry and non-collocated as well as distributed sensor-actuator configuration. The same concept is the basic framework of ASFEM developed in this chapter, and it accounts for axial-flexural wave coupling due to out of plane bending actuation and anisotropic electromagnetic properties across the beam thickness.

Active Local or Low Authority Control (LAC) of Waves

LAC implies that the control law is based only on information from the vicinity of the actuator. Typically, collocated actuators and sensors are used for LAC. This is because if the actuators and sensors are collocated, then the input-output transfer function is positive real, with an alternating pole-zero structure and phase bounded by $\pm 90^0$ [30-31]. If, in addition, the compensator applied to the structure is strictly positive real, then the closed-loop system is guaranteed to be stable and the compensator will add damping to the structure. Also, the corresponding modeling and analysis also become straight-forward. Once, stability is guaranteed in LAC, it makes a prefect compliment for global or high authority control (HAC). By providing broadband increase in damping, local controller(s) make the flexible modes robust in the roll-off region and improve performance at higher frequencies, where HAC is not designed to work [20 & 32]

However, collocated sensor/actuator configuration is not always the best option for LAC because of two reasons. First is that for control of dispersive waves at high frequencies, time delay in collocated feedback can be higher than the arrival time of wave packets. This is bound to produce phase difference and loss of coherence, and hence sensor-actuator collocation may not be always the best choice. Instead, such incoherence can be exploited to obtain optimal control performance. Second is that a non collocated sensor/actuator configuration may yield less power requirement for same broadband control objective. This would require analysis of the non collocated sensor/actuator configurations for LAC.

In the case studies carried out in this chapter, the control performance of axial-flexural coupled wave is modeled. Optimal performance that can be realized is also discussed. As suggested in Reference [33], the coupling between axial and flexural waves can be modeled by an equivalent asymmetric scattering termination for one dimensional structural member. An experimental study by them supporting this model reveals that control of one wave type (axial or flexural) results in standing waves in both types, if scattered waves from terminations are not effectively controlled. Since, exact coupling between various wave types was not studied in available literature, it was concluded in Reference [34] that effective control of all the waves would require at least one actuator per wave type. An experimental investigation supporting this conclusion can be found in Reference [35]. The experimental study by [1] also indicates that coupling between axial and flexural waves will be important due to presence of transverse dynamic loading along with the primary axial loads.

From the above study, we can conclude that the control of wave transmissions is essentially local or falls into the category of low authority control (LAC). Therefore, a general framework to develop computational model is required, which can deal with arbitrary sensor/actuator collocation. Distributed actuator dynamics with non collocated feedback is not tractable in modal method. In methods based on traveling waves and transfer function matrices, a number of complicated and application-specific algebraic manipulations make the analysis cumbersome. Further, optimization of such non collocated sensor/actuator configuration for broadband LAC applications is a challenging task

Active Material Systems and Devices

Use of bulk transducer material with piezoelectric, magnetostrictive, electro-opto-mechanical and electro-magnetic properties have long been exploited in ultrasonic, ferroelectric and optical devices. For structural vibration and wave related applications, they are comparatively new in modeling, design, fabrication and range of applications. The additional issues, namely the distributed actuator dynamics, actuation bandwidth, actuator authority etc. are some of the important aspects, which are being addressed for vibration and wave related applications. Also, issues in integration of these bulk materials with the host composite structures have provided a new dimension in design. Subsequently, accurate modeling, analysis and understanding of different actuation and sensing mechanisms play important role in cost-effective technology development. The most commonly used piezoelectric material is the

Lead Zirconium Titanate (PZT) with varying crystal structures [36 & 37]. At present several categories of PZT (hard and soft) are available in wafer, deposited thin-film, powder and fiber forms [38-39]. PZT wafers (typical thickness is of order of mm to μm) polarized with uniform surface electrodes or polarized with Inter-digital Electrodes (IDEs) can be used in surface-bonded or embedded form in laminated composite with ply cut-outs or resin pockets. PZT powder mixed with polymer matrix and reinforcing fibers can be manufactured using appropriate curing methods in presence of polarizing field. Piezoelectric fibers of various shapes and binder matrix with improved conductivity can be used to manufacture Piezoelectric Fiber Composite (PFC). In all these active composites, the effects of residual stress, low compaction, dielectric breakdown, electrical insulation are the major manufacturing issues that are intense areas of research, which will help in improving the performance of such active composite for broad-band, reliable sensing and control applications. Among magnetostrictive materials for structural sensing and actuation, the Terfenol-D ($Tb_x Dy_{(1-x)} Fe_2$) appears to be the best candidate having wide linear constitutive relation, smallest hysteresis and high actuator authority. However, for both the PZT and Terfenol-D, hysteresis and non-linear constitutive relation is common, and appropriate DC bias electric/magnetic field is required to achieve best results. Terfenol-D is available in the form of rod as well as powder. These powders can be mixed with polymer matrix in composite and hence has huge potential to be used for structural actuation.

Table 1. Comparison of solid state actuation materials with their bulk properties.

	PZT 5H	PVDF	PMN	Terfenol-D	Nitinol
Actuation Mechanism	piezo- (31) ceramic	piezo film	electro-strictive	magneto-strictive	SMA
Max. Strain	0.13%	0.07%	0.1%	0.2%	2%-8%
Modulus (GPa)	60.6	2	64.5	29.7	28(m), 90(a)
Density (kg/m³)	7500	1780	7800	9250	7100
Actuation Energy Density (J/kg)	6.83	0.28	4.13	6.42	252-4032
Hysteresis	10%	>10%	<1%	2%	High
Temperature Range	-20° to 200°C	Low	0° to 40°C	High	-
Bandwidth	100kHz	100kHz	100kHz	<10kHz	<5Hz

Table 1 shows comparison of of the properties of bulk *PZT 5H* (hard), *PVDF*, *PMN*, *Terfenol-D* and *Nitinol*. *PZT 5H* is commonly used for high actuator authority and high dielectric breakdown voltage in structural control. Polyvenyl difluoride (*PVDF*)is commonly used as thin-film sensor. Lead Magnesium Niobet (*PMN*) falls in the category of electrostrictive material with quadratic non-linearity in constitutive relation, and can be used for sensing and low strength actuation. *Terfenol-D* is commonly used in structural actuation with high actuator authority. However, excessive heating and weight due to its magnetic housing and magnetizing coil limit its wider application. *Nitinol* is the industrial name of the Nickel Titanium Alloy and shows the shape memory effect due to diffusion less transformation between its martensitic and austenitic microstructures. The temperature-induced transformation, which is primarily responsible for the shape-memory effect, is commonly exploited for static shape control and low-frequency vibration control. Beside, the stress-induced superelastic effect also helps providing a high passive damping feature of Shape Memory Alloys (*SMAs*) in vibration and wave control applications.

For sensing and active control of vibration and waves, frequency bandwidth of the transducer is one important parameter, which provides their operational limits in terms of frequency range of excitation. For vibration and wave sensing, any measurement at the higher end of the frequency bandwidth is likely to introduce high signal-to-noise ratio in the sensor output. On the other hand, for vibration and wave control applications, requirement of any actuation at the higher end of the bandwidth and above may cause control spill-over. As can be seen in Table 1, *Nitinol* is suitable for low frequency applications, whereas *Terfenol-D* and *PZT* are suitable for medium and high frequency applications, respectively.

Apart from these, packaged actuators are also used for vibration and wave control. Figure 1 show some typical configuration of magnetostrictive and piezoelectric stacked actuator. In such packaged transducers with housing, multi-modal control can be obtained using suitable block-force (provided by pre-stressing the core) and different directional mounts having a group of such transducers. Also, additional tuned mass can be used to provide passive damping features. However, for multi-modal control, one needs to have adaptive feed-forward or feed-back algorithm and error sensors in to guide these actuators.

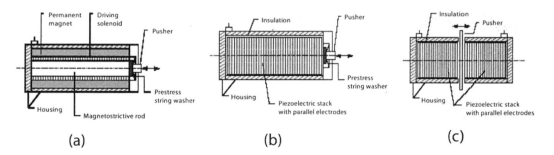

Figure 1. Configuration of the packaged magnetostrictive/piezoelectric transducers.
(a) Magnetostrictive rod actuator. (b) Piezoelectric stack actuator. (c) Piezoelectric stack pair.

More recently, PFC actuators have been explored for its use in high control authority actuation. However, their application for distributed structural actuation in vibration and wave control related applications are very few (e.g. see reference [40]). Also, there is destabilizing effect of such distributed but finite actuator dynamics on the control performance, especially for non-collocated feedback, which needs to be accurately dealt in the framework of distributed parameter model. In the present study, we consider the effect of integration of similar PFC actuators and sensors with host composite beam structures on their control performance as one of the key issue. We also consider broadband actuation and sensing capability (up to kHz range) of these PFCs to develop a linear wave mechanics based structure-control interaction model.

Modeling of Composite Structures Integrated with Transducer Devices

Before obtaining a distributed parameter model for integrated active/passive material system to be used for structural wave control applications, it is essential to incorporate the key features of these material assemblies accurately. Several literature and finite element models are at present available for carrying out detail 2D/3D analysis and design for smart transducers [41-43]. These transducers externally bonded or embedded inside the composite, when acting as sensors, generate electrical signal (with the help of appropriate circuit and signal-processors) due the induced mechanical strain. On the other hand, when they are designed as actuator, they induce mechanical strain in the host structure due to application of electro-magnetic field. In the present study, we consider only piezoelectric (e.g. *PZT*) and magnetostrictive materials (e.g. *Terfenol-D*) as the bulk transducer materials. Modeling of surface-bonded and embedded wafers or thin-film type structure as well as stacked actuators fitted in composite rod/beam segments are considered for modeling. A more complicated material system in the form of Inter-Digital Transducers (*IDEs*), which are often used in MEMS, ultrasonics and Surface Acoustic Wave (*SAW*) applications, are derived next. Use of Piezoelectric Fiber Composite (*PFC*), which is of more recent development in distributed structural actuation technology, has opened up the vast scope of using active fiber composite in various forms. Although simplified model of different *PFCs* based on homogenization have been developed [39], such constitutive model has not been applied for studying their distributed actuation capabilities for structural wave control or *SAW* type applications. At the end of the following section, a uniform field model of *PFCs* derived in [44] is studied later in the context of distributed actuator dynamics.

Figure 2 (a) and (b) show the active composite beam segments with piezostack and layers respectively. Similar configuration is also possible for Terfenol-D rod, stack and matrix patch with magnetostrictive particles. However, for Terfenol-D rod or stack, the magnetic coil assembly and housing can be modeled with appropriate additional mass and stiffness lumping. For matrix patch with magnetostrictive particles, the magnetic coil assembly can also be modeled in similar ways. Assuming perfect bonding between the stack or layers with the electrodes and the host structures, the beam kinematics can be defined as

$$u(x,z,t) = u^0(x,t) - z\frac{\partial w}{\partial x}, \quad w(x.z,t) = w(x,t) \tag{1}$$

for Euler-Bernoulii beam and

$$u(x,z,t) = u^0(x,t) - z\phi(x,t), \quad w(x.z,t) = w(x,t) + z\psi(x,t) \tag{2}$$

for shear deformable beam. Here, u^0 is the mid-plane axial displacement, w is the transverse displacement, ϕ is the slope of the beam and ψ is the parameter that defines the lateral contraction due to Poisson's ratio. For concise representation, we proceed with the constitutive model and equation of motion for the first order shear deformable beam, and these can be reduced to the Euler-Bernoulli (thin beam) model by making the shear rigidity to infinity and rotational inertia equal to zero. For sensing, we assume a single layer with output electric field $E(x,t)$ due to mechanical strain. For actuation with layered configuration (Figure 2}), we assume multiple layers with input electric field $E(x,t)$ in each layer. To arrive at the constitute model, we start with the constitutive model of the bulk active materials and then reduce it to a plane stress constitutive model. The derivations for piezoelectric actuators are presented in detail. The same procedure is also applicable to bulk magnetostrictive materials in rod or layered form.

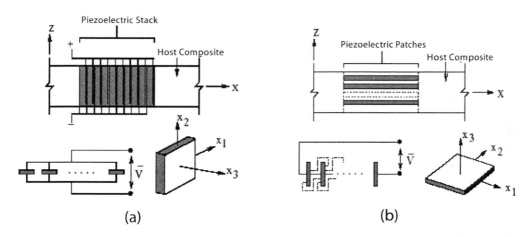

(a)   (b)

Figure 2. Configuration of the piezoelectric transducer integrated with the host composite beams. (a) Piezoelectric stack configuration with the equivalent parallel plate capacitor model. (b) Piezoelectric patches adhesively bonded or embedded in host laminated composite along and the equivalent parallel plate capacitor model.

Plane Stress Constitutive Model for Piezoelectric Material

Here, we assume the media to be orthotropic with the direction of the fibers in the composite representing the three planes of symmetry. We first establish the constitutive model at the lamina level, which are synthesized to get the laminate level constitutive model.

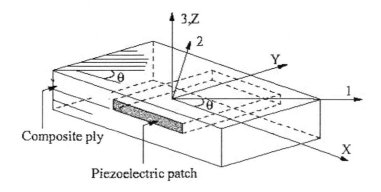

Figure 3. Local and Global Coordinate system for a lamina with embedded Piezoelectric Patch.

Consider a lamina with the piezoelectric layer as shown in Figure 3. The constitutive model in directions *1, 2,* and *3* for such a lamina can be written as

$$\begin{Bmatrix} \{\sigma\} \\ \{D\} \end{Bmatrix} = \begin{bmatrix} [C] & -[e] \\ -[e]^T & [\mu] \end{bmatrix} \begin{Bmatrix} \{\varepsilon\} \\ \{E\} \end{Bmatrix} \quad \text{or} \quad \{\bar{\sigma}\} = [\bar{C}]\{\bar{\varepsilon}\} \tag{3}$$

Expanding the above equation, we get

$$\begin{Bmatrix} \sigma_{11} \\ \sigma_{22} \\ \sigma_{33} \\ \sigma_{23} \\ \sigma_{31} \\ \sigma_{12} \\ D_1 \\ D_2 \\ D_3 \end{Bmatrix} = \begin{bmatrix} C_{11} & C_{12} & C_{13} & 0 & 0 & 0 & 0 & 0 & -e_{31} \\ C_{12} & C_{22} & C_{23} & 0 & 0 & 0 & 0 & 0 & -e_{32} \\ C_{13} & C_{23} & C_{33} & 0 & 0 & 0 & 0 & 0 & -e_{33} \\ 0 & 0 & 0 & C_{44} & 0 & 0 & 0 & -e_{24} & 0 \\ 0 & 0 & 0 & 0 & C_{55} & 0 & -e_{15} & 0 & 0 \\ 0 & 0 & 0 & 0 & 0 & C_{66} & 0 & 0 & 0 \\ 0 & 0 & 0 & 0 & e_{15} & 0 & \mu_{11} & 0 & 0 \\ 0 & 0 & 0 & e_{24} & 0 & 0 & 0 & \mu_{22} & 0 \\ e_{31} & e_{32} & e_{33} & 0 & 0 & 0 & 0 & 0 & \mu_{33} \end{bmatrix} \begin{Bmatrix} \varepsilon_{11} \\ \varepsilon_{22} \\ \varepsilon_{33} \\ \varepsilon_{23} \\ \varepsilon_{31} \\ \varepsilon_{12} \\ E_1 \\ E_2 \\ E_3 \end{Bmatrix}$$

Here, $E_i = -\nabla\Phi$, where Φ is the electric potential vector. The above constitutive model is then transformed to the global *x-y-z* coordinate system using the transformation matrix, which is given by

$$[T] = \begin{bmatrix} [T_{11}] & [0] \\ [0] & [T_{22}] \end{bmatrix} \tag{4}$$

where

$$[T_{11}] = \begin{bmatrix} C^2 & S^2 & 0 & 0 & 0 & -2CS \\ S^2 & C^2 & 0 & 0 & 0 & 2CS \\ 0 & 0 & 1 & 0 & 0 & 0 \\ 0 & 0 & 0 & C & S & 0 \\ 0 & 0 & 0 & S & C & 0 \\ CS & -CS & 0 & 0 & 0 & C^2 - S^2 \end{bmatrix}, \quad [T_{22}] = \begin{bmatrix} C^2 & S^2 & 0 \\ S^2 & C^2 & 0 \\ 0 & 0 & 1 \end{bmatrix}, C = \cos(\theta), S = \sin(\theta)$$

Here θ is the fiber orientation of the lamina. The constitutive model in the global *xyz* direction is then given by

$$\{\sigma\} = [T]^T \begin{bmatrix} \{C\} & -[e] \\ [e]^T & [\mu] \end{bmatrix} [T]\{\varepsilon\} = \begin{bmatrix} [\bar{\bar{C}}] & -[\bar{\bar{e}}] \\ [\bar{e}] & [\mu] \end{bmatrix} \{\varepsilon\}$$

In expanded form, the above equation becomes

$$\begin{Bmatrix} \sigma_{xx} \\ \sigma_{yy} \\ \sigma_{zz} \\ \sigma_{yz} \\ \sigma_{zx} \\ \sigma_{xy} \\ D_x \\ D_y \\ D_z \end{Bmatrix} = \begin{bmatrix} \bar{\bar{C}}_{11} & \bar{\bar{C}}_{12} & \bar{\bar{C}}_{13} & 0 & 0 & 0 & 0 & 0 & -\bar{\bar{e}}_{31} \\ \bar{\bar{C}}_{12} & \bar{\bar{C}}_{22} & \bar{\bar{C}}_{23} & 0 & 0 & 0 & 0 & 0 & -\bar{\bar{e}}_{32} \\ \bar{\bar{C}}_{13} & \bar{\bar{C}}_{23} & \bar{\bar{C}}_{33} & 0 & 0 & 0 & 0 & 0 & -\bar{\bar{e}}_{33} \\ 0 & 0 & 0 & \bar{\bar{C}}_{44} & 0 & 0 & 0 & -\bar{\bar{e}}_{24} & 0 \\ 0 & 0 & 0 & 0 & \bar{\bar{C}}_{55} & 0 & -\bar{\bar{e}}_{15} & 0 & 0 \\ 0 & 0 & 0 & 0 & 0 & \bar{\bar{C}}_{66} & 0 & 0 & 0 \\ 0 & 0 & 0 & 0 & \bar{\bar{e}}_{15} & 0 & \bar{\bar{\mu}}_{11} & 0 & 0 \\ 0 & 0 & 0 & \bar{\bar{e}}_{24} & 0 & 0 & 0 & \bar{\bar{\mu}}_{22} & 0 \\ \bar{\bar{e}}_{31} & \bar{\bar{e}}_{32} & \bar{\bar{e}}_{33} & 0 & 0 & 0 & 0 & 0 & \bar{\bar{\mu}}_{33} \end{bmatrix} \begin{Bmatrix} \varepsilon_{xx} \\ \varepsilon_{yy} \\ \varepsilon_{zz} \\ 2\varepsilon_{yz} \\ 2\varepsilon_{zx} \\ 2\varepsilon_{xy} \\ E_x \\ E_y \\ E_z \end{Bmatrix} \quad (5)$$

The elements of $[\bar{\bar{C}}]$ and $[\bar{\bar{e}}]$ are given by

$$\bar{\bar{C}}_{11} = 4C_{66}C^2S^2 + C^2(C_{11}C^2 + C_{12}S^2) + S^2(C_{12}C^2 + C_{22}S^2)$$

$$\bar{\bar{C}}_{12} = -4C_{66}C^2S^2 + S^2(C_{11}C^2 + C_{12}S^2) + C^2(C_{12}C^2 + C_{22}S^2), \quad \bar{\bar{C}}_{13} = C_{13}C^2 - C_{23}S^2$$

$$\bar{\bar{C}}_{16} = -2C_{66}CS(C^2 - S^2) + CS(C_{11}C^2 + C_{12}S^2) - CS(C_{12}C^2 + C_{22}S^2), \quad \bar{\bar{C}}_{21} = \bar{\bar{C}}_{12}$$

$$\bar{\bar{C}}_{22} = 4C_{66}C^2S^2 + S^2(C_{11}S^2 + C_{12}C^2) + C^2(C_{12}S^2 + C_{22}C^2), \quad \bar{\bar{C}}_{23} = C_{23}C^2 + C_{13}S^2$$

$$\bar{\bar{C}}_{26} = 2C_{66}CS(C^2 - S^2) + CS(C_{11}S^2 + C_{12}C^2) - CS(C_{12}S^2 + C_{22}C^2), \quad \bar{\bar{C}}_{31} = \bar{\bar{C}}_{13}, \bar{\bar{C}}_{32} = \bar{\bar{C}}_{23}$$

$$\bar{\bar{C}}_{33} = C_{33}, \quad \bar{\bar{C}}_{36} = CS(C_{13} - C_{23}), \quad \bar{\bar{C}}_{44} = C_{44}C^2 + C_{55}S^2, \quad \bar{\bar{C}}_{45} = CS(C_{55} - C_{44}),$$

$$\bar{\bar{C}}_{54} = \bar{\bar{C}}_{45}, \quad \bar{\bar{C}}_{55} = C_{44}S^2 + C_{55}C^2, \quad \bar{\bar{C}}_{61} = \bar{\bar{C}}_{16}, \quad \bar{\bar{C}}_{62} = \bar{\bar{C}}_{26}, \quad \bar{\bar{C}}_{63} = \bar{\bar{C}}_{36}$$

$$\bar{\bar{C}}_{66} = C_{66}(C^2 - S^2)^2 + C^2S^2(C_{11} - C_{12}) - C^2S^2(C_{12} - C_{22})$$

$$\bar{\bar{e}}_{31} = (e_{31}C^2 + e_{32}S^2), \quad \bar{\bar{e}}_{32} = (e_{31}S^2 + e_{32}C^2), \quad \bar{\bar{e}}_{33} = e_{33}, \quad \bar{\bar{e}}_{14} = (e_{15}C^2S + e_{24}CS^2),$$

$$\bar{\bar{e}}_{24} = (e_{24}C^3 + e_{15}S^3), \quad \bar{\bar{e}}_{15} = (e_{15}C^3 - e_{24}S^3), \bar{\bar{e}}_{25} = (e_{24}C^2S - e_{15}CS^2),$$

$$\bar{\bar{e}}_{36} = CS(e_{31} - e_{32}), \quad \bar{\bar{\mu}}_{11} = \mu_{11}C^4 + \mu_{22}S^4, \quad \bar{\bar{\mu}}_{12} = C^2S^2(\mu_{11} + \mu_{22}), \quad (6)$$

$$\bar{\bar{\mu}}_{22} = \mu_{22}C^4 + \mu_{11}S^4, \quad \bar{\bar{\mu}}_{33} = \mu_{33}$$

For 2-D analysis, we normally employ either plane stress or plane strain assumptions. For plane stress assumption in the x-z plane, we substitute $\sigma_{yy} = \sigma_{xy} = \sigma_{yz} = D_x = D_y = 0$ in the Equation (5). Simplifying this, we can write the constitutive model for a 2-D piezoelectric composite as

$$\left\{ \begin{matrix} \{\sigma\} \\ D_z \end{matrix} \right\} = \begin{bmatrix} [\hat{C}] & -[\hat{e}] \\ [\hat{e}]^T & \hat{\mu} \end{bmatrix} \left\{ \begin{matrix} \{\varepsilon\} \\ E_z \end{matrix} \right\} = \left\{ \begin{matrix} \sigma_{xx} \\ \sigma_{zz} \\ \sigma_{xz} \\ D_z \end{matrix} \right\} = \begin{bmatrix} \hat{C}_{11} & \hat{C}_{13} & 0 & -\hat{e}_{31} \\ \hat{C}_{13} & \hat{C}_{33} & 0 & -\hat{e}_{32} \\ 0 & 0 & \hat{C}_{55} & 0 \\ \hat{e}_{31} & \hat{e}_{32} & 0 & \hat{\mu}_{33} \end{bmatrix} \left\{ \begin{matrix} \varepsilon_{xx} \\ \varepsilon_{zz} \\ 2\varepsilon_{xz} \\ E_z \end{matrix} \right\} \qquad (7)$$

where,

$$\hat{C}_{11} = \bar{\bar{C}}_{11} + \frac{1}{\Delta}\left[\bar{\bar{C}}_{12}(\bar{\bar{C}}_{26}\bar{\bar{C}}_{16} - \bar{\bar{C}}_{66}\bar{\bar{C}}_{12}) + \bar{\bar{C}}_{16}(\bar{\bar{C}}_{26}\bar{\bar{C}}_{16} - \bar{\bar{C}}_{22}\bar{\bar{C}}_{16}) \right]$$

$$\hat{C}_{13} = \bar{\bar{C}}_{13} + \frac{1}{\Delta}\left[\bar{\bar{C}}_{12}(\bar{\bar{C}}_{26}\bar{\bar{C}}_{36} - \bar{\bar{C}}_{66}\bar{\bar{C}}_{23}) + \bar{\bar{C}}_{16}(\bar{\bar{C}}_{26}\bar{\bar{C}}_{23} - \bar{\bar{C}}_{22}\bar{\bar{C}}_{36}) \right]$$

$$\hat{C}_{33} = \bar{\bar{C}}_{33} + \frac{1}{\Delta}\left[\bar{\bar{C}}_{23}(\bar{\bar{C}}_{26}\bar{\bar{C}}_{36} - \bar{\bar{C}}_{66}\bar{\bar{C}}_{23}) + \bar{\bar{C}}_{36}(\bar{\bar{C}}_{26}\bar{\bar{C}}_{23} - \bar{\bar{C}}_{22}\bar{\bar{C}}_{36}) \right]$$

$$\hat{C}_{55} = \bar{\bar{C}}_{55} - \frac{\bar{\bar{C}}_{45}^2}{\bar{\bar{C}}_{44}} \qquad\qquad (8)$$

$$\hat{e}_{31} = \bar{\bar{e}}_{31} + \frac{1}{\Delta}\left[\bar{\bar{e}}_{32}(\bar{\bar{C}}_{12}\bar{\bar{C}}_{66} - \bar{\bar{C}}_{16}\bar{\bar{C}}_{26}) + \bar{\bar{e}}_{36}(\bar{\bar{C}}_{16}\bar{\bar{C}}_{22} - \bar{\bar{C}}_{12}\bar{\bar{C}}_{26}) \right]$$

$$\hat{e}_{32} = \bar{\bar{e}}_{32} + \frac{1}{\Delta}\left[\bar{\bar{e}}_{32}(\bar{\bar{C}}_{23}\bar{\bar{C}}_{66} - \bar{\bar{C}}_{26}\bar{\bar{C}}_{36}) + \bar{\bar{e}}_{36}(\bar{\bar{C}}_{36}\bar{\bar{C}}_{22} - \bar{\bar{C}}_{23}\bar{\bar{C}}_{26}) \right]$$

$$\hat{\mu}_{33} = \bar{\bar{\mu}}_{33} + \frac{1}{\Delta}\left[\bar{\bar{e}}_{36}(\bar{\bar{C}}_{22}\bar{\bar{C}}_{69} + \bar{\bar{C}}_{66}\bar{\bar{C}}_{29}) + \bar{\bar{e}}_{32}(\bar{\bar{C}}_{29}\bar{\bar{C}}_{66} - \bar{\bar{C}}_{26}\bar{\bar{C}}_{69}) \right]$$

For piezoelectric layered configuration (Figure 2(b))with assumed shear deformable kinematics in X-Z plane, the required constitutive model from Equation (7) finally becomes

$$\left\{ \begin{matrix} \sigma_{xx} \\ \sigma_{xz} \\ D_z \end{matrix} \right\} = \begin{bmatrix} \hat{C}_{11} & 0 & -\hat{e}_{31} \\ 0 & \hat{C}_{55} & 0 \\ \hat{e}_{31} & 0 & \hat{\mu}_{33} \end{bmatrix} \left\{ \begin{matrix} \varepsilon_{xx} \\ 2\varepsilon_{xz} \\ E_z \end{matrix} \right\} \qquad (9)$$

The above equation also holds for passive fiber reinforced polymer matrix composite layers with $\hat{e}_{31} = 0$ and $\hat{\mu}_{33}$ almost negligible compared to piezoelectric materials.

For piezoelectric stack configuration (Figure 2(a), the constitutive model becomes

$$\left\{ \begin{matrix} \sigma_{xx} \\ \sigma_{xz} \\ D_z \end{matrix} \right\} = \begin{bmatrix} \hat{C}_{33} & 0 & -\hat{e}_{32} \\ 0 & \hat{C}_{55} & 0 \\ \hat{e}_{32} & 0 & \hat{\mu}_{33} \end{bmatrix} \left\{ \begin{matrix} \varepsilon_{xx} \\ 2\varepsilon_{xz} \\ E_z \end{matrix} \right\} \qquad (10)$$

The actuation of the stack happens along x direction, which is same as the direction vector of the electric field, that is, x_3 as shown in Figure 2 (a) having material coordinate (x_1, x_2, x_3) for individual stack. For passive fiber reinforced polymer matrix composite, the constitutive model remains same as discussed in the context of Equation (9).

Constitutive Model for Piezo Fiber Composite (PFC)

To illustrate the derivation of the constitutive model for piezoelectric fiber composite (PFC) actuation, we consider rectangular packing square PZT fibers with matrix as shown in Figure.4. The rectangular cross-section of the fibers can provide maximum volume fraction of ceramic, which is preferable for actuation. The configuration can be obtained using fibers that have been tape cast and diced, extruded or cast into a mold.

Figure 4 shows an actuator element with its host composite structure in an arbitrary q^{th} actuator with local coordinate system as(X_a^q, Y_a^q, Z_a^q). The representative volume element (RVE), of the two-phase ceramic-matrix composite system is described by one quadrant axi-symmetric model about the x_3 axis. Here, h is the total depth a single PFC layer, p is the uniform spacing of the inter digitated electrodes spanning along x_1, and b is the width of each electrode. The constitutive relations for orthotropic active ceramic bulk form [45] can be represented as

$$\left\{\begin{array}{c} \sigma_{xx} \\ \sigma_{zz} \\ \sigma_{xz} \\ D_z \end{array}\right\} = \left[\begin{array}{cccc} C^E_{11} & C^E_{12} & C^E_{13} & -e_{31} \\ C^E_{12} & C^E_{22} & C^E_{23} & -e_{32} \\ C^E_{13} & C^E_{23} & C^E_{33} & -e_{33} \\ e_{31} & e_{32} & e_{33} & \mu^s_{33} \end{array}\right] \left\{\begin{array}{c} \varepsilon_{xx} \\ \varepsilon_{zz} \\ 2\varepsilon_{xz} \\ E_z \end{array}\right\} \qquad (11)$$

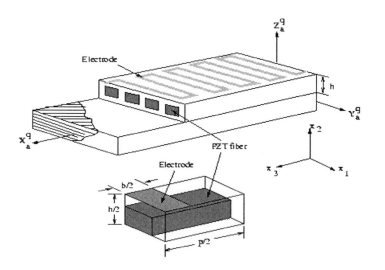

Figure 4. Configuration of Piezoelectric Fiber Composite (PFC) for composite beam actuation.

This is of very similar form as that of PZT actuator. For 1-D waveguide analysis, this requires reduction into single equivalent constitutive law by considering the volume fraction of the Piezo fibre (PZT) to the total volume of the laminate. For pure Piezoceramic, $C_{11}^E = C_{22}^E$, $C_{23}^E = C_{13}^E$, $e_{32} = e_{31}$. For matrix phase, all e_{ij} are zero, and their mechanical and dielectric properties are represented without superscripts. Assuming negligible distortion of the equipotent lines and electric fields beneath the electrodes, and imposing proper field continuity between the ceramic and matrix phases, the effective unidirectional constitutive law for a PFC beam structure can be expressed as [44]

$$\sigma_{zz} = C^{eff}{}_{33}\varepsilon_{zz} - e^{eff}{}_{33}E_z \tag{12}$$

where,

$$C_{33}{}^{eff} = (\bar{\bar{C}}_{33}V_1^p + C_{22}V_1^m) - \frac{V_1^p V_1^m (C_{12} - \bar{\bar{C}}_{13})^2}{C_{22}V_1^p + \bar{\bar{C}}_{11}V_1^m} \tag{13}$$

$$\bar{\bar{C}}_{11} = (\bar{C}_{11}V_2^p + C_{22}V_2^m) - \frac{V_2^p V_2^m (C_{12} - \bar{C}_{12})^2}{C_{11}V_2^p + \bar{C}_{12}V_2^m},$$

$$\bar{\bar{C}}_{13} = (\bar{C}_{13}V_2^p + C_{12}V_2^m) - \frac{V_2^p V_2^m (C_{12} - \bar{C}_{12})C_{12} - \bar{C}_{23})}{C_{11}V_2^p + \bar{C}_{22}V_2^m}, \tag{14}$$

$$\bar{\bar{C}}_{11} = (\bar{C}_{33}V_2^p + C_{11}V_2^m) - \frac{V_2^p V_2^m (C_{12} - \bar{C}_{23})^2}{C_{11}V_2^p + \bar{C}_{22}V_2^m},$$

$$e^{eff}{}_{33} = \bar{\bar{e}}_{33}V_1^p + \frac{\bar{\bar{e}}_{31}V_1^p V_1^m (C_{12} - \bar{\bar{C}}_{13})}{C_{22}V_1^p + \bar{\bar{C}}_{11}V_1^m} \tag{15}$$

$$\bar{\bar{e}}_{31} = \bar{e}_{31}V_2^p + \frac{\bar{e}_{32}V_2^p V_2^m (C_{12} - \bar{C}_{12})}{C_{11}V_2^p + \bar{C}_{22}V_2^m},$$

$$\bar{\bar{e}}_{33} = \bar{e}_{33}V_2^p + \frac{\bar{e}_{32}V_2^p V_2^m (C_{12} - \bar{C}_{23})}{C_{11}V_2^p + \bar{C}_{22}V_2^m}, \tag{16}$$

$$\bar{C}_{jk} = C_{jk}^E + \frac{V_3^m e_{3j}e_{3k}}{V_3^m \mu_{33} + V_3^m \mu^s{}_{33}}, \qquad \bar{e}_{3j} = \frac{\mu_{33}e_{3j}}{V_3^p \mu_{33}V_3^m \mu^s{}_{33}} \tag{17}$$

Here, v_i^p and v_i^m, for $i=1,2$ represents respectively the length fraction of ceramic and matrix phase along direction i and

$$V_3{}^p = \frac{\dfrac{p}{h}}{\dfrac{p}{h} + (1 - V_2{}^p)} \qquad b << p \tag{18}$$

represents the volume fraction of ceramic phase in RVE. Similar models for uniform packing circular fiber can be found in [39]. Essentially, these models provide dominant electro-mechanical coupling in direction 3, which can be aligned along the local host beam axis during bonding or embedding. This is unlikely in uniformly electroded PZT plate structure.

Design Steps for Broadband Control

Once the bulk material properties of *PFC* and the *IDE* geometry are available, one can compute minimum specific actuation length that can sustain the allowable shear stress due to assumed electro-mechanical loading. After obtaining the effective properties of the chosen PFC actuator as discussed in the previous section, a coupled broadband analysis considering the distributed PFC actuator dynamics needs to be carried out. Since a closed form solution for multiple modal control over broad frequency band is not possible, optimal PFC actuator configuration (length, placement and electrical input) can be obtained by extremizing one or more cost functions over broadband by iteration, which can be the most general strategy. However, preliminary information regarding optimal actuator configuration for static and steady-state requirement available in closed form [46] can be used as the initial configuration for optimal control. Also, any additional constraint due to actuator electrical saturation and limited power supply can be imposed for practical applications. Figure 5 represents the above design steps for PFC distributed actuation for broadband structural control in a nutshell.

In the following sections, first we will explain the general framework for Spectral Finite Element (SFEM) formulation and then go on to derive an Active Spectral Finite Element Model (ASFEM) for closed-loop broadband control of vibration and waves in composite skeletal structures. The basic framework of the ASFEM model is based on the SFEM. Next the effect of distributed PFC actuator dynamics for broadband control is studied using the developed ASFEM.

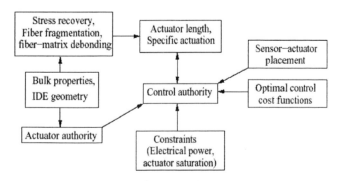

**Figure 5: Schematic diagram showing the design steps for PFC distributed
actuation for broadband structural control**

A case study on the active feedback control of multiple wave transmission in helicopter gearbox support-struts are carried out, which brings out various advantages of ASFEM while dealing with the various complexities related to finite and integrated strut-actuator dynamics and non-collocated feedback control of waves.

Spectral Finite Element Method (SFEM)

SFEM is FEM formulated in the frequency domain and wavenumber space. That is, these elements will have interpolating functions that are complex exponentials or Bessel functions. These interpolating functions are also functions of the wavenumbers. In this method, a governing partial 1-D wave equation, when transformed into frequency domain using DFT, removes the time derivative and reduces the PDE to a set of ODE, which have complex exponentials as solutions. In SFEM, we use these exact solutions as the interpolating functions. As a result, the mass is distributed exactly and hence, one single element is sufficient between any two discontinuities to get exact response irrespective of the frequency content of the exciting pulse. That is, one SFEM can replace hundreds of FEM normally required for wave propagation analysis. Hence, SFEM is an ideal candidate for multi-modal control of structures. In addition to smaller system sizes, other major advantages of the SFEM are the following

1. Since the formulation is based on frequency domain, system transfer functions are the direct by-product of the approach. As a result, one can perform inverse problems such as force identification/system identification in a straightforward manner.
2. The approach gives the dynamic stiffness matrix as function of frequency, directly from the formulation. Hence, we have to deal with only one element dynamic stiffness as apposed to two matrices in FEM
3. Since different normal modes have different amount of damping at various frequencies, by formulating the elements in the frequency domain, one can treat the complex damping mechanisms more realistically
4. The SFEM lets you formulate two sets of elements, one is the finite length element and the other is the infinite element or throw-off element. These throw-off element acts as a conduit of energy out of the system. There are various uses of this infinite throw-off element such as, adding maximum damping, obtaining good resolution of the responses in the time and frequency domains and also in modeling large lengths, which are computationally very expensive in FEM.
5. SFEM is probably the only technique that gives you responses both in time and frequency domain in the same analysis.

SFEM can be formulated in a similar manner as FEM by writing the weak form of the governing differential equation and substituting the assumed functions for displacements and integrating the resulting expression. Since the functions involved are much more complex, integration of these functions in closed form takes a longer time. In addition, by this approach, we cannot obtain the dynamic stiffness matrix of the throw-off element, as the throw off element is normally complex. Hence, we adopt an equilibrium approach of element formulation, which eliminates integration of the complex functions. The above approach is given in great detail in References [47-49]. In this chapter, we will concentrate more on the formulation of the ASFEM

Formulation of the spectral elements requires determination of the spectrum (the variation of wavenumber with frequency) relations and the dispersion relations (speed with frequency). For this, the governing PDE and the associated force boundary conditions are derived using Hamilton's principle. Next, the interpolating functions are written in terms of wave solutions, from which the dynamic (spectral) shape functions are derived. Using this, the element dynamic stiffness matrix is derived.

Active Spectral Finite Element Model (ASFEM)

In this model, the beam network is discretized and classified into three different classes of elements. These classes are(1) spectral element for finite beams with mechanical and passive properties (2) distributed or point sensors and (3) distributed or point actuators. A schematic diagram of a sensor-actuator element configuration is shown in Figure 6

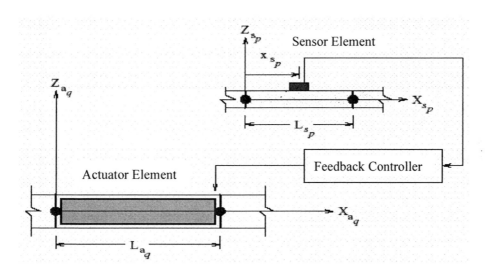

Figure 6. Sensor-actuator element configuration for Active Spectral Finite Element Model (ASFEM).

Here, it is assumed that that controller output for a single actuator can be designed based on a feedback signal constructed from a group of sensors. Furthermore, in Figure 6, the connectivity between the *pth* sensor and the *qth* actuator is also shown, where the sensor response is measured at the local coordinate system (X_s^p, Z_s^p) and the actuation force is provided at the local coordinate system (X_a^q, Z_a^q).

Spectral Element for Finite Beams

The Spectral Finite Element formulation for elementary and higher order laminated composite beams is given in References [47 & 48] and hence not explained in detail here. The nodal displacement vector is given by

$$\{\hat{u}\}^e = \{\hat{u}_1^0, \hat{w}_1, \hat{\theta}_1, \hat{u}_2^0, \hat{w}_2, \hat{\theta}_2\}^T$$

Here, $\hat{u}_1^0$ and $\hat{u}_2^0$ are the mid-plane axial displacements, $\hat{w}_1$ and $\hat{w}_2$ are the transverse displacements, $\hat{\theta}_1$ and $\hat{\theta}_2$ are the slopes at the two nodes of the spectral finite element. The nodal force vector can be written as

$$\{\hat{f}\}^e = \{\hat{N}_1, \hat{V}_1, \hat{M}_1, \hat{N}_2, \hat{V}_2, \hat{M}_2\}^T$$

where N, V, M represents the axial force, shear force and the bending moment at the two ends of the beam spectral element. As in conventional FEM, the displacement field can be expressed in terms of nodal displacement through shape functions. However, these shape functions, as apposed to polynomials in FEM, will contain terms involving complex exponentials. They can be written as

$$\{\hat{u}(x, \omega\} = [\aleph(x, \omega)]\{\hat{u}\}^e \tag{19}$$

In the above expression $[\aleph(x, \omega)]$ is a matrix of dynamic shape functions. Using these, one can relate the nodal forces and nodal displacement through a dynamic stiffness matrix $[\hat{K}]^e$ as given below

$$\{\hat{f}\}^e = [\hat{K}]^e\{\hat{u}\}^e \tag{20}$$

The stiffness matrix can be transformed to any desired coordinate system and assembled as in the case of conventional FEM

Sensor Element

For illustrative purposes, point sensor has been considered in the modeling. However, it should be noted that the formulation does allow for distributed sensors such as piezoelectric film sensors. The force balance equations for the sensor element are identical in form to that of finite beams explained earlier. Based on the response measured by a displacement sensor (s), which is located at (x_{s_p}, z_{s_p}) in p^{th} sensor element (denoted by subscript s_p), the actuator input spectrum can be expressed with the help of Equation (19) as

$$\hat{\eta}_u(x_{s_p},z_{s_p},\omega_n)=(i\omega_n)^m\,\alpha\left[\sum_{j=1}^{6}(\aleph_{1j}-z_{s_p}\aleph_{3j})\hat{u}_j^e\right] \tag{21}$$

when longitudinal displacement is measured, and

$$\hat{\eta}_w(x_{s_p},z_{s_p},\omega_n)=(i\omega_n)^m\,\alpha\left[\sum_{j=1}^{6}\aleph_{2j}\hat{u}_j^e\right] \tag{22}$$

when transverse displacement is measured. In the above equations, $i=\sqrt{-1}$, α is the sensitivity parameter and $m=0$, 1, and 2 for displacement, velocity and acceleration spectra. Similarly, if strain sensor is used, one can write the actuator input spectrum as

$$\hat{\eta}_\varepsilon(x_{s_p},z_{s_p},\omega_n)=(i\omega_n)^m\,\alpha\left[\sum_{j=1}^{6}\left(\frac{\partial\aleph_{1j}}{\partial x}-z_{s_p}\frac{\partial\aleph_{3j}}{\partial x}\right)\hat{u}_j^e\right] \tag{23}$$

Actuator Element

In the formulation of ASFEM presented here, we consider the PID (Proportional Integral Derivative) feedback control scheme [50]. Other types of frequency domain control schemes, such as feed-forward control can also be implemented. Controller output in the form of current spectrum ($\hat{I}$) (or voltage spectrum $\hat{\varphi}$) for the qth actuator, and the resulting field (magnetic field $\hat{H}$ for magnetostrictive material or electric field $\hat{E}$ for piezoelectric material) can be written as

$$\hat{I}=\sum_p\gamma\hat{\eta},\quad \hat{H}=\beta\hat{I},\quad \hat{\eta}=(\hat{\eta}_u,\hat{\eta}_w,\hat{\eta}_\varepsilon) \tag{24}$$

where, $\hat{\eta}$ is given by Equations (21-23). The constant γ is a scalar gain, and β is the actuator sensitivity parameter introduced to account for the actuator assembly and packaging properties (e.g. the solinoid configuration for packaged Terfenol-D rod actuator [51], voltage-to-electric field conversion factor for plane-polarized PZT wafers etc.). Next, after substituting for $\hat{H}$ from the magneto-mechanical (or electro-mechanical) force boundary condition into Equation (24) and following the same procedure as used for discretizing the purely mechanical domain using SFEM, the force balance equation for the qth actuator element (denoted by subscript a_q) in actuator local coordinate system can be obtained as

$$\{\hat{f}\}^e_{a_q} = [\hat{K}]^e_{a_q}\{\hat{u}\}^e_{a_q} + \left[A^{eff}_{33} \quad 0 \quad -B^{eff}_{33} \quad -A^m_{33} \quad 0 \quad B^m_{33} \right]\beta\gamma\hat{\eta} \qquad (25)$$

where

$$[A^{eff}_{33}, B^{eff}_{33}] = \int e^{eff}_{33}[1, z]dA$$

defines the equivalent mechanical stiffness due to effective magneto-mechanical (or electro-mechanical) coupling coefficient e^{eff}_{33} (see Equation (11) for PFC) for actuation in longitudinal mode. Similar vector with non-zero second and fifth elements in Equation (25) for actuation in shear mode can also be used. After substituting $\hat{\eta}$ in terms of the sensor element shape function matrix ($[\aleph]$) and the corresponding nodal displacement vector ($\{u\}^e$) from Equations (21-23), the Equation (25) can be re-written as

$$\{\hat{f}\}^e_{a_q} = [\hat{K}]^e_{a_q}\{\hat{u}\}^e_{a_q} + [\hat{K}]^e_{a_q \leftarrow s_p}\{\hat{u}\}^e_{s_q} \qquad (26)$$

where, the notation $[\hat{K}]^e_{a_q \leftarrow s_p}$ is introduced to represent the Sensor-Actuator Stiffness Influence Matrix (SASIM). The obtained stiffness matrix is then transformed at the actuator nodes from local coordinate system to the global coordinate system using the standard procedures as adopted in conventional FEM. This procedure leads to the final expression for the qth actuator element with the pth feedback sensor, which is given by

$$\{\hat{f}\}_{a_q} = [\Gamma]^T_{a_q}[\hat{K}]^e_{a_q}[\Gamma]_{a_q}\{\hat{U}\}_{a_q} + [\Gamma]^T_{a_q}[\hat{K}]^e_{a_q \leftarrow s_p}[\Gamma]_{s_p}\{\hat{U}\}^e_{s_p} \qquad (27)$$

Hence, the assembled closed-loop MIMO system with general sensor-actuator configuration in the ASFEM is obtained in the form

$$\begin{Bmatrix} \{\hat{f}_{s_p}\} \\ \cdot \\ \cdot \\ \cdot \\ \{\hat{f}_{a_q}\} \end{Bmatrix} = \begin{bmatrix} [\Gamma]^T_{s_p}[\hat{K}]^e_{s_p}[\Gamma]_{s_p} & \cdot\ \cdot\ \cdot & [0] \\ & \cdot & \cdot\ \cdot\ \cdot & \cdot \\ & \cdot & \cdot\ \cdot\ \cdot & \cdot \\ & \cdot & \cdot\ \cdot\ \cdot & \cdot \\ [\Gamma]^T_{a_q}[\hat{K}]^e_{a_q \leftarrow s_p}[\Gamma]_{a_q} & \cdot\ \cdot\ \cdot & [\Gamma]^T_{a_q}[\hat{K}]^e_{a_q}[\Gamma]_{a_q} \end{bmatrix} \begin{Bmatrix} \{\hat{U}_{s_p}\} \\ \cdot \\ \cdot \\ \cdot \\ \{\hat{U}_{a_q}\} \end{Bmatrix} \qquad (28)$$

As evident from the above derivation, restrictions are not placed on sensor and actuator locations. Also, in terms of computational cost and broadband analysis capabilities, the proposed ASFEM is a better option compared to conventional state-space model which are very high order and several accuracy related problems due to error in model-order reduction, modal truncation etc. needs to be addressed before they can be applied for broadband LAC. So far Low Authority Control (LAC) is concerned, after combining the displacement (or strain) field generated by the primary disturbance (external mechanical load) and the secondary sources (actuators), one can obtain the wave coefficient vector $\{\tilde{\boldsymbol{u}}\}$ for a sub domain Ω of interest as

$$\{\tilde{u}_\Omega\} = \hat{T}_\Omega^{-1}\left[[\hat{K}]+[\hat{K}(x_s)_{a\leftarrow s}]\right]^{-1}\{\hat{f}_\Omega\} \tag{29}$$

At this stage, if a transfer function based concept of wave cancellation is chosen for designing the controller. Equation (29) provides a direct way to carry out identification of appropriate control gains for known sensor and actuator locations that will reduce certain elements of $\{\tilde{u}_\Omega\}$ to zero, and hence the corresponding wave components can be controlled. However, analytical approach to achieve this is limited by the fact that one cannot obtain an explicit expression for the dependence of local wave components on sensor and actuator locations and other control parameters for a complex problem, which may have more than one discretized subdomain. Hence, a semi-automated scheme integrated with ASFEM is chosen to analyze the spatially rediscretized system by changing sensor locations or actuator locations on an iterative basis. This is feasible because of fast computation and small system size permitted by ASFEM.

Numerical Implementation

As the initial step, input time dependent forces or disturbances are decomposed into Fourier components by using the forward FFT. Note that all of the element-level operations as well as the global system-level operations are carried out at each discrete frequency ω. Except for this basic difference, the proposed program architecture is almost identical (for an open-loop configuration) to a finite element program in terms of features such as input, assemblage, solving of the system, and output. For a closed-loop system, we use Equation (28) to implement the explicit form of the global dynamic stiffness matrix at a particular frequency, which is in most of the cases, neither banded nor symmetric. Here, a non-symmetric sparse complex matrix inversion routine has been used as a part of the global system solver. After solving the closed-loop system at each frequency, time history of displacements, strains, stresses etc. are then post-processed using inverse FFT.

Effect of Broadband Actuator Dynamics

Here we consider a composite cantilever beam with surface-bonded PFC to study the effect of distributed actuator dynamics towards broadband control of transverse response of the cantilever tip under transverse impact type loading at the tip. Essentially the control point of interest for LAC in this case is the cantilever tip. This requires a feedback sensor to be placed at the cantilever tip. It is well known that optimal placement of actuator is dictated by the location of high average strain [46]. Hence, while controlling the static or first mode shape under tip loading on a cantilever beam, it is essential that the actuator be placed at the root of the cantilever beam. For velocity feedback with collocated sensor, [46] has shown that the damping of a particular vibration mode, while using surface-bonded PZT wafer, can be expressed as

$$\xi = \frac{e_{31}\gamma_{\dot{\omega}}h_B b}{2M\omega_0 L_B(6+\psi)}\Big[\hat{\phi}(x_2), x - \hat{\phi}(x_1), x\Big] \tag{30}$$

where $\gamma_{\dot{\omega}}$ is the velocity feedback gain, ω_0 is the natural frequency associated with the mode, M is the modal mass, L_B is the length of the beam, b is the width of the beam, h_b is the thickness of the beam, $\psi = C_B h_B / C_{PZT} h_B$ is the effective stiffness ratio for PZT wafer thickness of h_B. $\hat{\phi}(x_2), x - \hat{\phi}(x_1), x$, is the difference between the gradients of the strain mode shape at the two ends (x_1, x_2) of the PZT actuator. With the specified actuator-to-beam length scale, the maximum modal damping of a single mode is obtained for maximum feedback gain

$$(\gamma_{\dot{\omega}})_{\max} = \frac{E_{\max}h_{PZT}}{\omega_0\hat{w}_{\max}} \tag{31}$$

where $E_{\max}$ is the saturation electric field. Note that when the feedback sensor is placed non-collocated with the actuator, the sensor output signal will have phase difference with the modal strain at the actuator ends and the effect can be destabilizing for phase difference more than 180°. Similar consequence will also be evident while controlling more than one mode using the same configuration. For example, while controlling the second mode along with the first mode, one has to overcome the difficulty of almost zero average modal strain around the strain node (point of zero modal strain) at $x = 0.216L_B$. Two options are available to overcome this difficulty. One is to use segmented actuators, where one of the actuator located at $x < 0.216L_B$ must be driven 180° out of phase with a second actuator located at $x > 0.216L_B$. Obviously, control over more number of modes means more strain nodes and hence more number of segmented actuators. Also, the possibility of interaction between the controlled modes and modal spillover for multiple segmented actuators becomes evident. This necessitates the requirement of appropriate optimal control strategy. The second option for multi-modal control is to use single actuator at the root of the cantilever beam with optimal length of the actuator and frequency-weighted optimal gain while use a non-collocated sensor. Direct feedback from control point of interest is found to be more suitable when a large number of modes over broad frequency band are to be controlled. The fundamental behavior of this non-collocated sensor-actuator configuration for LAC resembles with that of the disturbance propagation in structural network [52].

In the following numerical simulation, the sensitivity of the PFC actuator length (actuator located at the cantilever root) while using velocity feedback from the sensor (located at the cantilever tip) is studied. The configuration is shown in Figure 7.

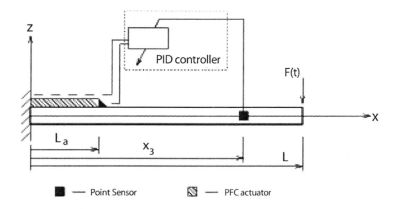

Figure 7. A composite cantilever beam with surface bonded PFC actuator and non-collocated velocity feedback sensor placed at $x = x_s$ for broadband local control at the tip.

The beam is of length $L = 1m$, and thickness $2\ cm$. AS/3501-6 graphite-epoxy material property with play-stacking sequence $[0_5^0 / 90_5^0]$ is considered. Assuming Euler-Bernoulli beam kinematics, the coupled electro-mechanical wave equation can be expressed as

$$-A_{33}\frac{\partial^2 u^0}{\partial x^2} + B_{33}\frac{\partial^3 w}{\partial x^3} + A_{33}^{\text{eff}}\frac{\partial E_3}{\partial x} = 0 \qquad (32)$$

$$\rho A \frac{\partial^2 w}{\partial t^2} - B_{33}\frac{\partial^3 u^0}{\partial x^3} + D_{33}\frac{\partial^4 w}{\partial x^4} + B_{33}^{\text{eff}}\frac{\partial^2 E_3}{\partial x^2} = 0 \qquad (33)$$

The force boundary conditions are given by

$$\frac{\partial u^0}{\partial x} - B_{33}\frac{\partial^2 w}{\partial x^2} - A_{33}^{\text{eff}}E_z = N_x,$$

$$\frac{\partial^2 u^0}{\partial x^2} - D_{33}\frac{\partial^3 w}{\partial x^3} - B_{33}^{\text{eff}}\frac{\partial E_3}{\partial x} = V_x \qquad (34)$$

$$B_{33}\frac{\partial u^0}{\partial x} + D_{33}\frac{\partial^2 w}{\partial x^2} + B_{33}^{\text{eff}}E_3 = M_x$$

where,

$$[A_{33}, B_{33}, D_{33}] = \int_A C_{33}^{\text{eff}}[1, z, z^2]dA, \quad [A_{33}^{\text{eff}}, B_{33}^{\text{eff}}] = \int_A e_{33}^{\text{eff}}[1, z]dA \quad (35)$$

In the above equation, N_x is the axial force, V_x is the shear force and M_x is the bending moment.

The beam is subjected to an impact loading of smoothed triangular profile, which is having a frequency content of 44 kHz, in the transverse direction at the cantilever tip. Note that under such loading, which is likely to excite many higher order modes, the control analysis becomes challenging because of additional axial-flexural coupling due to the unsymmetric ply-stacking sequence. Control of multiple spectral peaks in the frequency response of transverse tip displacement is considered as the local control objective. If satisfied, this requirement will also ensure the stability of close-loop system. Because all the resonant modes will be damped and hence, one can expect the poles to move to the left-half of the complex phase-plane. Also, the possibility of modal truncation over a sufficiently large frequency band can be eliminated. It is important to note that the waves that will travel from tip to the fixed-end of the beam will be of same order of magnitude as the incident impact. It is also necessary that the scattered axial and flexural waves from the fixed-end be suppressed. This is also one of the reason apart from those discussed in the context of Equations (30) and (31), for which the PFC actuator is placed adjacent to the fixed-end. The sensor is assumed at $x = x_s$, which is considered near the tip for direct velocity feedback to the actuator in advance.

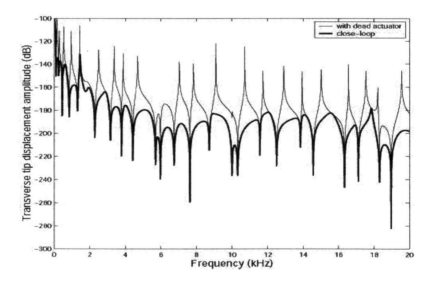

Figure 8. Close-loop transverse displacement at the cantilever tip under unit impulse excitation at the tip $x_s / L = 1, L_a / L = 0.25$.

A non-dimensional scalar feedback gain g is derived from the feedback gain γ (Equation (24)) to perform parametric study. These two quantities are related as $g = (c_0 \alpha \beta) \gamma / E_0$, where E_0 is a reference AC voltage and c_0 is the speed of sound in air. An optimal closed-loop performance which corresponds to $L_a = 0.25m$, $x_s = 1.0m$ (at the tip) and $g = 3.4 \times 10^6$ is shown in Figure 8. From Figure 8, the locations of the forced resonances and anti-resonances along the frequency axis and corresponding spectral amplitudes of transverse displacement at the tip are clearly seen. It can be seen that the configuration is able to suppress most of the resonant modes. Further, we study the effect of parametric variation on amplitude level over the frequency range of *20kHz* under consideration. First, it is assumed that the feedback gain *g* chosen above is optimal and is not sensitive to small variation in other parameters, such as L_a and x_s.

L_a is slowly varied from *0.15m* to *0.35m* corresponding to velocity feedback from various sensor locations x_s moving away from cantilever tip (Figure 7). Integral effect of the change in amplitude level of closed-loop response (transverse displacement at the tip) over whole frequency range is evaluated using the control cost function

$$\Pi = \sum_{n=1}^{N/2} 20.0 \left[\log_{10} \left| \hat{w}(\omega_n)^2_{open} \right| - \log_{10} \left| \hat{w}(\omega_n)^2_{close} \right| \right] \tag{36}$$

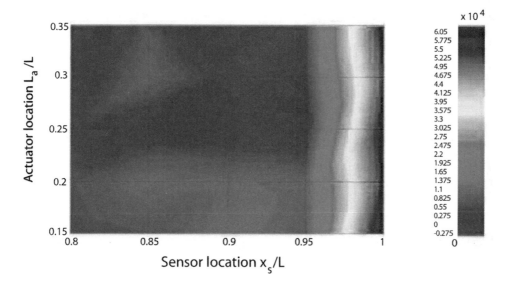

Figure 9. Performance of actuator and control point interaction by variation in total amplitude level of transverse tip response.

In Figure 9, the sensitivity of Π is shown by a two dimensional solution space involving the actuator length L_a and sensor location x_s. This plot confirms the result of Figure 8 that one optimum solution exists at $x_s / L = 1, L_a / L = 0.25$ and yields a total reduction of *6.025dB* in Π. Figure 9 also predicts that another solution exist at $x_s / L = 1, L_a / L = 0.15$. It is clear from the plot that sensor placed before *0.96m* never performs effectively over the significant resonant modes. Whereas, collocated configuration before *0.8m* may not produce stable performance.

Linear Quadratic Optimal Control Using Spectral Power

As a distributed measure of the controlled structural response, we consider frequency domain power flow at the spectral element nodes which is defined as the product of the nodal force vector and conjugate of the nodal velocity vector. In matrix notation, this is expressed as

$$\hat{P}(\omega_n)^e = \{\hat{f}^e\}^T (i\omega_n \{\hat{u}^e\})^* \tag{37}$$

Considering power expression for an element internal point (x,z), itcan be shown that the real part of the complex power is constant over an uniform element domain and spatial rate of change of total power is purely imaginary, i.e. $\mathbf{Re}(\partial P / \partial x) = \mathbf{0}$.This implies that real part of the complex power is the contribution from far field disturbance transmitted through the structural joints. The imaginary part amounts to the energy trapped in the element, which is essentially the near field effect on the sensor. Since, we emphasize on accuracy of the feedback signal from the sensor near the control point in LAC, one possible approach is to consider the power amplitude. Here, we choose a linear quadratic performance measure H_{ctrl} for the closed-loop system subjected to an upper-bound E_{drv} of the actuator input voltages. For multiple LAC objectives, the elements of the matrix measure to be achieved are expressed as

$$H_{ctrl} = \frac{\left|\hat{P}(\omega_n)_0\right| - \left|\hat{P}(\omega_n)_m\right|}{\left|\hat{P}(\omega_n)_0\right|} > 0, \quad E_{drv} = \frac{\left|\hat{E}(\omega_n)_{max}\right| - \left|\gamma(\omega_n)_m \hat{\eta}(\omega_n)_m\right|}{} > 0 \tag{38}$$

The subscripts *ctrl* and *drv* respectively stands for the control points (for LAC) and driving points (actuator element nodes), respectively. Amplitude of $\hat{E}(\omega_n)_{max}$ represents allowable peak voltage to the PFC actuator and the corresponding frequency range depends on the capacity of the voltage amplifier being used. It is to be noted that ω_n must be considered within the actuator bandwidth to match with numerically simulated performance. In ASFEM, all the required matrix computations for open-loop as well as closed-loop are performed at every discrete frequency ω_n.Equation (28) is solved for unit amplitude of the FFT signal of external mechanical disturbance. After solving the system, actual frequency response for displacement, force, power and controller input-output transfer function are obtained by convolving with the original mechanical loading spectrum. Constrained optimization of power flow at the control points, which is governed by Equation (38), is performed at ω_n. As shown in Equation (38), frequency weighting of the feedback gain $\gamma(\omega_n)$ at ω_n for a sensor S_p is obtained by maximizing H_{ctrl} with constraint $E_{drv} > 0$ using non-uniform iterative scheme. Here, subscript m represents the mth iteration at a particular ω_n and subscript 0 represents the uncontrolled system. In ASFEM, the spatially discretized model is much smaller in size compared to other MIMO models. Taking advantage of this aspect, a semi-automated scheme is integrated with ASFEM to achieve optimal location of sensors and actuators.

Broadband Control of a Three Member Composite Beam Network

In the following numerical simulations, a three member composite beam network is considered with the objective of controlling one of its nodal displacements over a frequency band of *10 kHz*. The Y-shaped network is made of three AS/3501-6 graphite-epoxy composite beam members connected through a common rigid joint (node 1) as shown in Figure 10. The free-end (node 11) of the network is subjected to the impact type excitation as in the previous example. The local control objective is framed as the simultaneous control of horizontal displacement *u* as well as transverse displacement *w* at the common joint (node 1) of the network with minimum number of PFC actuators of prescribed length.

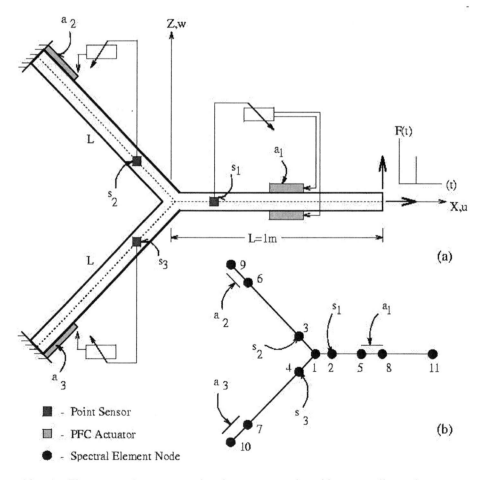

Figure 10. A Three member composite beam network with non-collocated sensor-actuator configuration to control waves at the network joint. (a) Schematic diagram of the feedback model (b) Spectral element configuration.

From the nature of the considered structural geometry and boundary conditions, it can be said that there will be multiple scattered waves which are axial-flexural coupled. Therefore, significant near field will exist on the sensors placed near node 1. However, at the initial stage of disturbance propagation, there will be only dominant transmitted wave component arriving at node 1 from free-end side. But, at latter stages of time, effect of multiple scattering amounting a major part of trapped energy in the fixed-end portions will increase. Based on this behavior, we assume four PFC actuators (each of length *0.15m*) in the locations shown in Figure 10. For direct velocity feedback to these actuators, we consider three point sensors near node 1(Figure 10) Sensor S_1 is used for transverse velocity $(i\omega_n\hat{w})$ feedback to the actuator pair a_1.

Sensors S_2 and S_3 are used for axial velocity feedback $(i\omega_n\hat{u})$ in $\pm 135°$ rotated coordinate systems which are local to the two inclined members. All of these three sensors are assumed at *0.1m* from node 1.

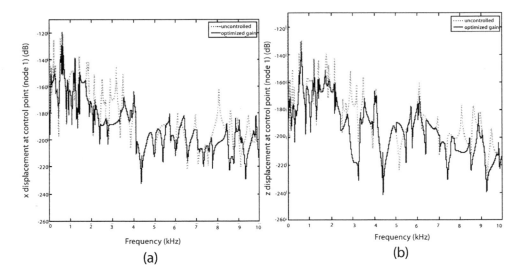

Figure 11. Optimized close-loop performance showing (a) Displacement amplitude in X direction and (b) Displacement amplitude at Z direction, at the control point (node 1 in Figure 10}.

Spectral element configuration of the above closed-loop system is schematically represented in Figure 10(b) showing the mechanical and active element nodes in SFEM. The spatially discretized system size is *27 x 27* including three sensor nodes and three actuator elements. Constrained optimization as discussed in last section for the control point (node 1) is carried out assuming allowable specific actuator input voltage of *9kV/cm* to each of the PFC actuators. To control the primary disturbance transmitted from node 11 towards the control point, the actuator pair a_1 is placed at *0.45m* from control point node 1. In Figure 11 (a) and (b), the horizontal and the transverse displacements corresponding to optimal closed-loop performance are plotted, respectively. Figure 12(a) shows the closed-loop constrained power amplitude spectrum at the control point. Frequency weighted feedback gain amplitude $\log|\gamma|$ for the three actuators are plotted in Figure 12(b). From Figures 11 (a),(b) and Figure

12(a),it can be noticed that controlled performance at different frequency bands over resonant modes are similar in the power amplitude spectrum and the displacement amplitude spectrum. Therefore, optimization of the constrained power flow proves to be an effective strategy for active distributed control of structural waves. However, in both the horizontal and transverse responses, performance near *0.6* and *4 kHz* may not be satisfactory. This can be attributed to the fact that only amplitude of the power was optimized. Real and imaginary components of the close-loop power (which constituted the far field and near field effects respectively, due to asynchronized phase) were not optimized separately. Therefore, the current optimization scheme has worked on an average basis by considering both the amplitude of near field and far field disturbances, and not by considering the associated phase information.

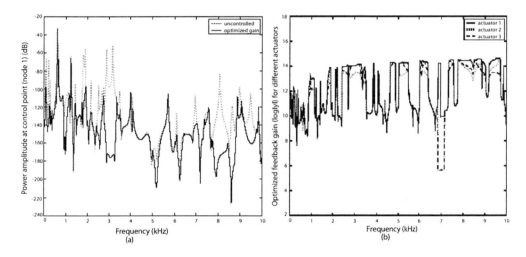

Figure 12. Optimized close-loop performance showing (a) Optimized power flow at the control point (node 1 in Figure 10) and (b) Feedback gain spectrum for three actuators.

Conclusion

This chapter is focused on the development of SFEM integrated with control algorithm for analysis and design of MIMO control system for structural vibration and broadband wave control applications. An overview of the problem complexity and modeling difficulties encountered in existing modeling techniques are presented. Some of the main issues in distributed parameter control system modeling, which are namely the control-spillover due to insufficient modal information, stability of the Low Authority Control (LAC) instead of global control, finite actuator dynamics, and single actuator multi-modal control etc. are discussed. Constitutive modeling of layered piezoelectric materials surface-bonded or embedded in host composite structures was presented. Modeling of Piezoelectric Fiber Composite (PFC) based on uniform field model is presented. Design criteria for such PFCs in surface-bonded or embedded form, which become an ideal candidate for distributed structural actuation is outlined considering the structural performance, actuator authority and overall control authority of the integrated system. Next, the Active Spectral Finite Model (ASFEM) is derived considering PID feedback sensors and distributed actuators in non-collocated form.

The resulting closed-loop model is shown to be very high dimensional one but having discretized system size as small as the total number of sensor nodes, actuator nodes and structural boundary nodes. This provides the suitability of ASFEM in broadband control where many vibration modes participates in the dynamics and conventional state-space model needs very large system size for accurate MIMO system realization and optimal control design. Numerical simulations are carried out to study the effect of distributed actuator dynamics. A case study on active control of multiple waves in a cantilever beam is presented next. The effect of sensor-actuator location in non-collocated feedback is studied. Also, the study shows a systematic approach towards development of single actuator multi-mode control system. To the end of the chapter, an optimal control algorithm based on ASFEM and spectral power flow is developed and implemented. Numerical simulations are presented to demonstrate the optimal control performance of a beam network with non-collocated point sensors and distributed PFC actuators

Acknowledgement

The author wishes to thank his former student and currently a post-doctoral research fellow at Wilfred Laurier University, Canada, Dr. Roy Mahapatra for his significant input and also his help in generating many of the results given in this chapter

References

[1]   Pelinescu, I. and Balachandran, B. Analytical and experimental investigations into active control of wave transmission through gearbox struts. *Proceedings of SPIE Smart Structures and Materials Conference on Smart Structures and Integrated Systems, Newport Beach*, CA, 2000, vol. l3985.

[2]   Wilke, P., Johnson, C., Grosserode, P. and Sciulli, D. Whole-spacecraft vibration isolation for broadband attenuation. *Proceedings of IEEE Aerospace Conference, Big Sky,* Montana, March 19-25, 2000.

[3]   Anderson, E.H., Evert, M.E., Glaese, R.M., Gooding, J.C., Pendleton, S.C.,Cobb, R.G., Erwin, R.S., Jensen, J., Camp, D., Fumo, J. and Jessen, M. Satellite Ultra quiet Isolation Technology Experiment(SUITE): electromechanical subsystems. *Industrial and Commercial Applications of Smart Structures and Technologies}, Newport Beach*, CA, March 2-4, 1999.

[4]   Leo, D.J. and Smith, C.A. Performance tradeoffs in active-passive vibration isolation. *11th Symposium on Structural Dynamics and Control}, Blacksburg, VA*, May 12-14, 1999

[5]   Anderson, E.H. and How, J.P. Active vibration isolation using adaptive feed forward control. *American Control Conference, Albuquerque*, NM, 1997, Paper I-97115B

[6]   Abe, M. and Fujino, Y. Dynamic characterization of multiple tuned mass dampers and some design formulas. *Earthquake Engineering and Structural Dynamics}*, 1994, 23(8), 813-835.

[7]   Abe, M. and Igusa, T. Tuned mass dampers for structures with closely spaced natural frequencies. *Earthquake Engineering and Structural Dynamics}*, 1995, 24(2), 247-261.

[8]   Inaudi, J.A. and Kelly, J.M. Experiments on tuned mass dampers using viscoelastic, frictional and shape-memory alloy materials. *Proceedings of First World Conference on Structural Control: International Association for Structural Control}*, Los Angeles, 3-5 August. G.W. Housner, et al, eds. Los Angeles: 1994, volume 2, pages TP3-127--TP3-136.

[9]   Lin, J.H., Zhang, W.S., Sun, D.K., He, Q., Lark, R.J. and Williams, F.W. Precise and efficient computation of complex structures with TMD devices. *Journal of Sound and Vibration}*, 1999, 223(5), 693-701.

[10] Clark, R.L., Saunders, W.R. and Gribbs, G.P. *Adaptive structures.* New York: Johm Wiley; 1998.

[11] Flotow, A.H. von. Disturbance propagation in structural network. *Journal of Sound and Vibration}*, 1986, 106(3), 433-450.

[12] Fujii, H., Ohtsuka, T. and Murayama, T. Wave-absorbing control for flexible structures with non collocated sensors and actuators. *Journal of Guidance Control and Dynamics}*, 1992, 15(2), 431-439.

[13] Tanaka, N. and Kikushima, Y. Optimal vibration feedback control of an Euler-Bernoulli beam: Toward realization of the active sink method. *Journal of Vibration and Acoustics}*, 1999, 121, 174-182.

[14] Elliott, S.J. and Billet, L. Adaptive control of flexural waves propagating in a beam. *Journal of Sound and Vibration.* 1993,163, 295-310.

[15] Roure, A. Self-adaptive broadband sound control system, *Journal of Sound and Vibration*, 1985, 101, 429-441.

[16] Mace, R.B. Active control of flexural vibrations. *Journal Sound and Vibration*, 1987, 114, 253-270.

[17] Pines, D.J. and Flotow, A.H. von. Active control of bending waves propagation at acoustic frequencies. *Journal of Sound and Vibration*, 1990, 142, 391-412.

[18] Kuo, S.M. and Morgan, D.R. Active noise control systems. New York: *John Wily*; 1996.

[19] Balas, M.J. Direct velocity feedback control of large space structures. *Journal of Guidance and Control}*, 1979, 2, 242-253.

[20] Auburn, J.N. Theory of control of structures by low-authority controllers. *Journal of Guidance and Control}*, 1980, 3, 444-451.

[21] Baumann, W.T. An adaptive feedback approach to structural vibration suppression. *Journal Sound and Vibration*, 1997, 205(1), 121-133.

[22] Makarenko, A.A. and Crawley, E.F. *Force and strain feedback for distributed actuation.* Technical Report, SSL $\#98-10$, Space Engineering Research Center, Massachusetts Institute of Technology, Cambridge, USA. 1998.

[23] McConnell, K.G. and Cappa, P. Transducer inertia and stringer stiffness effects on FRF measurements. *Mechanical Systems and Signal Processing*, 2000, 14(4), 625-636.

[24] Goh, C.J. and Caughey, T.K. On the stability problem caused by finite actuator dynamics in the collocated control of large space structures. *International Journal of Control*, 1985, 41(3), 787-802.

[25] Noyer, B. de M. and Hanagud, S. Single actuator and multi-mode acceleration feedback control. *Adaptive Structures and Material Systems*, ASME, 1997, 54, 227-235.

[26] Gopalakrishnan, S. and Roy Mahapatra, D. Active Control of Structure-Borne Noise in Helicopter Cabin Transmitted Through Gearbox Support Struts. IUTAM Symposium on

Designing for Quietness, Solid Mechanics and Its Applications, Edited by *Munjal, M.L.,* *Kluwer Academic Publishers*, Netherlands,2002, 99-125.

[27] Roy Mahapatra, D., and Gopalakrishnan, S. Optimal Spectral Control of Broadband Waves in Smart Composite Beams with Distributed Sensor-Actuator Configurations. Paper no: 4234-12, Proceedings of SPIE's 2000 *Symposium on Smart Materials and MEMS*, December 13-15, 2000, Melbourne, Australia.

[28] Roy Mahapatra, D., Gopalakrishnan, S., and Balachandran, B. Active Feedback Control of Multiple Waves in Helicopter Gearbox Support Struts. *Smart Structures and Materials*, 2001, 10, 1046-1058.

[29] Roy Mahapatra, D., and Gopalakrishnan, S. An Active Spectral Element Model for PID Feedback Control of Wave Propagation in Composite Beams. *CD-ROM Proceedings of First International Conference on Vibration Engineering and Technology of Machinery (VETOMAC- I)*, October 2000, IISc, Bangalore.

[30] Burke, S.E., Hubbard, J.E.J., Meyer, and J.E. Collocation: design constraints for distributed and discrete transducers. *Proceedings of 13th Biennial Conference on Mechanical Vibration and Noise*, Miami, Florida, Sept 22-25, 1991, DE Vol. 34, 75-81.

[31] Fleming, F.M. and Crawley, E.F. The zeroes of controlled structures: sensor/actuator attributes and structural modeling, *Proceedings of 32nd AIAA Structures, Structural Dynamics and Materials Conference}*, Baltimore, MD, *AIAA* paper 91-0984, 1991.

[32] Hall, S.R., Crawley, E.F., How, J. and Ward, B. A hierarchic control architecture for intelligent structures. *Journal of Guidance Control and Dynamics*, 1991, 14(3), 15-25.

[33] Gardonio, P. and Elliott, S.J. Active control of wave in a one-dimensional structure with scattering termination. *Journal of Sound and Vibration*, 1996, 192, 701-730.

[34] Sutton, T.J., Elliott, S.J., Brennan, M.J., Heron, K.H. and Jessop, D.A.C. Active isolation of multiple structural waves on a helicopter gearbox support strut. *Journal of Sound and Vibration*, 1997, 205(1), 81-101.

[35] Clark, R.L., Pan, J. and Hansen, C.H. An experimental study of multiple wave types in elastic beams, *Journal of Acoustical Society of America*, 1992, 89(1), 871-876

[36] Mason, W.P. *Piezoelectric crystals and their applications to ultrasonics*. Van Nostrand; 1950

[37] Cady, W.G. Piezoelectricity, *Dover Publications*; 1964.

[38] Bent, A., Hagood, N.W. and Rodgers, J.P. Anisotropic actuation with piezoelectric composites. *Journal of Intelligent Material Systems and Structures*, 1995, 6(3), 338-349.

[39] Bent, A.A. *Active fiber composites for structural actuation.* PhD Thesis. Massachusetts Institute of Technology, USA, 1995.

[40] Mahut, T., Agbossou, A. and Pastor, J. Dynamic analysis of piezoelectric fiber composite in an active beam using homogenization and finite element method. *Journal of Intelligent Material Systems and Structures*, 1998, 9, 1009-1015.

[41] Lerch, R. Simulation of piezoelectric devices by two- and three-dimensional finite elements. *IEEE Transactions on Ultrasonics, Ferroelectrics and Frequency Control*, 1990, 37(2), 233-247.

[42] Mackerle, J. Sensors and actuators: finite element and boundary element analyses and simulations. A bibliography (1997-1998), *Finite Elements in Analysis and Design,* 1999, 33, 209-220.

[43] Benjeddou, A. Advances in piezoelectric finite element modeling of adaptive structural elements: a survey. *Computers and Structures*, 2000, 76, 347-363.

[44] Roy Mahapatra, D. Development of Spectral Finite Element Models for Wave Propagation Studies Health Monitoring and Active Control of Waves in laminated Composite Structures, Ph.D. Thesis, Indian Institute of Science, India, 2003

[45] IEEE Std 176-1978}. IEEE Standard on Piezoelectricity. *The Institute of Electrical and Electronics Engineers*, 1978.

[46] Crawley, E.F. and Luis, J. de. Use of piezoelectric actuators as elements of intelligent structures. *AIAA Journal*, 1987, 25(10), 1373-1385.

[47] Roy Mahapatra, D., Gopalakrishnan, S., and Sankar, T.S. Spectral Element Based Solutions for Wave propagation Analysis of Multiply Connected Unsymmetric Laminated Composite Beams. *Journal of Sound and Vibration*, 2000, 237(5), 819-836.

[48] Roy Mahapatra, D. and Gopalakrishnan, S. Axial-Shear-Bending Coupled Wave Propagation in Thick Composite Beams. *Composite Structures*, 2003, 59(1), 67-88.

[49] Roy Mahapatra, D. and Gopalakrishnan, S. A Spectral Finite Element for Analysis of Wave Propagation in Uniform Composite Tubes. *Journal of Sound and Vibration*, 2003, 268 (3), 429-463.

[50] Anderson, B.D.O. and Moore, J.B. *Optimal Control*. Englewood Cliffs, New Jersey: Prentice Hall; 1990

[51] Butler, J.L. Application Manual for the Design of ETREMA Terfenol-D Magnetostrictive Transducers. Ames, *IA: EDGE Technologies Incorporated*, 1988.

[52] Miller, D.W., Hall, S.R. and Flotow, A. von. Optimal control of power flow in structural junctions. *Journal of Sound and Vibration*, 1990, 140(3), 475-497.

In: Progress in Smart Materials and Structures
Editor: Peter L. Reece, pp. 151-202

ISBN: 1-60021-106-2
© 2007 Nova Science Publishers, Inc.

Chapter 6

SMART COMPOSITE PLATFORMS FOR SATELLITES THRUST VECTOR CONTROL AND VIBRATION SUPPRESSION

Kougen Ma and Mehrdad N. Ghasemi-Nejhad[†]*

Intelligent and Composite Materials Laboratory,
Department of Mechanical Engineering University of Hawaii at Manoa,
Honolulu, HI

Abstract

Thrusters are commonly used in satellite attitude control systems to provide controllable external thrust force, and hence maintain or change satellite orbits. Ideally, the thrust vector of a satellite thruster should pass through its mass center. In reality, this thrust vector alignment is not always met during the lifetime of a satellite, and consequently resulting in a disturbance torque that could cause the losses of the satellite orientation and orbit keeping. To eliminate the effects of this disturbance torque in the presence of a large thrust level, a reaction control system (RCS), which spins the satellite in the opposite direction, is currently used. This RCS consists of several small auxiliary thrusters that impose a significant mass penalty and onboard fuel consumption, and hence increase the launch cost and shorten the satellite lifetime. In addition, the firing of thrusters in a satellite generates vibration that resonates throughout the entire satellite structure, and hence renders onboard sensitive devices non-operational.

In this chapter, a novel satellite thrust vector control and vibration suppression technology is introduced, including the following five aspects: 1) the concept of the novel technology, 2) the development of a two-degree-of-freedom smart composite platform, 3) the platform control, 4) satellite thrust vector control using the platform, and 5) the satellite structure vibration suppression using the platform.

The core of this novel technology is to place a smart structural interface ---- the UHM smart composite platform, between the satellite thruster and the satellite structure to make the satellite thruster steerable. By steering the thruster in real time the thrust vector of a satellite can always be pointed in the desired direction and a vibration suppression capability can also

---

[*]. E mail address: kougen@hawaii.edu, Tel: 808-956-5639; Fax: 808-956-2373
[†]. E mail address: nejhad@eng.hawaii.edu, Tel: 808-956-7560; Fax: 808-956-2373

be provided by the same smart platform simultaneously. To achieve this novel technology, the UHM smart composite platform employing the state-of-the-art smart structures science and technology is developed. This platform consists of smart composite panels, smart composite struts, and advanced control systems, and possesses simultaneous precision positioning and vibration suppression capabilities. The inverse kinematics of the platform is analyzed, and an adaptive nonlinear modeling method is proposed to model the platform kinematics. Advanced control strategies are investigated and applied to specially treat the existing nonlinearities, and two control strategies, namely local control strategy and global control strategy, are presented and compared experimentally.

To investigate the satellite thrust vector control, the platform is assembled onto a satellite structure. The satellite attitude dynamic model for thrust vector control is then built, and the satellite attitude controller and intelligent controller for the smart composite platform are designed. The successful performance of the thrust vector control employing the UHM smart composite platform is proven here. The results indicate that the smart composite platform can precisely achieve the thrust vector control, and the misalignment of the thrust vector of the satellite can be corrected effectively with satisfactory position accuracy of the thrust vector.

The vibration suppression capability of this novel technology is also assessed. The combined system dynamics of the satellite structure and the smart composite platform is analyzed, and the dominant modes of the satellite structure are determined. A MIMO adaptive control scheme is then developed to suppress the satellite structure vibration employing four PZT stack actuators in the three smart composite struts and the central support of the platform as well as three PZT patch actuator pairs in the platform device plate. A convergence factor vector concept is also introduced to ease the multi-channel convergent rate control. This vibration controller is adjusted based on the vibration information of the satellite structure and drives the smart composite platform to isolate the vibration transmission from the firing thruster to the satellite structure. Eleven vibration components of the structure and platform are controlled. The results demonstrate that the entire vibration of the structure at its dominant frequency can be suppressed to 7-10% of its uncontrolled value for various device plate position configurations.

Keywords: Precision platform, parallel robot, intelligent control, smart structures, thrust vector control, precision positioning, vibration suppression, local control, global control

Introduction

Since the ability to control the thrust vector of any propulsion system is extremely advantageous, thrust vector control (TVC) has been considered a key area of research in launch systems, upper stages, and vertical or short takeoff and vertical landing aircrafts. For example, a thrust vector controller for a space shuttle vehicle with multiple engines was presented in [1]. The controller maintained vehicle trajectory and thrust vector while minimizing risk and damage to each engine and to the propulsion system as a whole by independently controlling the thrust magnitude and exhaust cone gimbal angles of each engine. For rockets, missiles [2], and jet planes [3, 4] TVC becomes a crucial technology of controlling their trajectories and achieving high maneuverability. In satellite systems, TVC systems can be used to correct the misalignment of the thrust vector caused by the shift of the mass center of a satellite due to fuel consumption and to reduce the vibration induced by the thruster firing to improve the satellite attitude control and vibration environment of onboard sensitive devices.

The thrust vector of a satellite main thruster should ideally pass through the mass center of the satellite; however, it practically does not, which induces a so-called thrust vector misalignment. This thrust vector misalignment results in a disturbance torque that causes satellite rotation, long orbit-transfer time, extra requirements of attitude/orbit control systems, and large positioning error. To eliminate the effects of this disturbance torque in the presence of a large thrust level, satellites should be either spun in the opposite direction by auxiliary thrusters or equipped with thrust orientation control devices. In general, gimbals are used to provide coarse position correction for thrust vector alignment and as a means of reorienting the thruster in the intended thrust direction; however, this usually imposes a significant mass penalty. In addition to gimbals, reaction control system (RCS) thrusters, which are small auxiliary thrusters in addition to the main thruster, are conventionally used for thrust vector control in satellites. The RCS thrusters obviously consume satellite fuel thus shortening the life of the satellites.

Although the RCS thrusters can correct the satellite thrust vector misalignment, the firing of the thrusters inevitably generate vibration that resonates throughout the satellite structure, and the RCS cannot take care of the satellite structure vibration due to thruster firing. It is known that the thruster-firing-induced disturbance is the dominant source of satellite vibration, causing high vibration levels with long durations [5].

Instead of the RCS thrusters, a steerable main thruster is another alternative for the TVC of satellites. However, many issues related to using a steerable main thruster are still open such as how to steer the thruster, how to make the steering mechanism as light as possible, and how to control the steering mechanism.

The emerging smart composite structures technology provides a promising technical venue towards the TVC problems of satellites. Hexapod Stewart platforms with six degrees of freedom for the device plate [6-12] have been developed for precision positioning or vibration suppression as optical benches, although some of them may not possess considerable motion ranges or frequency bands. In the Intelligent and Composite Materials Laboratory (ICML) of the University of Hawaii at Manoa (UHM), a smart composite platform, namely UHM platform, is being developed. The UHM smart platform incorporates composites, sensors, actuators, and controllers as a means for fine-tuning position tolerances of the satellite thrust vector and suppressing satellite structure vibration during the main thruster firing, i.e., the UHM smart composite platform possesses simultaneous precision positioning and vibration suppression (SPPVS) capabilities [13-15]. The UHM smart composite platform is a smart structural interface placed in between the satellite main thruster and the satellite structure, and makes the satellite main thruster steeable.

The UHM smart platform is an assembly of active composite panels (ACPs) [16-21] and active composite struts (ACSs) [20-22] both with SPPVS capabilities. The ACSs with integrated sensors/actuators/controllers can function as both actuators and load carrying members. The preliminary studies [13, 14] demonstrated that the UHM smart platform could provide sufficient range and speed of motion for the thrust vector alignment such that the gimbals system for TVC application in satellites may be subsequently eliminated, and the positioning accuracy can meet the practical requirements. In addition, the UHM platform is driven by electricity from the solar panels of a satellite, resulting fuel saving, and hence lifetime prolongation of the satellite. The vibration suppression capability of the smart composite platform for satellite vibration reduction has also been demonstrated [15].

This chapter includes the innovative concept for satellites thrust vector control using the UHM smart composite platform, general configuration, kinematics and inverse kinematics, and precision position control of the UHM platform, thrust vector control of satellites using the UHM platform, and the vibration suppression of satellite structures employing the UHM platform. First, an innovative thrust vector control concept and the general configuration of the UHM platform are introduced in Sections 2 and 3, respectively. In Section 4, the inverse kinematics and singularity analysis of the platform are conducted, and an adaptive kinematic modeling method is proposed. In Section 5, two intelligent control strategies; namely, local control strategy and global control strategy, are proposed for the precision position control of the platform, and experiments and comparisons are performed. In Section 6, the configuration of the thrust vector control system is illustrated, the satellite mathematical model for thrust vector misalignment correction is built, the attitude controller is designed, and the thrust vector control of a satellite is conducted together with the intelligent controller of the UHM platform discussed in Section 5. In Section 7, satellite structure vibration reduction using the UHM platform is demonstrated and followed by the discussions. Section 8 presents conclusions.

Innovative Concept for Thrust Vector Control of Satellites

Figure 1 shows a schematic of a satellite. The main thruster of a satellite is presently hard-mounted to the satellite structure to provide the thrust for recovery of the satellite orbital altitude, and the RCS thrusters are also hard-mounted to perform satellite attitude control during orbit recovery. Ideally, the thrust vector of the main thruster is designed to pass through the mass center of the satellite to avoid the momentum buildup that could cause the satellite to be thrown off its orbit if not compensated. Practically, this arrangement will change along with missions such as the motions of solar array and antenna pointing as well as the fuel consumption. Therefore, the RCS thrusters are employed to maintain the thrust vector alignment. The drawbacks of this technology include (a) too many RCS thrusters increase the weight penalty of the satellite; (b) the RCS thrusters consume onboard fuel that is crucial to the lifetime of a satellite, and (c) when the main thruster and the RCS thrusters fire, the satellite structure vibrates.

The UHM smart platform provides an innovative thrust vector control concept for satellites. This concept is illustrated in Figure 2 where the UHM smart platform is placed in between the main thruster and the satellite structure. Therefore, the connection between the main thruster and the satellite structure is no longer rigid. The UHM smart platform is a structural interface between the main thruster and the satellite structure, and the main thruster is thus steerable. By controlling the tip/tilt of the UHM smart platform, the thrust vector of the main thruster can be controlled, and furthermore the attitude of the satellite is controlled as well. This concept results in four innovations. First, the UHM smart platform takes the place of the RCS thrusters for thrust vector control. Therefore, the weight penalty of the RCS thrusters is removed, and the vibration due to the RCS thrusters no longer exists. Second, the UHM smart platform is driven by electricity from solar energy through solar panels, and hence the onboard fuel is saved and the satellite lifetime is prolonged. Third, the smart platform provides the vibration suppression capability in addition to TVC of the misalignment, and the disturbance of the main thruster can be actively isolated. Fourth, due to the large motion range of the UHM smart platform (i.e., $\pm 15^0$), the gimbals system could potentially be eliminated, further simplifying and lightening the entire satellite.

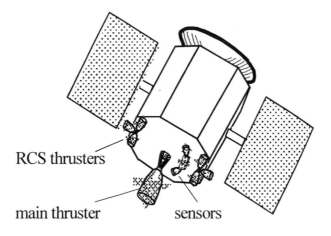

RCS thrusters

main thruster   sensors

Figure 1. Schematic of satellite.

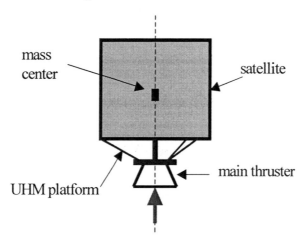

mass
center   satellite

main thruster

UHM platform

Figure 2. Satellite with UHM smart platform.

Smart Composite Platform

Figure 3 shows the UHM smart composite platform with two degrees of freedom. This platform is the updated version of the previous platform [13, 14]. It consists of a base, a circular device plate, three active composite struts, and a central support. The three active struts are connected to the vertices of the base and device plate through joints. The device plate is constrained to pivot about the central joint of the central support as the active struts extend or retract to allow precision control of the angular displacement of an axial main thruster mounted on the device plate. The central support carries the majority of the thruster force. There are a DC precision locomotion motor with an encoder, a piezoelectric stack actuator, and a load cell in each strut, and a heavy-duty piezoelectric stack actuator and a load cell in the central support. The struts possess simultaneous precision position and vibration suppression capabilities due to the serial configuration of the precision locomotion motor and the piezoelectric stack actuator. The central support can provide vibration suppression. The device plate and the base are made of woven carbon/epoxy pregreg composite. Especially, the device plate is a smart composite plate with six embedded piezoelectric patch actuators and three embedded piezoelectric patch sensors using active fiber composite piezoelectric patches [19]. The optimal configuration of the piezoelectric patch actuators in the device plate was investigated using different optimization technologies [21, 23] and the final configuration is illustrated in Figure 3(b). By controlling the motion of each strut, the device plate can be driven to a given position to correct the misalignment of the thrust vector as stated earlier, and vibration suppression can be achieved by controlling the piezoelectric stack actuators in the active composite struts and the central support as well as the piezoelectric patches in the smart composite device plate.

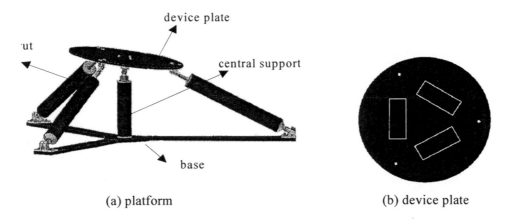

(a) platform   (b) device plate

Figure 3. UHM 2-DOF smart composite platform.

The dimensions of this platform are as follows: the diameters of the device plate and the base are 406 mm and 1,220 mm, respectively, and its height is 300 mm. The DC precision locomotion motor has high resolution and low power consumption, which is required for precision positioning in this application. This motor maximum velocity is greater than 30 mm/s, its unidirectional repeatability is 0.5 μm, its maximum stroke is 50 mm, and its maximum push/pull force is about 100 N. The push force capability, pull force capability, resolution, maximum stroke, and maximum input voltage of the piezoelectric stack actuator in the three struts are 800 N, 300 N, 0.9 nm, 90 μm, and 100 V, respectively. The maximum stroke and push/pull force-carrying capability of the piezoelectric stack actuator in the central support are 80 μm, and 30,000 N/3,500 N, respectively. The high-load carrying capability of the piezoelectric stack actuator in the central support ensures the thrust transmission of the main thruster to the satellite in this application of the UHM platform. The device plate has 18 layers of W3F 282 42'' F593 graphite plain-weave fabric prepreg with epoxy resin, making it 6 mm in thickness. The six piezoelectric patch actuators are embedded into the device plate, and they are back to back and a layer beneath the skin in both sides of the device plate. The piezoelectric patch actuators are active fiber composite piezoelectric ceramic patches [19] with the dimension of 0.33 mm in thickness, 55 mm in width, and 135 mm in length. Three narrow active fiber composite piezoelectric patch sensors are also embedded into the device plate, which are placed parallel and adjacent to the piezoelectric patch actuators in the bottom side of the device plate. These patches employ PZT-5A [24, 25] piezoelectric material and perform as extension motors.

Inverse Kinematics, Sigularity and Adaptive Modeling of Kinematics

Inverse Kinematics

Kinematic analysis is an important step in the design and control of the platform. Figure 4 shows the UHM platform as it would be attached to the satellite and is illustrated schematically in Figure 2. In Figure 4, two coordinate systems are assumed. $X_1Y_1Z_1O$ is fixed to the device plate and $X_2Y_2Z_2O$ is fixed to the base of the platform. The angular rotations of the device plate are assumed as α and β about the axes of X_2 and Y_2, respectively. Therefore, the transformation between the coordinates $X_2Y_2Z_2O$ and $X_1Y_1Z_1O$ is as follows:

$$\begin{Bmatrix} X_2 \\ Y_2 \\ Z_2 \end{Bmatrix} = \begin{bmatrix} \cos\beta & \sin\alpha\sin\beta & \cos\alpha\sin\beta \\ 0 & \cos\alpha & -\sin\alpha \\ -\sin\beta & \sin\alpha\cos\beta & \cos\alpha\cos\beta \end{bmatrix} \begin{Bmatrix} X_1 \\ Y_1 \\ Z_1 \end{Bmatrix} \tag{1}$$

The positions of the joint points A, B, and C in the coordinate $X_1Y_1Z_1O$ are as follows:

$$\begin{Bmatrix} X_1 \\ Y_1 \\ Z_1 \end{Bmatrix}_A = \begin{Bmatrix} -r \\ 0 \\ -h_1 \end{Bmatrix} \quad \begin{Bmatrix} X_1 \\ Y_1 \\ Z_1 \end{Bmatrix}_B = \begin{Bmatrix} r/2 \\ r\sqrt{3}/2 \\ -h_1 \end{Bmatrix} \quad \begin{Bmatrix} X_1 \\ Y_1 \\ Z_1 \end{Bmatrix}_C = \begin{Bmatrix} r/2 \\ -r\sqrt{3}/2 \\ -h_1 \end{Bmatrix} \tag{2}$$

where r is the radius of the device plate and h_1 is the distance from the pivot point O to the center of the device plate. According to Eq. (1), the positions of the points A, B, and C in the coordinate $X_2Y_2Z_2O$ can be derived as the following:

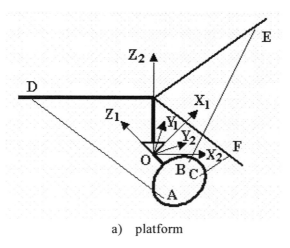

a) platform

Figure 4. Continued on next page.

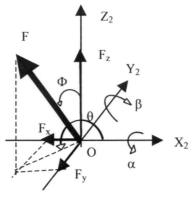

(b) coordinate systems

Figure 4. Coordinate systems of platform.

$$
\begin{Bmatrix} X_{2A} \\ Y_{2A} \\ Z_{2A} \end{Bmatrix} = \begin{Bmatrix} -r\cos\beta - h_1\cos\alpha\sin\beta \\ h_1\sin\alpha \\ r\sin\beta - h_1\cos\alpha\cos\beta \end{Bmatrix}
\tag{3a}
$$

$$
\begin{Bmatrix} X_{2B} \\ Y_{2B} \\ Z_{2B} \end{Bmatrix} = \begin{Bmatrix} (r\cos\beta)/2 + (r\sqrt{3}\sin\alpha\sin\beta)/2 - h_1\cos\alpha\sin\beta \\ (r\sqrt{3}\cos\alpha)/2 + h_1\sin\alpha \\ -(r\sin\beta)/2 + (r\sqrt{3}\sin\alpha\cos\beta)/2 - h_1\cos\alpha\cos\beta \end{Bmatrix}
\tag{3b}
$$

$$
\begin{Bmatrix} X_{2C} \\ Y_{2C} \\ Z_{2C} \end{Bmatrix} = \begin{Bmatrix} (r\cos\beta)/2 - (r\sqrt{3}\sin\alpha\sin\beta)/2 - h_1\cos\alpha\sin\beta \\ -(r\sqrt{3}\cos\alpha)/2 + h_1\sin\alpha \\ -(r\sin\beta)/2 - (r\sqrt{3}\sin\alpha\cos\beta)/2 - h_1\cos\alpha\cos\beta \end{Bmatrix}
\tag{3c}
$$

It is easy to find the positions of the points D, E, and F on the base in the coordinate $X_2Y_2Z_2O$:

$$
\begin{Bmatrix} X_{2D} \\ Y_{2D} \\ Z_{2D} \end{Bmatrix} = \begin{Bmatrix} -R \\ 0 \\ h_2 \end{Bmatrix} \quad \begin{Bmatrix} X_{2E} \\ Y_{2E} \\ Z_{2E} \end{Bmatrix} = \begin{Bmatrix} R/2 \\ R\sqrt{3}/2 \\ h_2 \end{Bmatrix} \quad \begin{Bmatrix} X_{2F} \\ Y_{2F} \\ Z_{2F} \end{Bmatrix} = \begin{Bmatrix} R/2 \\ -R\sqrt{3}/2 \\ h_2 \end{Bmatrix}
\tag{4}
$$

in which R is the radius of the base and h_2 is the distance from the pivot point O to the center of the base. According to Eqs. (3) and (4), the lengths of the three struts can be calculated as:

$$L_{AD}^2 = (X_{2A} - X_{2D})^2 + (Y_{2A} - Y_{2D})^2 + (Z_{2A} - Z_{2D})^2$$
$$= R^2 + r^2 + h_1^2 + \mathrm{h}_2^2 - 2Rr\cos\beta - 2rh_2\sin\beta - 2h_1\cos\alpha(R\sin\beta - h_2\cos\beta) \tag{5a}$$

$$L_{BE}^2 = (X_{2B} - X_{2E})^2 + (Y_{2B} - Y_{2E})^2 + (Z_{2B} - Z_{2E})^2$$
$$= R^2 + r^2 + h_1^2 + h_2^2 + rh_2(\sin\beta - \sqrt{3}\sin\alpha\cos\beta) + 2h_1h_2\cos\alpha\cos\beta \tag{5b}$$
$$-0.5Rr(\cos\beta + 3\cos\alpha + \sqrt{3}\sin\alpha\sin\beta) + h_1 R(\cos\alpha\sin\beta - \sqrt{3}\sin\alpha)$$

$$L_{CF}^2 = (X_{2C} - X_{2F})^2 + (Y_{2C} - Y_{2F})^2 + (Z_{2C} - Z_{2F})^2$$
$$= R^2 + r^2 + h_1^2 + h_2^2 + rh_2(\sin\beta + \sqrt{3}\sin\alpha\cos\beta) + 2h_1h_2\cos\alpha\cos\beta \tag{5c}$$
$$-0.5Rr(\cos\beta + 3\cos\alpha - \sqrt{3}\sin\alpha\sin\beta) + h_1 R(\cos\alpha\sin\beta + \sqrt{3}\sin\alpha)$$

Assuming that $\alpha = \beta = 0$, the initial lengths of all the three struts are

$$L_{AD}^0 = L_{BE}^0 = L_{CF}^0 = \sqrt{(R-r)^2 + (h_1 + h_2)^2} \tag{6}$$

Therefore, the length change of each strut (i.e., strut motion) can be obtained from:

$$\Delta L_{AD} = L_{AD} - L_{AD}^0, \ \Delta L_{BE} = L_{BE} - L_{BE}^0, \ \Delta L_{CF} = L_{CF} - L_{CF}^0 \tag{7}$$

Figure 5 shows the length changes of the three struts with respect to varying tilt angles α and β, demonstrating that (a) the tilt angle about Y_2 axis, i.e., β, is mainly controlled by the strut AD, and this fact can also be deduced from the platform architecture in that the strut AD is always almost normal to Y_2 axis; (b) for one degree tip/tilt of the device plate, the three struts have to move 2-3 mm.

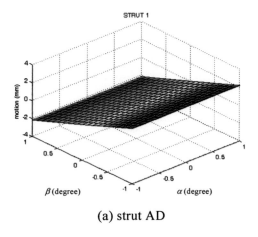

(a) strut AD

Figure 5. Continued on next page.

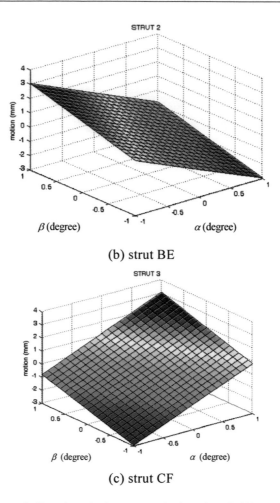

(b) strut BE

(c) strut CF

Figure 5. Strut length changes vs. device plate tip/tilt angles.

In practice, it is easier to determine the desired angle ranges of Φ and θ in Figure 4(b) instead of the tilt angles α and β, though the angles of Φ and θ cannot be measured directly. Φ is the angle between the thrust vector (i.e., Z_1) and the Z_2 axis, and θ is the angle between the X_2 axis and the projection of the thrust vector in the X_2-Y_2 plane. Normally, the desired angle range of Φ is ± 1 degree for satellite TVC. The gimbals can potentially be eliminated if Φ is about ± 6 degrees. The relations between α, β, and Φ θ are easily known as the following:

$$\cos\Phi = \cos\alpha\cos\beta \qquad (8a)$$

$$\sin\theta = -\sin\alpha/\sin\Phi \qquad (8b)$$

Figure 6 is the transformation between α, β, and Φ. Figure 7 illustrates the struts motions versus tilt angles Φ and θ for achieving one degree tilt angle of Φ, indicating that the required maximum struts motions are about 2.2 mm, but the extension is +2.2154 mm and the contraction is -2.1653 mm. The difference in extension and contraction represents the weak non-linearity of the platform kinematical characteristics. In addition, all the three struts have the same motion tracks along the angle θ varying from 0 to 360 degrees, although their phases are different.

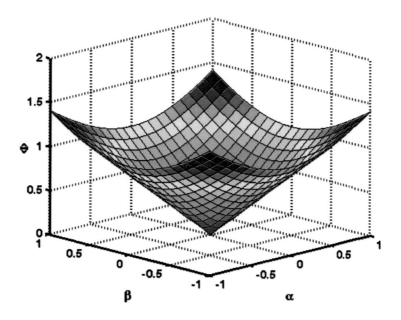

Figure 6. Transformation between α, β and Φ.

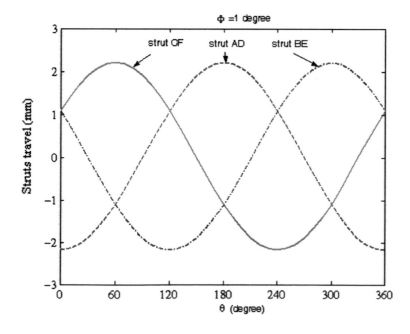

Figure 7. Strut motions vs. angle θ ($\Phi = 1$).

Singularity

According to Eq. (7), the relationships between the tilt angular velocities and the strut velocities are the following:

$$\begin{Bmatrix} \Delta \dot{L}_{AD} \\ \Delta \dot{L}_{BE} \\ \Delta \dot{L}_{CF} \end{Bmatrix} = \begin{bmatrix} J_{AD}^{\alpha}(\alpha,\beta) & J_{AD}^{\beta}(\alpha,\beta) \\ J_{BE}^{\alpha}(\alpha,\beta) & J_{BE}^{\beta}(\alpha,\beta) \\ J_{CF}^{\alpha}(\alpha,\beta) & J_{CF}^{\beta}(\alpha,\beta) \end{bmatrix} \begin{Bmatrix} \dot{\alpha} \\ \dot{\beta} \end{Bmatrix} = \mathbf{J} \begin{Bmatrix} \dot{\alpha} \\ \dot{\beta} \end{Bmatrix} \tag{9}$$

where $\Delta \dot{L}_{AD}$, $\Delta \dot{L}_{BE}$, and $\Delta \dot{L}_{CF}$ are the velocities of the three struts; $\dot{\alpha}$ and $\dot{\beta}$ are the tilt angular velocities of the device plate about OX_2 and OY_2 axes, respectively; $J_{AD}^{\alpha}(\alpha,\beta)$, ..., $J_{CF}^{\beta}(\alpha,\beta)$ are the sensitivity functions of the length changes of the three struts with respect to the tilt angle α or β, i.e., $J_{AD}^{\alpha}(\alpha,\beta) = \partial\Delta L_{AD}/\partial\alpha$, $J_{AD}^{\beta}(\alpha,\beta) = \partial\Delta L_{AD}/\partial\beta$, etc.; and $\mathbf{J}$ is the Jacobian matrix.

To avoid kinematic singularity, the Jacobian matrix $\mathbf{J}$ must be non-singular, i.e., the condition number of the Jacobian matrix $\mathbf{J}$ has to be near one and not too large. The infinite condition number of the Jacobian matrix means singularity of the smart platform. Figure 8 illustrates the condition number of the Jacobian matrix $\mathbf{J}$ of the UHM smart platform in the range of $\alpha = [-1, 1]$ degree and $\beta = [-1, 1]$ degree, depicting that the condition number of the Jacobian matrix $\mathbf{J}$ is about 2, which means that the Jacobian matrix of the smart platform is well-conditioned and the smart platform has no singularity problem.

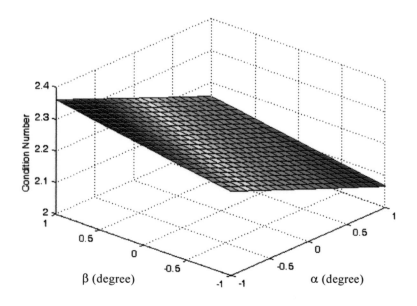

Figure 8. Condition number of the Jacobian matrix.

Adaptive Modeling of Kinematics

Kinematic model is necessary in kinematic analysis as well as controller synthesis and realization. However, kinematic models usually cannot be derived in a straight-forward manner. A number of kinematic modeling methods have been investigated in the literature. Here, an adaptive kinematic modeling is presented which is based on the inverse kinematics and adaptive least mean square (LMS) method. Figure 9 illustrates the block-diagram of the adaptive kinematic modeling.

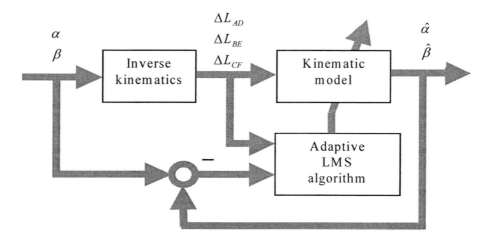

Figure 9. Adaptive kinematic modeling.

The kinematic model of the smart composite platform can be written in the following form:

$$\hat{\alpha} = \mathbf{A}^T \mathbf{V}$$
$$\hat{\beta} = \mathbf{B}^T \mathbf{V}$$

(10)

where $\hat{\alpha}$ and $\hat{\beta}$ are the output of the kinematic model, i.e., the estimated tilt angles of the smart composite platform. $\mathbf{A}$ and $\mathbf{B}$ are coefficients that need to be determined, $\mathbf{A}=[a_1 \ a_2 \ a_3 \ a_4 \ a_5 \ a_6]^T$, $\mathbf{B}=[b_1 \ b_2 \ b_3 \ b_4 \ b_5 \ b_6]^T$, V is the input vector consisting of the output of the inverse kinematics, i.e., ΔL_{AD}, ΔL_{BE}, and ΔL_{CF}. Due to the slight nonlinearity, as shown in Eq. (5) and Figure 7, of the inverse kinematics of the smart composite platform, the input vector $\mathbf{V}=[\Delta L_{AD} \ \Delta L_{AD}^2 \ \Delta L_{BE} \ \Delta L_{BE}^2 \ \Delta L_{CF} \ \Delta L_{CF}^2]^T$ is selected, which is a sum of two-order polynomials of the displacements of the three struts. Now, let us define the estimated errors of the two tilt angles as the following:

$$e_{\alpha} = \alpha - \hat{\alpha}$$
$$e_{\beta} = \beta - \hat{\beta}$$

(11)

and define the cost as:

$$J_k = e_\alpha^2 + e_\beta^2 \tag{12}$$

the coefficients $\mathbf{A}$ and $\mathbf{B}$ can then be iterated by using the following adaptive recursive algorithm:

$$\begin{aligned}
\mathbf{A}(k+1) &= \mathbf{A}(k) + 2\mu_\alpha e_\alpha \mathbf{V} \\
\mathbf{B}(k+1) &= \mathbf{B}(k) + 2\mu_\beta e_\beta \mathbf{V}
\end{aligned} \tag{13}$$

where μ_α and μ_β are the convergence factors that control the convergent rate of the adaptive algorithm; and k represents the time steps.

For the UHM smart composite platform, the final values of $\mathbf{A}$ and $\mathbf{B}$ are:

$$\mathbf{A}^T = [\text{-0.149921 -0.003923 -71.046372 -18.729984 71.198952 19.375486}] \times 10^{-4}$$
$$\mathbf{B}^T = [\text{-82.123047 -20.672060 41.063303 11.361680 41.059139 11.804007}] \times 10^{-4}$$

From the final values of $\mathbf{A}$ and $\mathbf{B}$, the following points, which practically reflect the characteristics of the smart composite platform, can easily be observed:

(a) the absolute values of a_1 and a_2 are much smaller than those of the rest elements in $\mathbf{A}$, meaning that the struts BE and CF have much higher control authority over the tilt angle α than the strut AD has;

(b) a_3 and a_5 are very close in their absolute values but with opposite signs, so are a_4 and a_6, meaning that the struts BE and CF have to always move in opposite direction to control the tilt angle α ;

(c) b_1 and b_2 are in opposite sign of b_4 to b_6, meaning that in order to control the tilt angle β the struts BE and CF always move in the same direction, and the strut AD has to move in opposite direction of the struts BE and CF; and

(d) the elements in $\mathbf{B}$ are in the same order, meaning that the control authority of the struts BE and CF over β is nearly the same as that of the strut AD.

Smart Platform Control

Control Strategies

Based on the architecture of the UHM smart platform, two control strategies are proposed: local control strategy (LCS) and global control strategy (GCS). Figure 10 shows the schematic of the local control strategy in which each active composite strut (ACS) is controlled individually, and for the UHM smart composite platform N=3 (see Figure 3). Employing the desired tilt angles α and β as well as Eq. (7), the motion of each strut can be calculated. Next, the calculated motion of each strut becomes the command to each strut position control loop. The motion of each strut is measured by the integrated encoder and compared with the command from the calculation. The position error is used to correct the position of each strut through a strut controller. The tilt angle accuracy of the platform device plate in this control strategy depends on the understanding of the platform kinematics since a mapping between the tilt angles of the platform device plate and the strut motions is needed and no direct tilt angle information of the device plate is used for this control strategy.

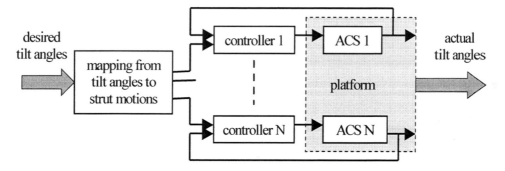

Figure 10. Local control strategy (LCS).

Figure 11 shows the schematic of the global control strategy in which the three active composite struts (ACSs) are controlled globally. A tilt sensor is located on the device plate of the smart platform to measure its tip/tilt angles, i.e., α and β. The tilt sensor signals are fed back to the global controller to make the three control signals available for the three active struts. The merit of this control strategy is that the tilt angle accuracy of the platform device plate is controlled directly.

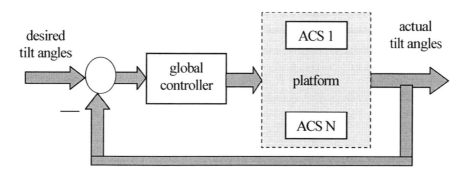

Figure 11. Global control strategy (GCS).

Intelligent Precision Position Control of Platform

As mentioned earlier, each active strut of the smart platform includes a DC precision locomotion motor for precision positioning and a piezoelectric stack actuator for vibration suppression. The DC locomotion motor behaves nonlinearly, as illustrated in Figure 12 where a dead zone exists in the curve of the motor velocity versus the motor input voltage.

To finely control the position of the smart platform, fuzzy logic control (FLC) is used here due to the fact that the FLC can be employed in developing both linear and non-linear systems. The FLC is also capable of providing the high degree of accuracy required by high-performance systems, without the need for detailed mathematical models [26-29].

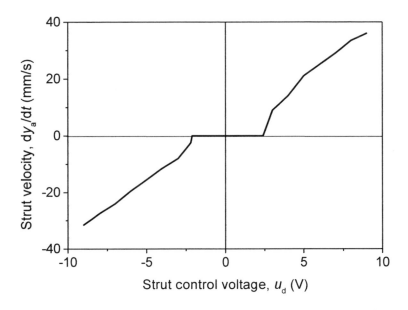

Figure 12. Strut input voltage vs. strut velocity.

The FLC uses a form of quantification of imprecise information (input fuzzy sets) to generate an inference scheme, which is based on a knowledge base of control signals to be applied to the system. The benefit of this quantification is that the fuzzy sets can be represented by a unique linguistic expression, such as small, medium, or large. The linguistic representation of a fuzzy set is known as a term, and a collection of such terms defines a term-set, or a library of fuzzy sets. A fuzzy controller converts a linguistic control strategy, typically based on expert knowledge, into an automatic control strategy. The FLC therefore is a rule-based controller. A FLC is comprised of four primary components: (1) fuzzifier that fuzzifies the values of the state variables monitored during the process into fuzzy linguistic terms, (2) a knowledge base that contains fuzzy IF-THEN rules and membership functions, (3) fuzzy reasoning whose result is a fuzzy output for each rule, and (4) defuzzification that converts a fuzzy quantity, represented by a membership function, into a precise or crisp value.

The first step in designing a FLC is to decide which state variables of the controlled system can be taken as the input signals to the controller. In the local control strategy, the error of each strut displacement is selected as the input. In the global control strategy, the errors of tilt angles of the platform device plate are used as the inputs of the global controller. The controller outputs are the voltages to the struts. Further, choosing the fuzzy sets to formulate the fuzzy control rules are also significant factors in the performance of the FLC. Empirical knowledge and engineering intuition play important roles in choosing fuzzy sets and their corresponding membership functions. After choosing proper fuzzy variables as inputs and outputs of a FLC, one must decide on the fuzzy sets. These sets transform the numerical values of the inputs to fuzzy quantities. The number of these fuzzy sets specifies the quality of the control, which can be achieved using FLCs. As the number of the fuzzy sets increases, the management of the rules is more involved and the tuning of the controller is less straightforward. Accordingly, a tradeoff between the quality of control and computational time is required to choose the number of fuzzy sets. A number of fuzzy sets for each of the input variables are normally chosen to represent the system under study. Here, seven fuzzy sets are defined for the position error and five fuzzy sets are defined for the output of each controller. The seven input fuzzy sets are Negative Large (NL), Negative Medium (NM), Negative Small (NS), Zero (Z), Positive Small (PS), Positive Medium (PM), and Positive Large (PL), and the five output fuzzy sets are Negative Large (NL), Negative Small (NS), Zero (Z), Positive Small (PS), and Positive Large (PL). After specifying the fuzzy sets, their membership functions are then determined. In this paper, sigmoidal-shaped, trapezoidal-shaped, and Gaussian-shaped functions are employed to fix the limits of the membership functions.

The third step of designing a FLC is the choice of the fuzzy rules. The decision-making logic is how the controller output is generated. The decision-making uses input fuzzy sets, and the decision is governed by the values of the inputs. Furthermore, the knowledge base consists of knowledge of the application domain and the attendant control goals. It includes a database and a fuzzy control rule base.

After the input and output values are assigned to define fuzzy sets, each possible input condition must be mapped onto an output condition. Such mapping is commonly expressed as IF-THEN rules. These rules may be as simple as the followings:

IF (input is (NL or NM)) THEN (output is NL);
IF (input is NS) THEN (output is NS);
IF (input is Z) THEN (output is Z);
IF (input is PS) THEN (output is PS); and
IF (input is (PL or PM)) THEN (output is PL).

For FLCs with several inputs, the fuzzy rule base can be illustrated as a look-up table. The more the inputs are, the more complex the fuzzy rule base is, and the longer it takes for making the decision.

The last step is the Defuzzification. Commonly used strategies may be described as maximum criterion, mean of maximum, and centroid methods. In this study, the centroid method is used to defuzzify outputs into crisp values.

Simulations and Comparisons

The two control strategies proposed in Figures 10 and 11 are verified in this section by employing Matlab Simulink [27].

The block-diagram of the local control strategy is shown in Figure 13. It consists of the desired tilt angle generators, the mapping from the desired angles to the motion commands of the three struts (inverse kinematics), strut control loops, backlash models, the mapping from the positions of the three struts to the actual tilt angles of the device plate of the smart composite platform (kinematics), and scopes. The kinematic model comes from the adaptive kinematic modeling explained in Section 4. It should be noted that the kinematic model here is solely for the purpose of comparing the desired and actual tilt angles and is not necessarily used in the real control system, since each strut displacement measured by the integrated displacement sensor in the strut is used for the local controller. The existing joints backlash is 10 μ m. Figure 13(b) shows the details of each strut control loop including the FLC with the input for each active strut position error and the motor model with its dead zone shown in Figure 12.

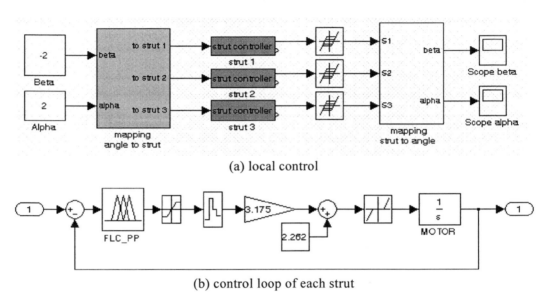

(a) local control

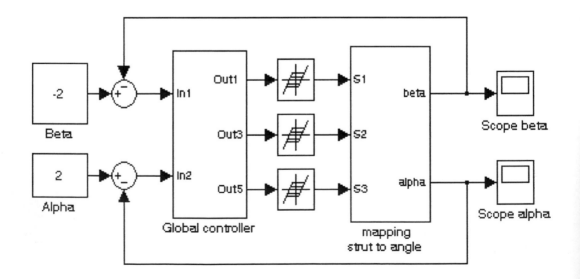

(b) control loop of each strut

Figure 13. Local control strategy block-diagram.

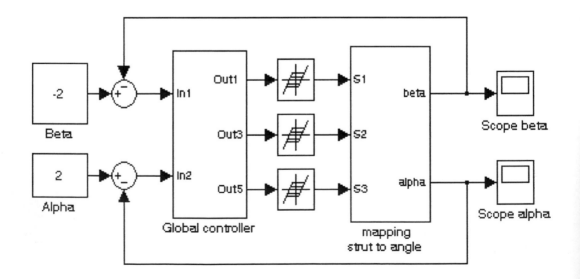

Figure 14. Global control strategy block-diagram.

The block-diagram of the global control strategy is shown in Figure 14. Similar to Figure 13, two generators generate the desired tilt angles. The main difference between Figures 13 and 14 is the mapping of the desired tilt angles into the motion commands of the three struts, i.e., the inverse kinematics in Figure 13 is eliminated from Figure 14. In the GCS, the actual tilt angles of the device plate, measured by a tip/tilt sensor, are directly fed back to the GCS controllers. The tilt angle errors become the input of the global controller. Similar to Figure 13, the mapping from the positions of the three struts to the actual tilt angles of the device plate of the platform (i.e., the kinematics) in Figure 14 is not necessarily used for the real control system, since the tilt angles can be measured by the tip/tilt sensor on the device plate directly.

Figures 15 and 16 illustrate the membership functions for the input and output fuzzy sets of the fuzzy logic controllers, respectively, employed in both the LCS and GCS. The detailed values of these membership functions are tabulated in Tables 1 and 2 for the input and output, respectively. The employed membership functions are defined as the followings:

Gaussian membership function:

$$GAUSS\_MF(x,\sigma,c) = e^{-\frac{(x-c)^2}{2\sigma^2}} \tag{14}$$

Trapezoidal membership function:

$$TRAP\_MF(x,a,b,c,d) = \max(\min(\frac{x-a}{b-a} \quad 1 \quad \frac{d-x}{d-c}),0) \tag{15}$$

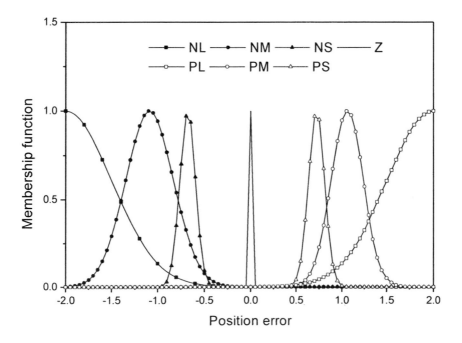

Figure 15. Input membership functions.

Sigmoidal membership function:

$$f(x,a,c) = \frac{1}{1+e^{-a(x-c)}} \tag{16}$$

$$DSIG\_MF(x,a_1,c_1,a_2,c_2) = f(x,a_1,c_1) - f(x,a_2,c_2) \tag{17}$$

where a, b, c, d, and σ are given constants, and x is a variable. It should be noted that the *DSIG_MF* membership function in the output fuzzy set Z in Figure 16 easily overcomes the nonlinearity of the active struts in Figure 12.

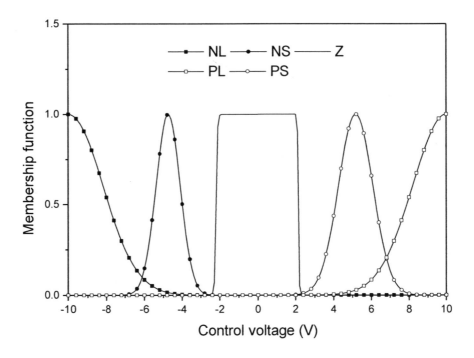

Figure 16. Output membership functions.

Table 1. Membership functions of input fuzzy sets.

Fuzzy Sets	Membership Functions
NL	GAUSS_MF $(x, 0.5, -2)$
NM	GAUSS_MF $(x, 0.261, -1.09)$
NS	GAUSS_MF $(x, 0.083, -0.68)$
Z	TRAP_MF $(x, -0.00529, -0.000529, 0.000529, 0.00476)$
PS	GAUSS_MF $(x, 0.088, 0.722)$
PM	GAUSS_MF $(x, 0.18, 1.06)$
PL	GAUSS_MF $(x, 0.5, 2)$

Table 2. Membership functions of output fuzzy sets.

Fuzzy Sets	Membership Functions
NL	GAUSS_MF (x, 1.8, -10)
NS	GAUSS_MF (x, 0.639, -4.75)
Z	DSIG_MF (x, 26, -2.2,51.8, 2.143)
PS	GAUSS_MF (x, 0.908, 5.17)
PL	GAUSS_MF (x, 1.8, 10)

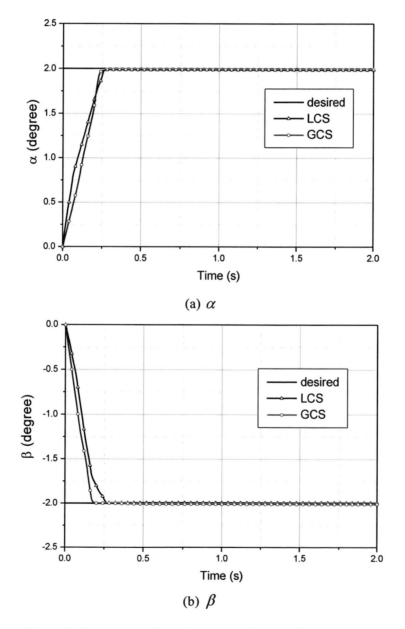

(a) α

(b) β

Figure 17. Device plate tilt angles using LCS and GCS (simulation).

Figure 17 shows the device plate tilt angles of the smart platform. The desired tilt angles are α =2 degrees and β =-2 degrees. For the LCS, it takes about 2.61 seconds to reach the desired α and 0.257 seconds to reach the desired β, both α and β with 1% steady-state error tolerance. For the GLS, these settling times become 0.245 and 0.179 seconds, respectively. The steady-state errors of α and β are 0.0102 degrees and 0.0055 degrees for the LCS and 0.0048 degrees and 0.0047 degrees for the GCS.

The effect of the joint backlashes on the controlled performance of the smart composite platform is also investigated and the results are listed in Table 3, depicting that the GCS can take care of the existing backlash/friction in the joints better than the LCS can. The control performance of the GCS does not change no matter the backlashes are included or not; however, the controlled performance of the LCS deteriorates when the backlashes are included.

Table 3. Performance Comparisons.

		LCS		GLS	
		α	β	α	β
Steady-state error (degree)	No backlashes	0.0075	0.004	0.0047	0.0047
	With backlashes	0.0102	0.0055	0.0047	0.0047
Settling time (second) (1% tolerance)	No backlashes	0.261	0.254	0.245	0.179
	With backlashes	0.262	0.257	0.245	0.179

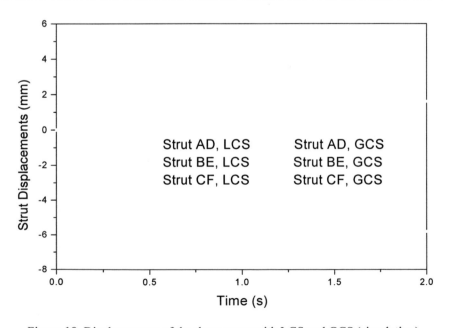

Figure 18. Displacements of the three struts with LCS and GCS (simulation).

Figure 18 illustrates the motions of the three struts. It shows that the struts AD and CF have to extend 4.459 and 1.563 mm, respectively, and the strut BE has to contract 5.81 mm to bring the device plate of the smart composite platform to the desired tilt angles. These motions of the three struts are consistent with those obtained from the inverse kinematics analysis shown in Figure 5.

The experimental studies of the smart composite platform control have been performed. The results reported here are similar to those for the first generation UHM smart platform [13]. Figure 19 illustrates the experimental tilt angles of the device plate of the smart composite platform by using the global control strategy. It depicts that relative steady state errors are 2.0% and 0.45% for α and β, respectively, and the rising times are about 1 second. These experimental results given in Figure 19 are similar to the simulated results given in Figure 17.

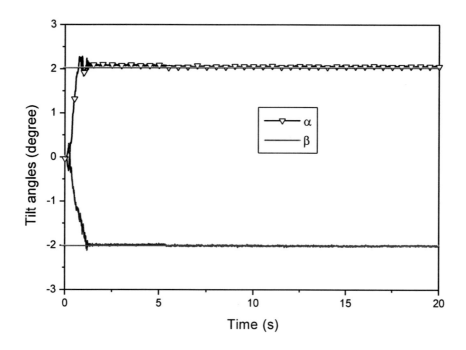

Figure 19. Device plate tilt angles using GCS (experiment).

Thrust Vector Control System of Satellites

Satellite Model with Smart Platform

Figure 20 shows the schematic of the thrust vector control system of satellites employing the UHM smart composite platform. To analyze and design a thrust vector controller, a satellite attitude dynamic model is needed.

The coordinate system $XYZO'$ in Figure 20 is fixed, O' is the mass center of the satellite. Φ_x, Φ_y, and Φ_z are rational angular displacements of the satellite about X, Y, and Z axes. O is the center of the device plate of the UHM smart platform and the point at which the thrust is applied. The offsets between O' and O in X, Y, and Z directions are also shown in Figure 20. Assume that the thrust level is F, the tilt angles of the smart composite platform are Φ and θ as shown in Figure 4(b), then the thrust components in the X_2, Y_2, and Z_2 axes are as the followings:

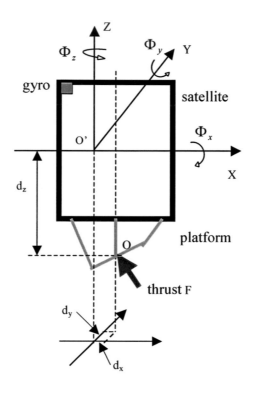

Figure 20. Satellite with smart composite platform.

$$F_x = F \sin \Phi \cos \theta = F \cos \alpha \sin \beta \qquad (18)$$

$$F_y = F \sin \Phi \sin \theta = -F \sin \alpha \qquad (19)$$

$$F_z = F \cos \Phi = F \cos \alpha \cos \beta \qquad (20)$$

where the coordinate system $X_2 Y_2 Z_2 O$ is parallel to the coordinate system $XYZO'$; F_x, F_y, and F_z are the components of the thrust F in X, Y, and Z axes, respectively; α and β are the tilt angles of the UHM smart platform device plate about X_2 and Y_2 axes, respectively; θ is the angle between the thrust vector and Z_2 axis; and Θ is the azimuth of the thrust orientation (also, see Figure 4).

From Figure 20, the thrust generated torques on the satellite about the axes X, Y, and Z, i.e., M_x, M_y, and M_z, can then be written as:

$$M_x = -F_y d_z + F_z d_y \tag{21}$$

$$M_y = -F_x d_z + F_z d_x \tag{22}$$

$$M_z = -F_x d_y + F_y d_x \tag{23}$$

where d_x, d_y, and d_z are the distances between the mass center of the satellite (O') and the center of the device plate of the smart platform, i.e. point O where the thrust is applied, in the *XYZO* coordinate system in Figure 20. Hence, the attitude dynamic equations of the satellites for thrust vector control are:

$$I_{xx} \ddot{\Phi}_x = M_x \tag{24}$$

$$I_{yy} \ddot{\Phi}_y = M_y \tag{25}$$

$$I_{zz} \ddot{\Phi}_z = M_z \tag{26}$$

where I_{xx}, I_{yy}, and I_{zz} are inertias of the satellites about *X*, *Y*, and *Z* axes; and $\ddot{\Phi}_x$, $\ddot{\Phi}_y$, and $\ddot{\Phi}_z$ are rational angular accelerations. Here, the satellite is considered as a rigid body.

For thrust vector misalignment correction, Φ_x, Φ_y, and Φ_z have to be controlled to zeros, i.e.,

$$M_x = M_y = M_z = 0 \tag{27}$$

According to Eqs. (18)-(23) and (27), the final tilt angles of the device plate of the smart composite platform can then be determined by the following equations:

$$\tan \beta = \frac{d_x}{d_z} \tag{28}$$

$$\tan \alpha = -\frac{d_y}{d_z} \cos \beta \tag{29}$$

Control Systems and Simulations

Figure 21(a) shows the block-diagram of the thrust vector control system of satellites employing the UHM smart composite platform. This system includes the satellite attitude controller, the smart platform controller, the UHM smart platform, and the satellite. For thrust vector control, the errors between the real attitudes of the satellite and the attitude commands will drive the attitude controller to generate the strut displacement commands of the smart platform. The UHM smart composite platform controller then adjusts the tilt angles of the platform to change the direction of the thrust vector of the main thruster that is mounted on the device plate of the platform to keep the thrust vector orientation, and further resulting in the attitude change of the satellite. There are two sensors in this block diagram. One is the gyro that is included in the satellite to measure the attitude rates of the satellite, and the other one is the tilt sensor that is included in the device plate of the smart platform to measure the platform tilt angles. Hence, there are practically two control loops in Figure 21. The inner-loop is for the smart platform control and its control strategy can be the local control strategy or the global control strategy introduced in Section 5. The outer-loop is for the thrust vector control of the satellite. For the thrust vector misalignment correction, the control system works the same as that in thrust vector control except the attitude commands are zeros. Figure 21(b) shows the block-diagram of the thrust vector control of satellites for simulations.

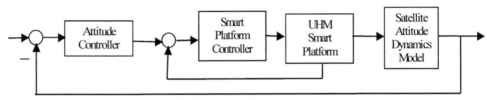

(a) block diagram of thrust vector control

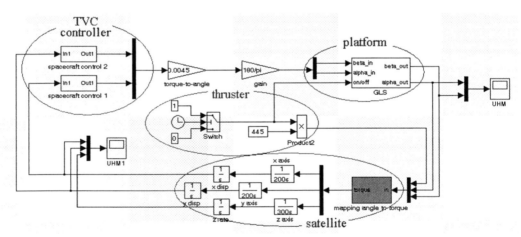

(b) Matlab Simulink block diagram of thrust vector control

Figure 21. Block-diagram of thrust vector control of satellites.

The parameters of the smart composite platform are given in Section 3. The inertias of the satellite are $I_{xx}=I_{yy}=200$ kg-m^2 and $I_{zz}=300$ kg-m^2. The thrust level is 445 N and the thruster fires at the 5th second. The offsets of the center of mass are: d_z=-500 mm and $d_x=d_y=10$ mm. That is, the smart platform should tilt about 1.146 degrees in α and -1.146 degrees in β, or 1.621 degrees in Φ according to Eq. (8), to make the thrust vector of the main thruster pass through the mass center of the satellite according to Eqs. (28) and (29).

The satellite attitude rates is integrated and filtered to obtain the satellite attitudes. The filter is a 2nd-order low-pass filter with the damping ratio of 0.707 and the resonance at 2 Hz.

The TVC controllers are PID controllers with the proportional, integral, and derivative gains of 800, 400, and 600, respectively.

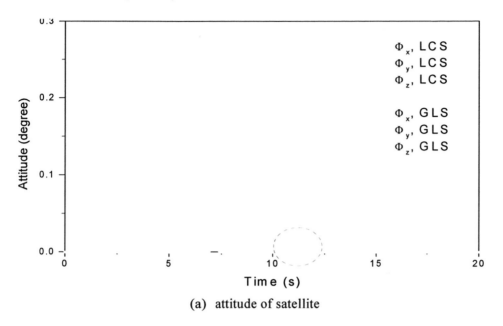

(a) attitude of satellite

Figure 22. Continued on next page.

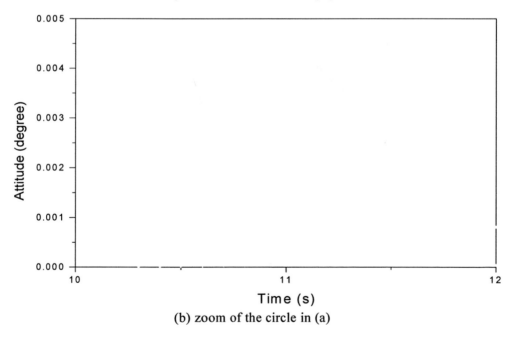

(b) zoom of the circle in (a)

Figure 22. Attitude of satellite.

Figures 22 to 24 depict the satellite attitudes, rates, and moments, respectively. The TVC controllers are the PID controllers and the smart platform controller employs either the local control strategy or the global control strategy. At the beginning, the satellite does not rotate because the thruster does not fire. At the 5th second, the thruster starts firing, resulting in applying torques on the satellite and rotating the satellite due to the fact that the thrust vector of the main thruster does not pass through the mass center of satellite, i.e., due to the misalignment. The satellite attitudes, rates, and moments increase while the thruster starts firing. Shortly, they become smaller and finally approach zeros due to the adjustment of the smart platform. The smart platform regulates the thrust vector in X and Y axes effectively, in the meanwhile, brings very tiny effect on the axial (Z axis) direction, which is in fact expected. Finally, the satellite attitude is maintained. Figure 22 shows that Φ_z is very small, Φ_x decreases to 0.005 degrees at 10.655 seconds with the GCS and 10.983 seconds with the

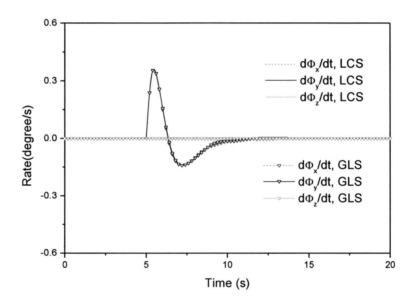

Figure 23. Rates of satellite.

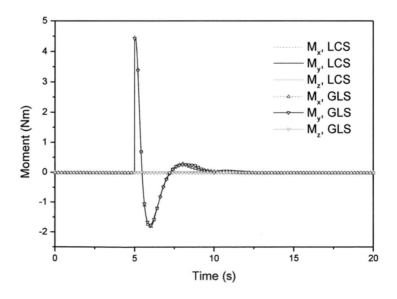

Figure 24. Satellite torques.

LCS. For Φ_y, these times become 10.503 seconds with the GCS and 10.655 seconds with the LCS, demonstrating that the GCS performs somewhat better than the LCS does. Figure 23 illustrates that the rates of the satellite are about -0.012 degree/s at the 10th second and the moments are about 0.04 N-m at the 10th second as shown in Figure 24.

Figure 25 shows the tilt angles of the smart platform, demonstrating that the smart platform responds to the thruster firing quickly, and the final position of the smart platform device plate is 1.142 degrees in α and -1.149 degrees in β, or 1.62 degrees in Φ according to Eq. (8). The accuracy of the thrust vector therefore is 0.01 degrees. The displacements of the struts are 2.538 mm for strut AD, 0.917 mm for strut BE, and -3.359 mm for strut CF to correct the given thrust vector misalignment.

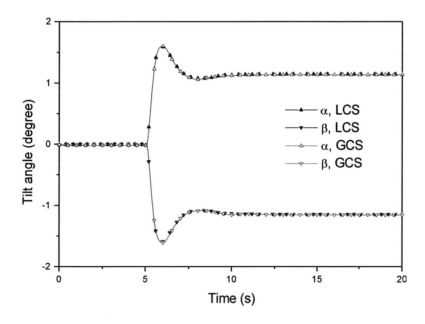

Figure 25. Tilt angles of smart platform.

Satellite Vibration Control Using Smart Platform

Dynamics of Combined Satellite Structure and Smart Platform

As mentioned earlier, the main thruster of a satellite is presently hard-mounted onto the satellite structure. When the main thruster fires, the body of the satellite vibrates, therefore, a vibration control system is needed to suppress the vibration. The UHM smart composite platform has both precision positioning and vibration suppression capabilities. It can function as a structural interface between the main thruster and the satellite, as shown in Figure 26, to provide simultaneous thrust vector misalignment correction and vibration suppression. As a result, the connection between the main thruster and the satellite body is no longer rigid. The satellite thrust vector misalignment control employing the UHM smart platform has been demonstrated in Section 6. In this section, all piezoelectric actuators including piezoelectric stack actuators and patch actuators in the UHM smart platform are expected to realize the vibration control of the satellite body and the thruster. The satellite structure is made of aluminum in the shape of an octagon with 480 mm long each side and the height of 1,070 mm. A parametric finite element model was developed and the dynamics of the combined structure was analyzed [30, 31]. The dynamic analysis implies that the first three modal frequencies of the combined structure are 38, 48, and 80 Hz, and the dominant mode of the satellite structure is at 48 Hz.

Figure 27 shows a typical thrust spectrum of satellite thrusters, implying that the dynamic thruster force is at a specific frequency. Therefore, the satellite structure vibration induced by the thruster firing is a harmonic vibration at this specific frequency, and the satellite structure will vibrate severely, if this specific frequency is equal to or close to one of the satellite modal frequencies.

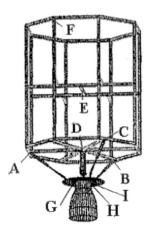

Figure 26. Satellite structure with UHM smart platform.

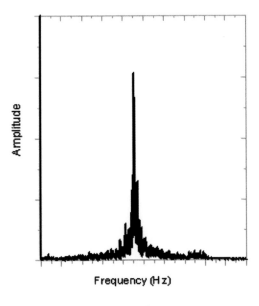

Figure 27. Typical thrust spectrum.

MIMO Adaptive Vibration Control

As mentioned earlier, the UHM smart platform has four piezoelectric stack actuators in the three struts as well as the central support and three pairs of piezoelectric patch actuators in the device plate. To control the vibration of the combined structure, displacements at nine points are measured resulting in a multi-input-multi-output control problem. These points are the connecting points, A, B, C, and D, of the UHM smart platform and the satellite structure, the mass center of the combined structure, E, the bottom of the satellite structure, F, and the three connecting points, G, H, and I, of the thruster and the device plate of the UHM smart platform, as marked in Figure 26. Adaptive feedforward vibration control [32-36] is chosen here to suppress the combined structure vibration. The basic idea of the adaptive feedforward control is "cancellation", i.e., the adaptive controller tries to adjust its parameters in real time such that the vibration due to the controller tends to be the same in magnitude as the vibration due to the external disturbance but with opposite phase, thus the vibration due to the external disturbance is cancelled by the vibration due to the control. The block diagram of adaptive feedforward control is shown in Figure 28.

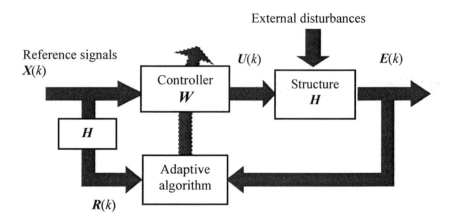

Figure 28. Multi-channel filtered-x adaptive algorithm.

For a structure with M actuators and L sensors and excited by a continuous disturbance, the vibration can be represented as the following:

$$e_l(k) = d_l(k) + \sum_{m=1}^{M} \sum_{j=1}^{J} h_{lm}(j) u_m(k - j + 1) \qquad (30)$$

where $d_l(k)$ and $e_l(k)$ are the uncontrolled vibration and the controlled vibration at the time step k, respectively; $u_m(k)$ is the the m-th actuator control signal at the time step k; h_{lm} is the finite impulse response function of the controlled structure between the l-th sensor and the m-th actuator; and J is the length of the finite impulse response of the controlled structures. Assuming that $\mathbf{E} = \begin{bmatrix} e_1 & e_2 & \cdots & e_L \end{bmatrix}^T$, $\mathbf{D} = \begin{bmatrix} d_1 & d_2 & \cdots & d_L \end{bmatrix}^T$,

$$\mathbf{H}_{lm} = \begin{bmatrix} h_{lm}(1) & h_{lm}(2) & \cdots & h_{ji}(J) \end{bmatrix}, \qquad \mathbf{U}_m(k) = \begin{bmatrix} u_m(k) & u_m(k-1) & \cdots & u_m(k-J+1) \end{bmatrix},$$

$$\mathbf{H} = \begin{bmatrix} \mathbf{H}_{lm} \end{bmatrix}_{l=1,2,\cdots,L\ m=1,2,\cdots,M}, \text{ and } \mathbf{U} = \begin{bmatrix} \mathbf{U}_1^T & \mathbf{U}_2^T & \cdots & \mathbf{U}_M^T \end{bmatrix}^T, \text{ then}$$

$$\mathbf{E}(k) = \mathbf{D}(k) + \mathbf{H}\mathbf{U}(k) \tag{31}$$

and further assuming that

$$u_m(k) = \sum_{i=1}^{I} w_{mi} x(k-i+1) = \mathbf{W}_m^T \mathbf{X}_I(k) \tag{32}$$

in which $\mathbf{W}_m = \begin{bmatrix} w_{m1} & w_{m2} & \cdots & w_{mI} \end{bmatrix}^T$, $\mathbf{X}_I(k) = \begin{bmatrix} x(k) & x(k-1) & \cdots & x(k-I+1) \end{bmatrix}^T$ is called the reference signal, and I is the length of the finite impulse response of the controller.

Defining the objective as shown in Eq. (33) and using the steepest descent method yield:

$$J_e(k) = \frac{1}{2} \mathbf{E}^T(k) \mathbf{E}(k) \tag{33}$$

$$\mathbf{W}_m(k+1) = \mathbf{W}_m(k) + \mu \frac{\partial J_e(k)}{\partial \mathbf{W}_m} \tag{34}$$

where μ is the convergence factor that rules the adaptation rate of the controller. According to Eqs. (32) and (33) and defining that

$$\mathbf{W}(k) = \begin{bmatrix} \mathbf{W}_1(k)^T & \mathbf{W}_2(k)^T & \cdots & \mathbf{W}_M(k)^T \end{bmatrix}^T \tag{35}$$

$$\mathbf{R}(k) = \begin{bmatrix} \mathbf{r}_{11}^T(k) & \mathbf{r}_{12}^T(k) & \cdots & \mathbf{r}_{1M}^T(k) \\ \mathbf{r}_{21}^T(k) & \mathbf{r}_{22}^T(k) & \cdots & \mathbf{r}_{2M}^T(k) \\ \vdots & \vdots & & \vdots \\ \mathbf{r}_{L1}^T(k) & \mathbf{r}_{L2}^T(k) & \cdots & \mathbf{r}_{LM}^T(k) \end{bmatrix}^T \tag{36}$$

$$\mathbf{r}_{lm}(k) = \begin{bmatrix} r_{lm}(k) & r_{lm}(k-1) & \cdots & r_{lm}(k-I+1) \end{bmatrix}^T \tag{37}$$

$$r_{lm}(k) = \mathbf{H}_{lm}^T \mathbf{X}_J(k) \tag{38}$$

$$\mathbf{X}_J(k) = \begin{bmatrix} x(k) & x(k-1) & \cdots & x(k-J+1) \end{bmatrix}^T \tag{39}$$

then Eq. (34) can be rewritten as the following:

$$\mathbf{W}(k+1) = \mathbf{W}(k) + 2\mu\mathbf{R}(k)\mathbf{E}(k) \tag{40}$$

It can be proved that the algorithm described above is stable when the value of the convergence factor μ is in the range of $0 < \mu < \lambda_{\max}^{-1}$, where $\lambda_{\max}$ is maximum eigenvalue of the filtered autocorrelation matrix, given by $E[\mathbf{R}(k)\mathbf{R}^T(k)]$, and the adaptation converges slower with the increase of the number of the sensors and actuators [34]. Since the eigenvalues of this matrix are rarely known, the convergence factor is usually selected manually.

During the adaptive control process, the estimated gradient vector given by $\mathbf{R}(k)\mathbf{E}(k)$ may not always coincide with the steepest gradient in which the objective is reduced with the fastest speed. However, the controller still converges towards its optimal values as long as the estimated gradient vector has a component towards this direction. This implies that an accurate estimate of the finite impulse response of the controlled structure is not required, i.e., this adaptive control scheme is tolerant of error in the estimation of finite impulse response of the controlled structures.

Besides the filtered-x LMS algorithm presented earlier, other adaptive filtering algorithms, such as Recursive Least Square (RLS), Fast Transversal Filtering (FTF), and Gradient-based Least Square (GLS) [35] can also be considered. The choice of algorithms depends on controlled structures, disturbances, the complexity and characteristics of algorithms, and the hardware. The LMS algorithm has smallest calculation, but the normal LMS algorithm converges slowly. The RLS algorithm converges faster, but its computation complexity is high, especially while the length of the filter is large. The FTF is complex in calculation, and the GLS has almost the same computation as the LMS but converges faster. Other means to expedite the adaptation process were also explored such as employing combined feedback and adaptive feedforward control [35, 36]. Another important issue in adaptive feedforward control is the selection of the reference signal. The selection depends on the availability of the external disturbance, the nature of the external disturbance, and the dynamics of the controlled structures. Generally, reference signals must correlate with the uncontrolled response of structures. The external disturbance that causes the structural vibration is the best reference signal, since the disturbance strongly correlates with uncontrolled responses. Unfortunately, it is hard to obtain the disturbance in many situations. One of the alternatives is a so-called virtual signal, which contains the characteristics of the disturbance. For example, for rotor vibration control, the predominant frequencies of uncontrolled response spectra are the rotating rate. Therefore, a virtual signal can be constructed using the rotating rate that can easily be measured. For wind-induced building vibration control, the virtual sinusoidal signal with the first natural frequency of the controlled building can be used as the reference signal, since the frequency bandwidth of the wind is very low and the dominant frequency in the building response is its first natural frequency [37]. The virtual reference signal was also used to suppress nonlinear vibration of smart structures [38]. Another method is to use estimated uncontrolled responses by employing an internal model [18]. Among these choices, the first one (i.e., the external disturbance) is direct and efficient. The second one (i.e., the virtual signals) is also very good in some cases. The last choice (i.e., the estimated uncontrolled response) is normally used with care, since it may cause the control system to become unstable.

Simulations and Discussions

First of all, the response of the combined test structure under 10 lbs dynamic thrust at 48 Hz
in the Z-direction has been carried out. The results show that the satellite structures vibration
in Y-direction is neglectable compared to its vibrations in X-and Z- directions, and the
vibration of the device plate is dominant in Z-direction. Therefore, the following vibrations
are measured: X- and Z-directions of the points A, B, and C, and Z-direction of the points D,
E, F, G, H, and I in Figure 26. Therefore, the objective is defined as the following:

$$J_e(k) = J_e^1(k) + J_e^2(k) \tag{41}$$

$$J_e^1(k) = \frac{1}{2}[e_{Ax}^2(k) + e_{Az}^2(k) + e_{Bx}^2(k) + e_{Bz}^2(k) + e_{Cx}^2(k) + e_{Cz}^2(k) + e_{Dz}^2(k) + e_{Ez}^2(k) + e_{Fz}^2(k)] \tag{42}$$

$$J_e^2(k) = \frac{1}{2}[e_{Gz}(k) - e_{Hz}(k)]^2 + \frac{1}{2}[e_{Gz}(k) - e_{Iz}(k)]^2 \tag{43}$$

where $J_e^1(k)$ and $J_e^2(k)$ are different. $J_e^1(k)$ is targeted at the vibrations at the points A, B,
C, D, E, and F, and $J_e^2(k)$ is targeted at the difference between the vibrations of the points
G, H, and I in Z-direction. This is due to the fact that the thrust vector of the main thruster
must be accurately kept in a given direction. According to the objective in Eqs. (42) and (43),
there are 11 outputs in this control system.

The control system also has seven inputs, i.e., the three piezoelectric stack actuators in
the three struts, the piezoelectric stack actuator in the central support, and the three
piezoelectric patch pairs in the device plate of the smart platform. Therefore, the system is a
seven-inputs-eleven-outputs system.

Prior to the simulations, the harmonic analyses of the combined structure are performed
to collect the dynamics of each control patch, i.e., **H** in Eq. (31). Then, the simulations are
conducted. In the simulations, three thrust position are considered: Case 1, the thrust angle of
$\Phi = 0$ with the azimuth of $\theta = 0$ in Figure 26; Case 2, $\Phi = 6^0$ and $\theta = 0$; and Case 3, $\Phi = 6^0$
and $\theta = 180^0$. As mentioned earlier, the dominant modal frequency of the satellite structure
is at 48 Hz; therefore, a sinusoidal excitation at 48 Hz is chosen as the reference signal. The
length of the finite impulse response of the controller is 2 and the sampling rate is 10,000 Hz.

Since there are large differences in the dynamics of each control path, a multi-
convergence factor method is proposed here. Instead of having a convergence factor in Eq.
(40), the multi-convergence factor method introduces a convergence factor vector. The
convergence factor vector is $[5 \times 10^7, 2 \times 10^6, 2 \times 10^6, 2 \times 10^6, 1 \times 10^5, 1 \times 10^5, 1 \times 10^5]$.
The first element in the convergence factor vector is associated with the piezoelectric stack
actuator in the central support, the second to the fourth elements are with the piezoelectric
stack actuators in the three struts, and the rest are with the three piezoelectric patch actuator
pairs in the device plate.

Table 4. The percent of controlled displacement over uncontrolled displacement after 300 seconds.

	Case 1 $\Phi = 0$, $\theta = 0$ % (dB)	Case 2 $\Phi = 6^0$, $\theta = 0$ % (dB)	Case 3 $\Phi = 6^0$, $\theta = 180^0$ % (dB)
J_e	0.79	0.6	0.69
e_{Az}	6.2 (-24,15)	5.7 (-24.88)	7.1 (-22.98)
e_{Ax}	8.6 (-21.30)	5.5 (-25.19)	6.1 (-24.29)
e_{Bz}	11.3 (-18.94)	5.9 (-24.58)	6.1 (-24.29)
e_{Bx}	8.3 (-21.62)	6.9 (-23.22)	5.6 (-25.04)
e_{Cz}	11.7 (-18.64)	6.7 (-23.48)	6.9 (-23.22)
e_{Cx}	7.7 (-22.27)	5.6 (-25.04)	5.7 (-24.85)
$e_{Gz} - e_{Hz}$	9.2 (-20.72)	5.8 (-24.73)	7.0 (-23.10)
$e_{Gz} - e_{Iz}$	9.3 (-20.63)	6.7 (-23.48)	6.8 (-23.35)
e_{Ez}	10.2 (-19.83)	7.8 (-22.16)	8.5 (-21.41)
e_{Fz}	8.6 (-21.31)	6.2 (-24.15)	5.9 (-24.58)
e_{Dz}	19.4 (-14.24)	18.3 (-14.75)	6.8 (-23.35)

Table 5. Amplitudes of control signals after 300 seconds (in Volts).

	Case 1 $\Phi = 0$, $\theta = 0$	Case 2 $\Phi = 6^0$, $\theta = 0$	Case 3 $\Phi = 6^0$, $\theta = 180^0$
Actuator in central support	267.83	289.45	212.13
Actuator in strut linked to A	3.15	61.41	18.69
Actuator in strut linked to B	14.38	4.28	51.52
Actuator in strut linked to C	21.14	7.56	53.29
Patch actuator in top plate 1	1811.76	2057.36	1263.47
Patch actuator in top plate 2	1682.05	1711.44	1075
Patch actuator in top plate 3	1623.23	1782.36	1118.19

Figures 29, 30, and 31 illustrate the vibration suppression and objective of the combined test structure in the three cases, respectively. Table 4 tabulates the percentage of the controlled displacements after 300 seconds over the uncontrolled displacements, demonstrating that the objective decreases to 0.7% of its uncontrolled value on average. The vibration of the test structure is reduced by 20.33 dB for Case 1, 23.24 dB for Case 2, and 23.67 dB for Case 3. Table 5 lists the control voltage for each actuator.

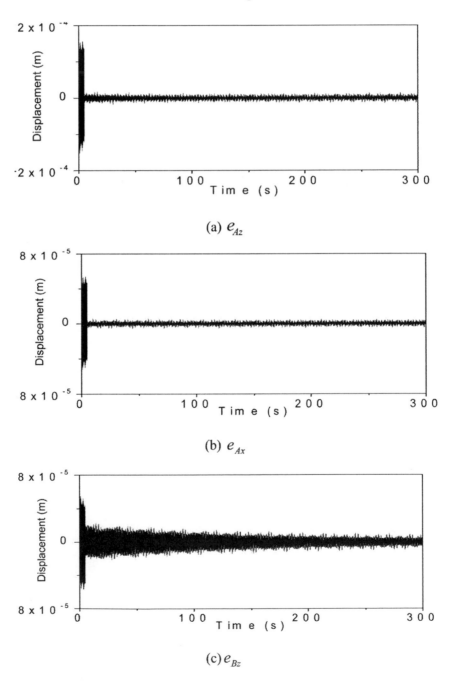

(a) e_{Az}

(b) e_{Ax}

(c) e_{Bz}

Figure 29. Continued on next page

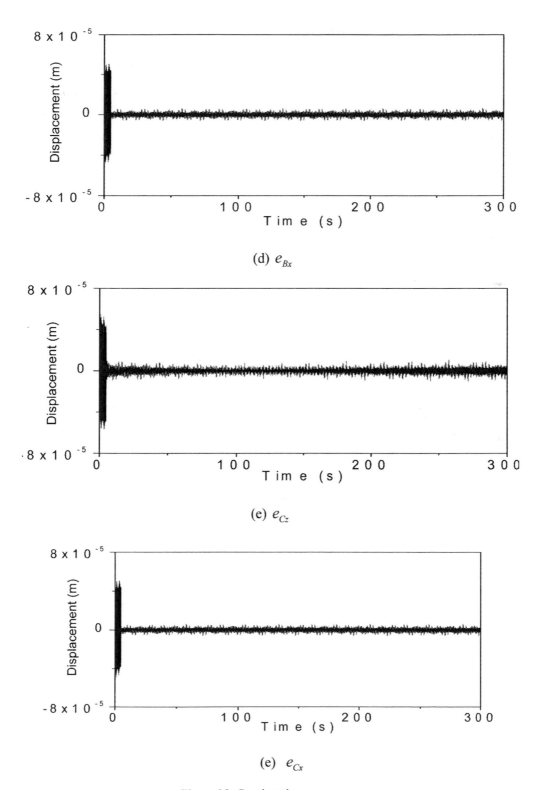

(d) e_{Bx}

(e) e_{Cz}

(e) e_{Cx}

Figure 29. Continued on next page.

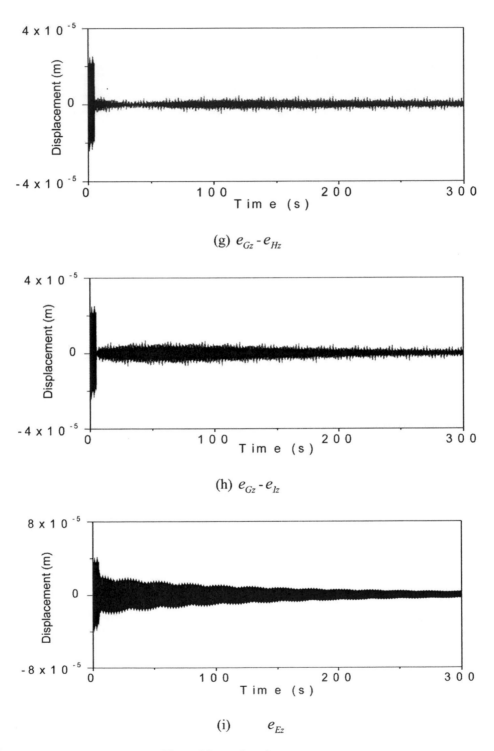

(g) $e_{Gz} - e_{Hz}$

(h) $e_{Gz} - e_{Iz}$

(i) e_{Ez}

Figure 29. continued on next page

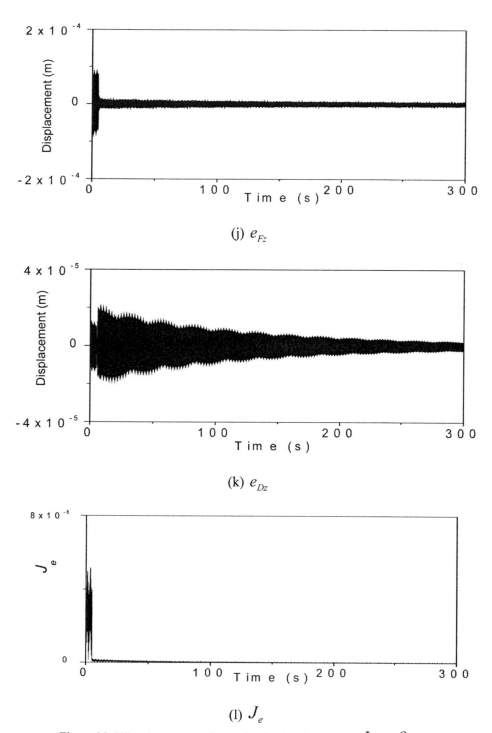

(j) e_{Fz}

(k) e_{Dz}

(l) J_e

Figure 29. Vibration suppression and objective for Case 1: $\Phi = 0$, $\theta = 0$.

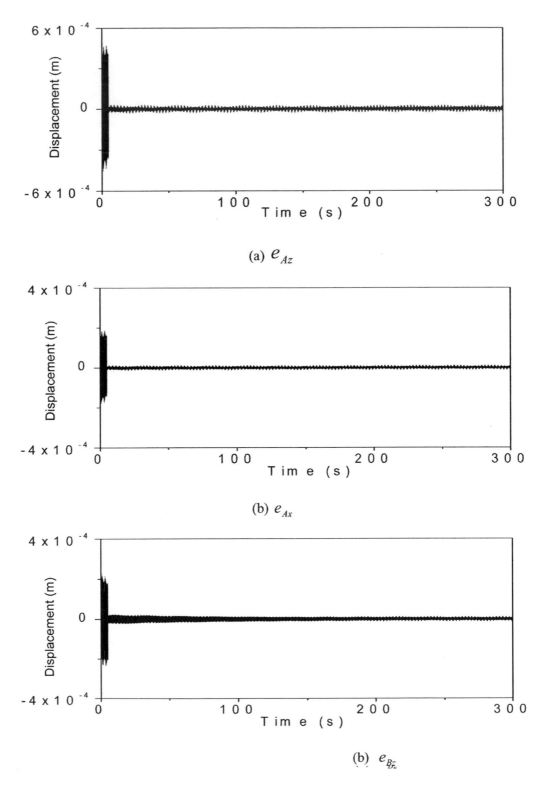

(a) e_{Az}

(b) e_{Ax}

(b) e_{Bz}

Figure 30. Continued on next page.

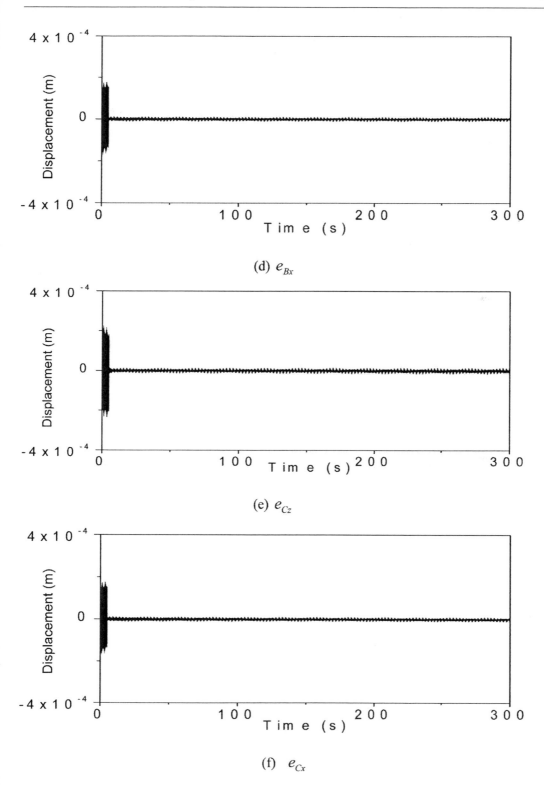

(d) e_{Bx}

(e) e_{Cz}

(f) e_{Cx}

Figure 30. Continued on next page.

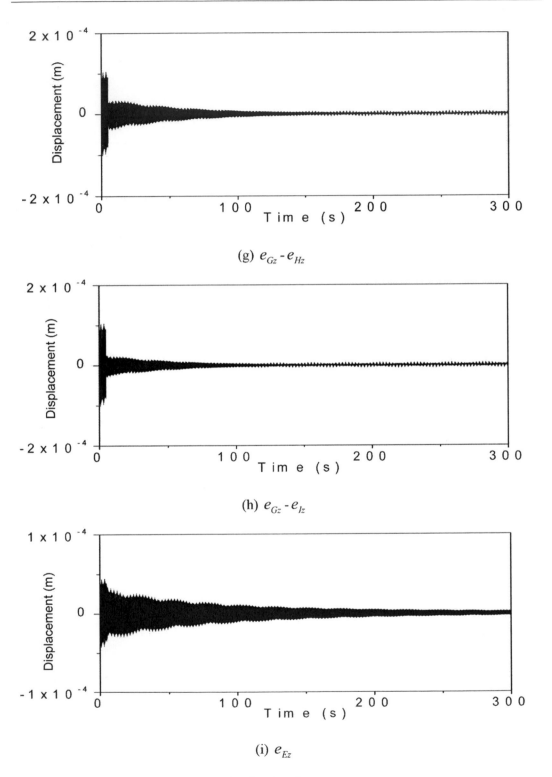

(g) $e_{Gz} - e_{Hz}$

(h) $e_{Gz} - e_{Iz}$

(i) e_{Ez}

Figure 30. Continued on next page.

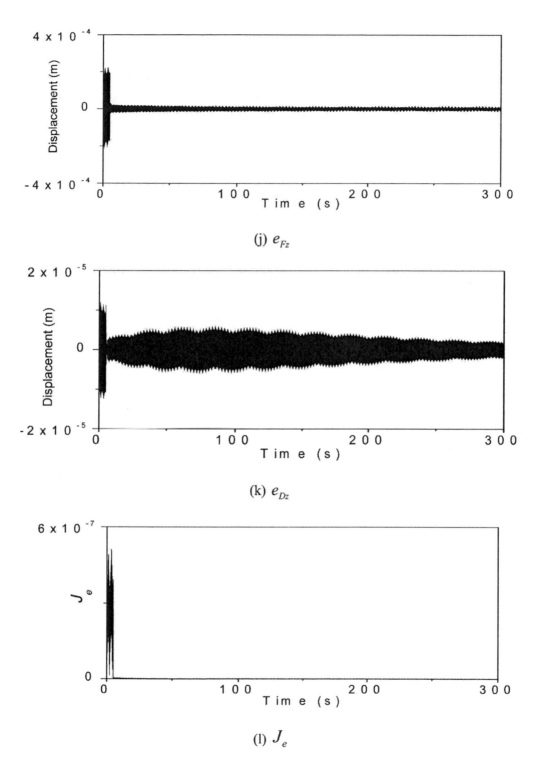

(j) e_{Fz}

(k) e_{Dz}

(l) J_e

Figure 30. Vibration suppression and objective for Case 2: $\Phi = 6^0$, $\theta = 0$.

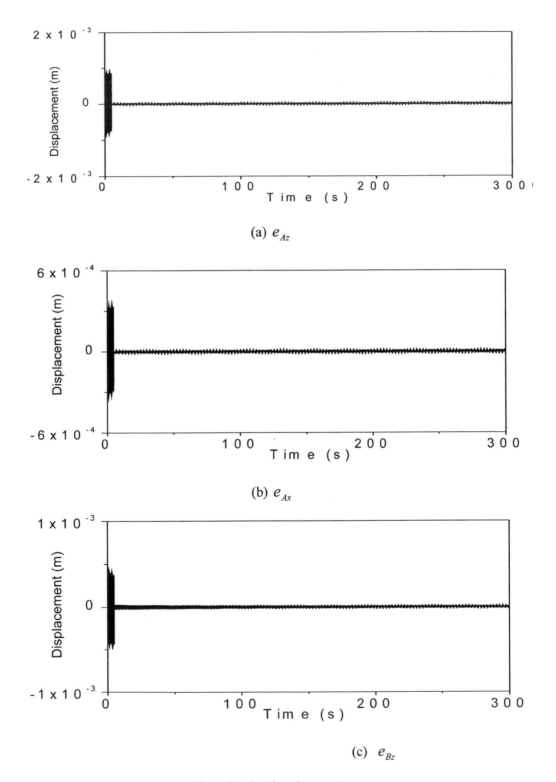

(a) e_{Az}

(b) e_{Ax}

(c) e_{Bz}

Figure 31. Continued on next page.

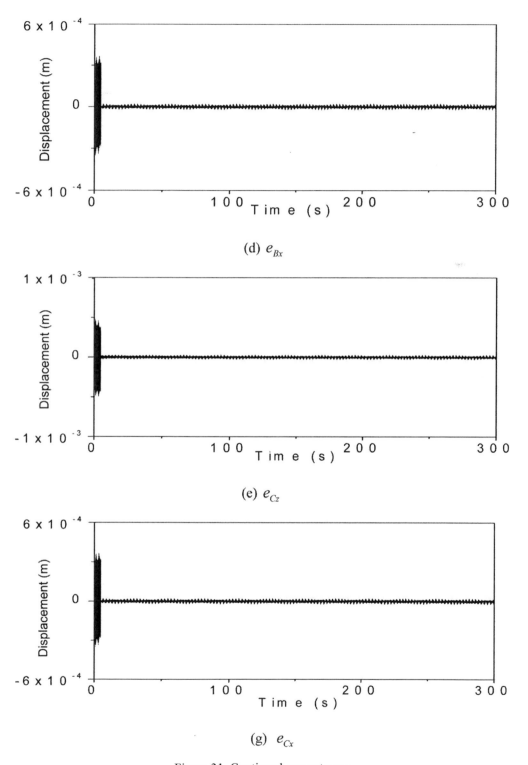

(d) e_{Bx}

(e) e_{Cz}

(g) e_{Cx}

Figure 31. Continued on next page.

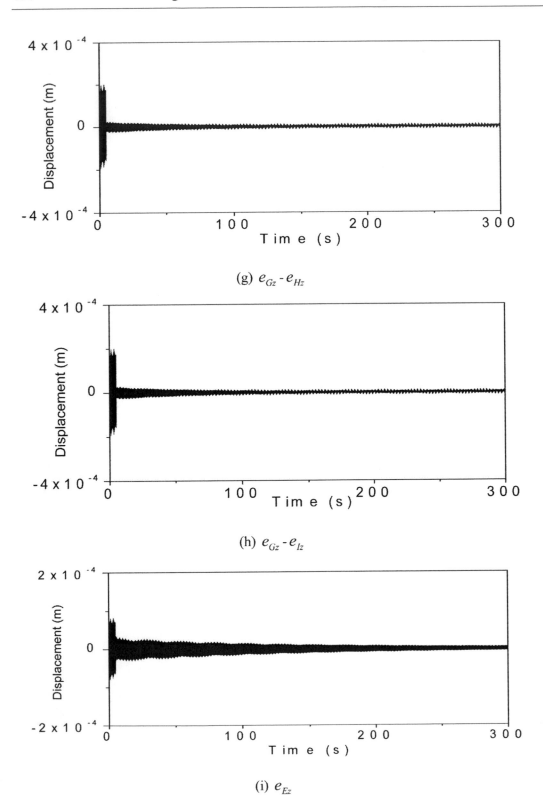

(g) $e_{Gz} - e_{Hz}$

(h) $e_{Gz} - e_{Iz}$

(i) e_{Ez}

Figure 31. Continued on next page.

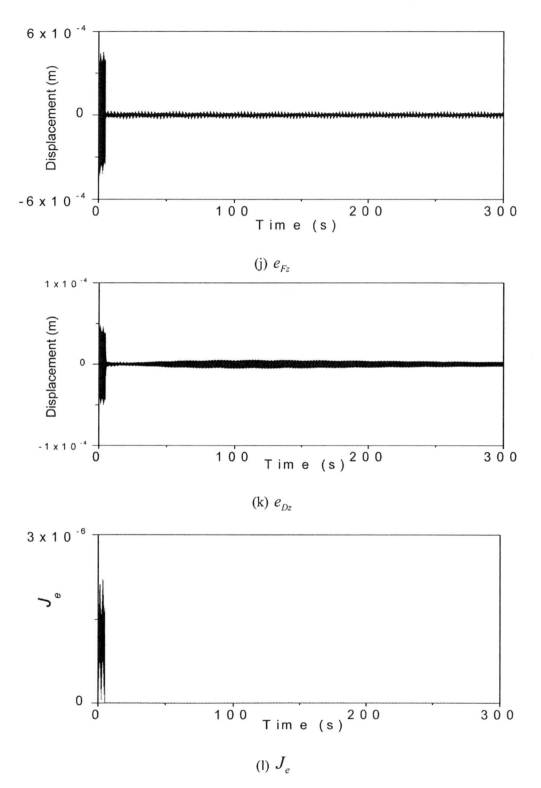

(j) e_{Fz}

(k) e_{Dz}

(l) J_e

Figure 31. Vibration suppression and objective for Case 3: $\Phi = 6^0$, $\theta = 180^0$.

Conclusion

Thrust vector control of satellites is a crucial problem in space. The RCS thrusters are traditionally used for the misalignment correction of the thrust vector, but there is a weight penalty and they consume the fuel of the satellite. The thruster-firing-induced vibration suppression is not solved yet.

In this chapter, a novel satellite thrust vector control and vibration suppression technology is introduced. The core of this novel technology is to place a smart composite platform between the satellite thruster and the satellite structure to make the satellite thruster steerable. By steering the thruster in real time, the thrust vector of a satellite can always be pointed in the desired direction and a vibration suppression capability can also be provided simultaneously by such platform. This chapter includes the configuration of the UHM smart composite platform, inverse kinematics and adaptive kinematics modeling, smart platform control, thrust vector misalignment correction of satellites, and vibration suppression of satellites.

The UHM smart composite platform employs the state-of-the-art smart structures technique. This platform consists of smart composite panels, smart composite struts, and advanced control systems, and possesses simultaneous precision positioning and vibration suppression (SPPVS) capabilities. The UHM platform inverse kinematics is derived and its kinematic singularity is also analyzed indicating that there is no kinematic singularity for the proposed design. In addition, an adaptive kinematic modeling method is introduced in which the nonlinearity of kinematics is considered.

Two control strategies; namely, local control strategy and global control strategy are presented for the smart platform control, and fuzzy logic control is employed for both control strategies. The local control strategy cannot take care of the existing joint tolerances and frictions, and the control system is more complex. The global control strategy, on the other hand, overcomes the drawbacks of the local control strategy and provides more accurate positioning of the device plate of the smart platform.

To investigate the satellite thrust vector control, the UHM platform is assembled onto a satellite structure. The satellite attitude dynamic model for thrust vector control is then built, and the satellite attitude controller and intelligent controller for the smart composite platform are designed. The successful performance of the thrust vector control employing the UHM smart platform is demonstrated. The results indicate that the smart composite platform can precisely achieve thrust vector control, the misalignment of the trust vector of the satellite can be corrected effectively with satisfactory position accuracy of the thrust vector. In addition, the thrust vector accuracy is not affected by the accuracy of the inner smart platform control loop, i.e., the local or control strategy, since there is a feedback outer loop.

The vibration suppression capability of this novel technology is also assessed. The combined system dynamics of the satellite structure and the smart composite platform is analyzed, and the dominant modes of the satellite structure are determined. A MIMO adaptive control scheme is then developed to suppress the vibration of the satellite structure employing four PZT stack actuators in the platform three smart composite struts and the central support, and three PZT patch actuator pairs in the platform device plate. A convergence factor vector concept is also introduced to ease the multi-channel convergent rate control. This vibration controller is adjusted based on the vibration information of the satellite structure and drives the smart composite platform to isolate the vibration transmission from the firing thruster to the satellite structure. Eleven vibration components of the structure and platform are controlled. Therefore, this is a seven-input-eleven-output system. The results demonstrate that the entire vibration of the combined structure, including the satellite structure, the UHM smart platform, and the thruster, at the dominant frequency of the combined structure can be decreased to 7-10% of its uncontrolled value in various device plate positions of the smart platform.

Acknowledgements

The authors acknowledge the financial support of the Office of Naval Research for the Adaptive Damping and Positioning using Intelligent Composite Active Structures (ADPICAS) project under the government grant number of N00014-00-1-0692 and N00014-05-1-0586, and the program officer Dr. Kam W. Ng. The authors also thank Mr. H. Edward Senasack, Jr., and Dr. Glenn Creamer of the Naval Research Laboratory for useful discussions.

References

[1]   Redmill, K., Ozguner, U., Musgrave, J., and Merrill, W. (1994). Intelligent hierarchical thrust vector control for a space shuttle. *IEEE Control Systems Magazine*, **4**(3), 13-23.

[2]   Yeh, F. K., Cheng, K., and Fu, L. (2002). Variable structure based nonlinear missile guidance and autopilot design for a direct hit with thrust vector control. In: *Proceedings of the 41st IEEE Conference on Decision and Control* (pp. 1275-1280). Piscataway, NJ: the Institute of Electrical and Electronic Engineers, Inc.

[3]   GlobalSecurity. Su-37 "Super Flanker" [online]. 2006 [cited January 1, 2006]. Available from URL: http://www.globalsecurity.org/military/world/russia/su-37.htm.

[4]   NASATech. Experiment on quasi-tailless flight of an X-31A airplane [online]. 2006 [cited January 1, 2006]. Available from URL: http://www.nasatech.com/Briefs/ May99/ DRC9612.html.

[5]   Dyne, S. J. C., Tunbridge, D. E. L., and Collins, P. P. (1993). The vibration environment on a satellite in orbit. In: *Proceedings of IEE Colloquium on High Accuracy Platform Control in Space* (pp.12/1-12/6). London: the Institution of Electrical Engineers.

[6]   Anderson, E. H., Moore, D. M., and Fanson, J. L. (1990). Development of an active truss element for control of precision structures. *Optical Engineering*, **29**(11), 1333-1341.

[7]   Boyd, J., T., Tupper Hyde, Osterberg D., and Davis T. (2001). Performance of a launch and on-orbit isolator. In: L. P. Davis (Ed), *Proceedings of SPIE* vol. 4327, *Smart Materials and Structures 2001: Smart Structures and Integrated Systems* (pp. 433-440). Bellingham, WA: the SPIE- the International Society for Optical Engineering.

[8]   Li, X., Hamann, J. C., and McInroy, J. E. (2001) Simultaneous vibration isolation and positioning control of flexure jointed Hexapods. ," In: L. P. Davis (Ed), *Proceedings of SPIE* vol. 4327, *Smart Materials and Structures 2001: Smart Structures and Integrated*

Systems (pp. 99-109). Bellingham, WA: the SPIE- the International Society for Optical Engineering.

[9] Quenon, D., Boyd, J., Buchele, P., Self, R., Davis, T., Hintz, T., and Jacobs, J. (2001) Miniature vibration isolation system for space applications. In: Anna-Maria R. McGowan (Ed), *Proceedings of SPIE* vol. 4332, *Smart Materials and Structures 2001: Industrial and Commercial Applications of Smart Structures Technologies* (pp. 159-170). Bellingham, WA: the SPIE- the International Society for Optical Engineering.

[10] McInroy, J. E., and Jafari, F. (2002). The state-of-the-art and open problems in stabilizing platforms for pointing and tracking. In: L. P. Davis (Ed), *Proceedings of SPIE* vol. 4701, *Smart Materials and Structures 2002: Smart Structures and Integrated Systems* (pp. 177-188). Bellingham, WA: the SPIE- the International Society for Optical Engineering.

[11] Physik Instrumente GmbH & Co. (2001). MicroPositioning, NanoPositioning, NanoAutomation: Solutions for cutting-edge Technology. Karlsruhe, Germany: Physik Instrumente GmbH & Co.

[12] Stewart, D. (1965). A platform with six degrees of freedom. *Proceedings of Institute of Mechanical Engineers,* London, 180(15), 371-386.

[13] Ma, K., and Ghasemi-Nejhad, M. N. (2005). Precision positioning of a parallel manipulator for spacecraft thrust vector control. *AIAA Journal of Guidance, Control, and Dynamics,* **28** (1), 185-188.

[14] Ma, K., and Ghasemi-Nejhad, M. N. (2004). 2-DOF precision platform for spacecraft thrust vector control: strategies and simulations. In: A. B. Flatau (Ed), *Proceedings of SPIE* vol. 5390, *Smart Materials and Structures 2004: Smart Structures and Integrated Systems* (pp. 46-55). Bellingham, WA: the SPIE- the International Society for Optical Engineering.

[15] Ma, K., and Ghasemi-Nejhad, M. N. (2005). MIMO adaptive control of thruster-firing-induced vibration of satellites using multifunctional platforms." In: R. C. Smith (Ed), *Proceedings of SPIE* vol. 5757, *Smart Materials and Structures 2005: Modeling, Signal Processing, and Control* (pp. 459-470). Bellingham, WA: the SPIE- the International Society for Optical Engineering.

[16] Ma, K., and Ghasemi-Nejhad, M. N. (2004). Frequency-weighted hybrid adaptive control for simultaneous precision positioning and vibration suppression of intelligent composite structures. *Smart Materials and Structures,* **13**(5), 1143-1154.

[17] Ma, K., and Ghasemi-Nejhad, M. N. (2005). Adaptive simultaneous precision positioning and vibration control of intelligent structures. *Journal of Intelligent Material Systems and Structures,* **16**(2), pp. 163-174.

[18] Ma, K., and Ghasemi-Nejhad, M. N. (2003). Simultaneous precision positioning and vibration suppression of smart structures ---- adaptive control methods and comparison. In: *Proceedings of the 42nd IEEE Conference on Decision and Control* (pp. 6386-6391). Piscataway, NJ: the Institute of Electrical and Electronic Engineers, Inc.

[19] Ghasemi-Nejhad, M. N., Russ, R., and Pourjalali, S. (2005). Manufacturing and testing of active composite panels with embedded piezoelectric sensors and actuators. *Journal of Intelligent Material Systems and Structures,* **16** (4), 319-334.

[20] Ghasemi-Nejhad, M. N. (2002). Active composite panels and active composite struts as building blocks of adaptive structures. In: *Proceedings of JSME/ASME International*

Conference on Materials and Processing (vol.1, pp.432-440). Tokyo, Japan: the Japan Society of Mechanical Engineers.

[21] Ghasemi-Nejhad, M. N., and Ma, K. (2004). Adaptive damping and positioning using intelligent composite active structures (ADPICAS). In: *Proceedings of ACTIVE 04* (paper no. A04_019). Ames, IA: the Institute of Noise Control Engineering of the USA, Inc.

[22] Ma, K., and Ghasemi-Nejhad, M. N. (2005). Simultaneous precision positioning and vibration suppression of reciprocating flexible manipulators. *Smart Structures and Systems*, **1**(1), 13-27.

[23] Yan, S., and Ghasemi-Nejhad, M. N. (2004) Modeling and genetic algorithms based piezoelectric actuator configuration optimization of an adaptive circular composite plate. In: R. C. Smith (Ed), *Proceedings of SPIE* vol. 5383, *Smart Materials and Structures 2004: Modeling, Signal Processing, and Control* (pp. 255-264). Bellingham, WA: the SPIE- the International Society for Optical Engineering.

[24] Ghasemi-Nejhad, M. N., Pourjalali, S., Uyema, M., and Yousefpour, A. (2006). Finite element method for active vibration suppression of smart composite structures using piezoelectric materials. *Journal of Thermoplastic Composite Materials*, **19** (in press).

[25] Ghasemi-Nejhad, M. N., Russ, R., and Ma, K. (2006). Finite element charts and active vibration suppression schemes for smart structures design. *Journal of Intelligent Material Systems and Structures*, **17** (in press).

[26] Li, T.-H. S., Lin, I-Fong, and Hung, T.-M, (2002). Behavior-based fuzzy logic control for a one-on-one robot soccer competition. In: *Proceedings of the 2002 IEEE International Conference on Fuzzy Systems* (vol.1, pp.470-475). Piscataway, NJ: the Institute of Electrical and Electronic Engineers, Inc.

[27] Mathworks, Inc. (2001). *Fuzzy Logic Toolbox*. Natick, MA: The Mathworks, Inc.

[28] Yen, J., and Langari, R. (1999). *Fuzzy Logic: Intelligence, Control, and Information*. Upper Saddle River, NJ: Prentice Hall.

[29] Verbruggen, H. B., and Babuška, R. (1999). *Fuzzy Logic Control: Advances in Applications*. Singapore: World Scientific Publishing Company.

[30] Antin, N. (2004). Design and analysis of an intelligent composite platform for thrust vector control. Master's thesis. Honolulu, HI: Department of Mechanical Engineering, University of Hawaii at Manoa.

[31] Antin, N., and Ghasemi-Nejhad, M. N. (2005). Active vibration suppression of a satellite frame using an adaptive composite thruster platform. In: A. B. Flatau (Ed), *Proceedings of SPIE* vol. 5764, *Smart Materials and Structures 2005: Smart Structures and Integrated Systems* (pp. 390-401). Bellingham, WA: the SPIE- the International Society for Optical Engineering.

[32] Widrow, B., Glover, J. R., McCool, J. M., Kaunitz, J., Williams, C. S., Hearn, R. H., Zeidler, J. R., Dong, E., and Goodlin, R. C. (1975). Adaptive noise canceling: principles and application. *Proceedings of the IEEE*, **63** (12), 1692-1716.

[33] Nelson, P. A., and Elliott, S. J. (1992). *Active Control of Sound*, New York: Academic.

[34] Vipperman, J. S., Burdisso, R. A., and Fuller, C. R. (1993). Active control of broadband structural vibration using the adaptive LMS algorithm. *Journal of Sound and Vibration*, **166** (2), 283-299.

[35] Ma, K. (2003). Vibration control of smart structures with bonded PZT patches: novel adaptive filtering algorithm and hybrid control scheme. *Smart Materials and Structures*, **12**(3), 473-482.

[36] Ma, K., and Melcher, J. (2003). Adaptive control of structural acoustics using intelligent structures with embedded PZT patches. *Journal of Vibration and Control*, **9**(11), 1285-1302.

[37] Ma, K., Chen X., and Gu Z. (1998) Adaptive control of wind-induced vibration of flexible structures---methods and experiments." *Journal of Vibration Engineering*, **11**(2), 131-137.

[38] Ma, K. (2003). Adaptive nonlinear control of a rectangular-clamped plate with PZT patches. *Journal of Sound and Vibration*, **264**(4), 835-850.

In: Progress in Smart Materials and Structures
Editor: Peter L. Reece, pp. 203-225

ISBN: 1-60021-106-2
© 2007 Nova Science Publishers, Inc.

Chapter 7

TEMPERATURE-INSENSITIVE SENSORS WITH CHIRP-TUNED FIBER BRAGG GRATINGS

Xinyong Dong[1], P. Shum[1] and C.C. Chan[2]*

[1]Network Technology Research Centre, Nanyang Technological University, Singapore
[2]School of Chemical and Biomedical Engineering, Nanyang Technological University,
Singapore

Abstract

Optical fiber grating sensor technology has attracted considerable interests of research and development in the last decade. The normal optical fiber Bragg grating (FBG) sensor systems, in which the measurand is related to the Bragg (or resonant) wavelength of one or several FBGs, usually require wavelength interrogators (or modulators) and temperature compensators when they are used practically. Therefore the cost may be increased and the construction of the system may become complex. Another FBG sensor prototype based on chirp-tuned FBGs may avoid these problems. In this chapter, we study the basic sensing principle of and, based on this study, propose two novel sensor designs for displacement measurement and tilt measurement, respectively. In this sensor prototype, the measurand is related to the bandwidth (or chirp) of the involved FBG(s) so that it can be sensed by direct measurement of the reflected optical power from the FBG(s). Due to this optical power-encoding property, the proposed sensors show great advantages including simple construction (no need of wavelength interrogator) and inherently insensitive to temperature thus eliminating the need for temperature compensation. The proposed sensor designs may have potential applications in the optical fiber sensor area.

Introduction

Following the realization of low loss optical waveguides in the 1960s, optical fibers have been developed to the point where they are synonymous with the modern telecommunication and optical sensor networks. A large amount of optical fiber-based passive and active devices, such as the optical fiber coupler, optical fiber attenuator, optical fiber amplifier and laser etc

---

* E mail address : exydong@ntu.edu.sg, Tel : 65-67904681; Fax : 65-67926894

have emerged in the market since then. With the significant discovery of photosensitivity in optical fibers in 1978 [1], a new class of fiber component, called the otpical fiber Bragg grating (FBG) has been developed and obtained broadly applications in optical fiber devices and subsystems. This small and robust optical component can perform many primary functions such as reflection and filtering, in a highly efficient and low loss manner. This is a comparatively simple device because it only consists of a periodic modulation of the refractive index along the fiber core within the grating-inscripted region. It can be easily fabricated by the ultraviolet laser inscription from side through phase mask [2]. By varying the period, tailoring the amplitude and/or introducing phase shifts to the modulation function of the refractive index of fiber core, fiber gratings may produce various reflection and transmission spectra and dispersion properties. These kinds of properties are very useful in many applications in optical fiber communications, such as the wavelength locking, add/drop multiplexing, gain flattening of erbium-doped fiber amplifiers, chromatic dispersion compensation and so on [3,4].

FBGs have also attracted considerable interests in various fiber-optic sensor implementations for the last decade due to their very sensitive strain and thermal responses [5-7]. The reliable linear response of the Bragg (or resonant) wavelength of an FBG to the applied axial strain has facilitated various transducer designs that can deal with a lot of measurands such as pressure [8,9], acceleration [10,11], displacement [12,13], bending [14], vibration [15], electrical current [16], underwater acoustic wave [17], etc. In addition to the well-known advantages of fiber-optic sensors such as electrically passive operation, immunity to radio frequency interference and electromagnetic interference, high sensitivity, compact size, light weight, corrosion resistance, and potentially low cost, FBG-based sensor is one of the promising candidates for sensing applications with its own advantages such as the inherent self-referencing capability and being easily multiplexed in a serial fashion along a single fiber. Therefore, FBG-based sensors and sensor systems have been widely studied especially in the area of smart materials and structures, where a large number of distributed and embedded sensor elements are needed to make the material or structure "smart" [18]. For example, many FBG sensor systems were developed and installed to monitor the health of structures such like bridge, high-level building, dam, mine, wing of aircraft, oil and gas transport pipelines, etc. In these cases, the health of structures can be assessed and tracked on a quasi real-time basis by wavelength scanning method to detect variations of Bragg wavelengths of sensing FBGs.

Generally, FBG sensors are based on the principle that the measured information is encoded into the shift of the Bragg wavelength. Therefore, a wavelength interrogator is necessary in order to get the wavelength shift information continuously [19]. A simple and straightforward method to realize this is to use a spectrometer or an optical spectrum analyzer, but neither is attractive in some practical applications due to big size, limited resolution capability and low cost-effectiveness. In most sensor systems, the wavelength shift of the FBG is converted into some easily measured parameters, such as amplitude, phase, or frequency, by using different wavelength interrogators, which may be based on various techniques such as linearly wavelength-dependent optical filters, scanning Fabry-Perot filters, tunable acousto-optic filters, matched fiber grating pairs, and unbalanced Mach-Zehnder interferometers [19]. The use of any of these wavelength interrogators may increase the complexity and cost of the whole sensing system significantly. In addition, for the use of FBG-based sensor in some practical situations, it is always necessary to distinguish between the strain effect and the thermal/temperature effect, because the Bragg wavelength of FBG is intrinsically sensitive to both factors. As a result, FBG sensing systems have to include an additional temperature compensation scheme, such as the reference FBG method, reflection-peak-split method, superstructure FBG method, and the FBGs with different responses method [5-7]. These methods may also increase the cost and reduce the multiplexing capability of FBG sensors for multi-point measurement.

Since wavelength interrogator and temperature compensator are indispensable in wavelength-encoded FBG sensor systems, it is natural that one looks into a non-wavelength-encoded sensing technique to reduce the cost and to simplify the system. In our previous study [20], bandwidth of chirped-FBG was used to track the bending state of a flexible beam on which the FBG was attached and chirped by strain filed produced on the beam lateral side. This method showed the advantage of temperature insensitive. Bandwidth of the chirped-FBG was not changed by temperature because every part of the grating had equal response to temperature variations. However, measuring the change of bandwidth or a wavelength interrogator was still necessary in that system. A much better method was reported by using the FBG-reflected optical power as the tracking signal [21,22]. In this method, measurand was related to chirp of the grating sensor in various ways so that sensing can be realized by directly measuring the optical power of the reflected light from the grating. The advantage of temperature insensitive is remained and the wavelength interrogator is replaced by a photodetector or an optical power meter for the measurement of optical power. This type of sensors, more than most optical fiber sensors, will be truly "smart", at least in that an aspect of self-compensation of temperature, eliminating the need for additional instrumentation. However, the previous reports demonstrated only their specific sensor design and results, didn't give a theoretical analysis of the basic sensing principle and the requirements for FBGs. In this chapter, the basic working principle is analyzed theoretically and the sensor performance is studied in terms of various FBG parameters. Additionally, two novel sensor designs based on this method are proposed and demonstrated experimentally.

Sensing Principle

The proposed sensor prototype consists of one or more FBGs with initially uniform periods, which are chirped by nonuniform strain fields related to the measurand so that their bandwidths, reflectivities and therefore reflected optical powers can respond to the variation of measurand. By detecting the reflected optical power from the grating using a photodetector or an optical power meter, measurement of the measurand can be realized.

Figure 1 shows the schematic diagram of the interrogation system of the proposed sensor prototype. Light from a broadband source is injected into the chirp-tuned FBG through an optical coupler. Reflected light from the grating is then guided via the same coupler and measured by a photo-detector (or power meter). The detected optical power can be expressed as

$$P = V \int_{\lambda} \rho(\lambda) R(\lambda) d\lambda \qquad (1)$$

where V is a factor that describes the total loss of the reflected light from the source to the detector, including the losses arising from fiber splicing and connection, insertion loss of the coupler (the detected light passes the coupler twice before being measured), etc, but excluding transmission loss of the grating. $\rho(\lambda)$ and $R(\lambda)$ are the power spectral density of the broadband light source and the reflectivity of the grating, respectively. Both are functions of wavelength.

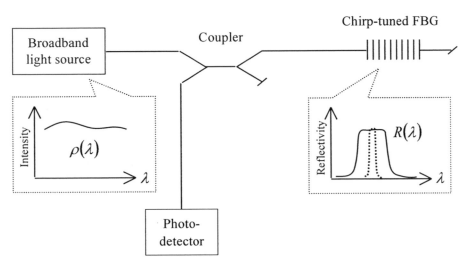

Figure 1. Interrogation system of the proposed sensor prototype.Schematic diagram and experimental setup of the proposed FBG-based displacement sensor: ASE, gain-flattened erbium-doped fiber amplified spontaneous emission source; OSA, optical spectrum analyzer; PD, photodetector.

It is notable that the function $R(\lambda)$, which also includes the information of the grating, will vary with the measurand. $\rho(\lambda)$ is supposed to be a fixed function, i.e. not changeable in the sensing process. Furthermore, in order to achieve linear response and temperature-insensitive measurement $\rho(\lambda)$ is expected to be a constant, which means that it does not change with wavelength. Such flat-output, broadband light source can be realized by using a super-luminescent light emitted diode or an output-flattened amplified spontaneous emission source with erbium-doped fiber. And for most cases, the reflection band of the chirped-FBGs can be regarded as of rectangular shapes. With the above assumptions, Eq. (1) can be therefore rewritten as

$$P = V\rho R \Delta\lambda_{bw} \tag{2}$$

where ρ is the constant power spectral density of the broadband light source, R is the average reflectivity of the reflection band of the grating, and $\Delta\lambda_{bw}$ is the 3-dB bandwidth of the grating. Since temperature variation cannot change the reflection profile but the central wavelength of the sensing grating, the detected optical power will inherently insensitive to temperature.

Theoretical Study

Fiber Bragg gratings are produced usually by exposing an optical fiber to a spatially periodical pattern of ultraviolet intensity. Due to the photosensitivity of material of the fiber core, a perturbation to the effective refractive index of the guided mode(s) of interest will be formed. It can be described by [2]

$$\delta n_{eff}(z) = \overline{\delta n}_{eff}(z)\left\{1 + v\cos\left[\frac{2\pi}{\Lambda}z + \phi(z)\right]\right\} \tag{3}$$

where $\overline{\delta n}_{eff}$ is the refractive index change spatially averaged over a single grating period of the grating, v is the fringe visibility of the index modulation, Λ is the grating period (i.e. the period of the index perturbation), and $\varphi(z)$ is the arbitrary spatially varying phase change which may describe the grating chirp.

Based on the coupled mode theory, a uniform FBG usually reflects light within a very narrow wavelength band related to the index modulation dept and grating length. The Bragg wavelength or resonant condition is usually given by the following expression

$$\lambda_B = 2n_{eff}\Lambda, \tag{4}$$

where n_{eff} is the effective refractive index of the fiber core. As is known, the strain response of FBG arising from both the physical elongation of the grating, corresponding to the fractional change in grating period, $\Delta\Lambda$, and the change in fiber index, Δn_{eff} due to photoelastic effects, can be expressed using

$$\frac{\Delta\lambda_B}{\lambda_B} = \frac{\Delta n_{eff}}{n_{eff}} + \frac{\Delta\Lambda}{\Lambda} = (1 - p_e)\varepsilon_{ax}, \tag{5}$$

where $\Delta\lambda_B$ is the change of the Bragg wavelength, $\varepsilon_{ax} = \Delta\Lambda/\Lambda$ is the axially applied strain along the FBG, and p_e is the effective photoelastic constant (~0.22) of the optical fiber material.

It can be seen from Eq (5) that the variation in Bragg wavelength of an FBG is directly proportional to the applied axial strain. Different level of strain will cause different change in Bragg wavelength. Therefore, if we apply a linearly varying strain field along the length of the FBG and assume that this FBG is composed of many small grating segments, the change in Bragg wavelength of each small grating segment will be different, depending on the local strain level. That introduces a chirp to the grating period with a chirp rate depending on the gradient of the strain field. The reflection band will be broadened and the bandwidth will be decided by the magnitude of the introduced chirp rate. For the convenience of analysis, here we express the linearly varying strain field as

$$\varepsilon_{ax}(z) = \varepsilon_0 + K_\varepsilon z, \quad (-0.5L_g \leq z \leq 0.5L_g), \tag{6}$$

where $\varepsilon_0 = \varepsilon_{ax}(0)$, is the strain applied at the center of the fiber grating, K_ε is the gradient of the strain field, and L_g is the length of the fiber grating. By substituting Eq. (6) into Eq. (5), we can get

$$\Delta\lambda_B(z) = \Delta\lambda_{B0} + \lambda_B K_\varepsilon (1 - p_e)z, \tag{7}$$

where $\Delta\lambda_{B0} = \Delta\lambda_B(0) = \lambda_B(1 - p_e)\varepsilon_0$, is the Bragg wavelength shift at the center of the grating. The strain-induced chirp rate in Bragg wavelength and the bandwidth of the grating reflection hence can be given respectively by

$$R_{ch} = \frac{d(\Delta\lambda_B)}{dz} = \lambda_B K_\varepsilon (1 - p_e), \tag{8}$$

$$\Delta\lambda_{bw} = \Delta\lambda_{bw0} + R_{ch}L_g = \Delta\lambda_{bw0} + \lambda_B K_\varepsilon L_g (1 - p_e) \tag{9}$$

where $\Delta\lambda_{bwo}$ is the initial bandwidth of the grating reflection. So the bandwidth of the strain-tuned uniform FBG is changed linearly with the gradient of the applied strain field. But to evaluate the reflected light power from the FBG, we also need to know details of the reflectivity, which may be reduced after the grating is chirped. Following we numerically study the reflection spectrum of the chirped-FBG by using a transfer matrix method, and study the relationship between reflected light power and reflection bandwidth.

In the transfer matrix method, the chirp-tuned FBG is divided into N grating segments. N is a large number so that each grating segment is very short and its period can be regarded as uniform. For each such uniform Bragg grating, we can use a 2 × 2 matrix to describe it. By multiplying all these 2 × 2 matrices together, we can obtain a single 2 × 2 matrix that describes the whole grating. We define A_i and B_i to be the field amplitudes of the co-propagating mode and the identical counter-propagating mode respectively after traversing the i-th grating segment. By starting with $A_0 = A(L_g/2) = 1$ and $B_0 = B(L_g/2) = 0$, the output amplitudes, $A_N = A(-L_g/2)$ and $B_N = B(-L_g/2)$ can be described as

$$\begin{bmatrix} A_N \\ B_N \end{bmatrix} = M \begin{bmatrix} A_0 \\ B_0 \end{bmatrix}; \quad M = M_N \cdot M_{N-1} \cdot ... \cdot M_i \cdot ... \cdot M_1, \tag{10}$$

where

$$M_i = \begin{bmatrix} M_{11} & M_{12} \\ M_{21} & M_{22} \end{bmatrix} \tag{11}$$

is the matrix that describes the i-th grating segment. Matrix elements are defined as follows [2]

$$M_{11} = \cosh(\alpha\Delta z) - i\frac{\delta}{\alpha}\sinh(\alpha\Delta z) \qquad (12)$$

$$M_{12} = -i\frac{\kappa}{\alpha}\sinh(\alpha\Delta z) \qquad (13)$$

$$M_{21} = \cosh(\alpha\Delta z) + i\frac{\delta}{\alpha}\sinh(\alpha\Delta z) \qquad (14)$$

$$M_{12} = i\frac{\kappa}{\alpha}\sinh(\alpha\Delta z), \qquad (15)$$

where Δz is the length of the i-th grating segment, δ and κ are "dc" and "ac" coupling coefficients, which are defined respectively as

$$\delta = 2\pi n_{eff}\left(\frac{1}{\lambda} - \frac{1}{\lambda_B}\right) + \frac{2\pi}{\lambda}\overline{\delta n}_{eff}(z) - \frac{1}{2}\frac{d\phi}{dz}, \qquad (16)$$

$$\kappa = \frac{\pi}{\lambda}v\overline{\delta n}_{eff}(z). \qquad (17)$$

α is defined as

$$\alpha = \sqrt{\kappa^2 - \delta^2}. \qquad (18)$$

In order to show the situation that an initially uniform FBG is chirped by a linearly varying strain field, all grating-related parameters, except for the phase term that describes the chirp rate, are set to fixed values. Following are the values used: $n_{eff} = 1.45$, $v\overline{\delta n}_{eff} = 4 \times 10^{-4}$, $L_g = 4$ cm, and $\lambda_B = 1550$ nm (central Bragg wavelength). The phase term for a linear chirp can be given by

$$\frac{1}{2}\frac{d\phi}{dz} = -\frac{4\pi n_{eff} z}{\lambda_B^2}\frac{d\lambda_B}{dz}. \qquad (19)$$

We change the value of $d\lambda_B/dz$ to get different chirp rates and reflection bandwidths.

Figure 2 shows several calculated reflection spectra of FBGs with different chirp rates. The narrowest reflection spectrum is achieved at $d\lambda_B / dz = 0$, corresponding to the originally uniform FBG. In this case, the 3-dB bandwidth is 0.5 nm, and the top-band reflectivity is 1. With the chirp rate and bandwidth being increased, the top-band reflectivity is reduced; however, the reduction is very small (less than 0.005) until the bandwidth is broadened to more than 4.2 nm. The details of average top-band reflectivity against 3-dB bandwidth are shown in Figure 3.

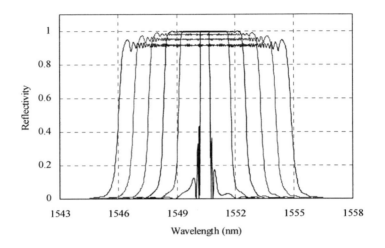

Figure 2. Simulated reflection spectra of chirp-tuned FBG: $n_{eff} = 1.45$, $v\overline{\delta n}_{eff} = 4 \times 10^{-4}$, $L_g = 4$ cm, and $\lambda_B = 1550$ nm (central Bragg wavelength).

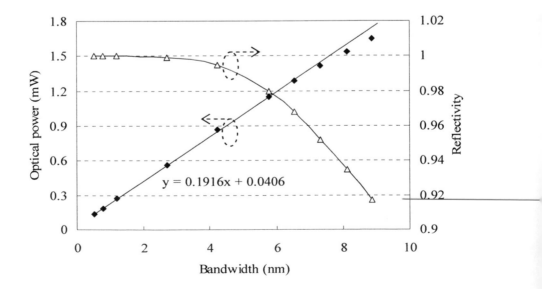

Figure 3. Calculated average top-band reflectivity and detectable optical power against 3-dB bandwidth of chirp-tuned FBG: $n_{eff} = 1.45$, $v\overline{\delta n}_{eff} = 4 \times 10^{-4}$, $L_g = 4$ cm, $\lambda_B = 1550$ nm, $\rho = 1$ mW/nm, and $V = 0.2$.

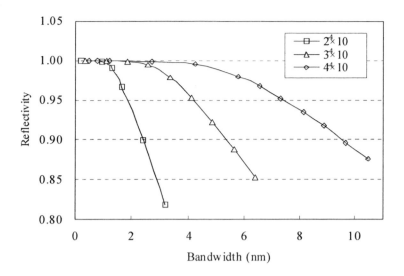

Figure 4. Calculated top-band reflectivity against 3-dB bandwidth for chirp-tuned FBG with various $\overline{v\delta n}_{eff}$ of 2, 3, and 4 × 10^{-4}: n_{eff} = 1.45 , L_g = 4 cm, and λ_B = 1550 nm.

By assuming a constant power spectral density of $\rho = 1$ mW/nm and a loss factor of $V = 0.2$, optical powers can also be calculated out by using Eq. (2). The calculated optical power data against 3-dB bandwidth are also shown in Figure 3. A good linear response of 0.19 mW/nm can be maintained till a wide bandwidth of 7.3 nm. Over this bandwidth, the response of optical power becomes nonlinear due to the obvious and fast reduction in reflectivity of the reflective band.

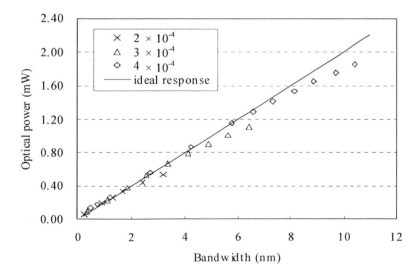

Figure 5. Calculated detectable optical power against 3-dB bandwidth for chirp-tuned FBG with various $\overline{v\delta n}_{eff}$ of 2, 3, and 4 × 10^{-4}: n_{eff} = 1.45 , L_g = 4 cm, λ_B = 1550 nm (central Bragg wavelength), $\rho = 1$ mW/nm, and $V = 0.2$.

To study the effects of the average refractive index change due to grating inscription, as well as the grating length, on the reflectivity and reflected optical power, we simulated the chirp-tuned FBG under different situations. Figure 4 shows curves of calculated top-band reflectivity against 3-dB bandwidth for the chirp-tuned FBG with various $v\overline{\delta n}_{eff}$ of 2, 3, and 4×10^{-4}. Other parameters are the same as those used in the aforementioned case. It shows that, with increasing the bandwidth, the grating reflectivity reduces more slowly in the case with a larger $v\overline{\delta n}_{eff}$ than that with a smaller one. In above three cases, high reflectivity of over 0.95 can be maintained till bandwidths are broadened to 1.9, 4.1, and 7.3 nm, respectively. It is easy to understand because larger is the value of $v\overline{\delta n}_{eff}$, stronger is the reflection of the grating. Corresponding to that, the good linear response of optical power to bandwidth can be maintained to a higher value of bandwidth in the case of a larger $v\overline{\delta n}_{eff}$ than that with a smaller $v\overline{\delta n}_{eff}$, as shown in Figure 5.

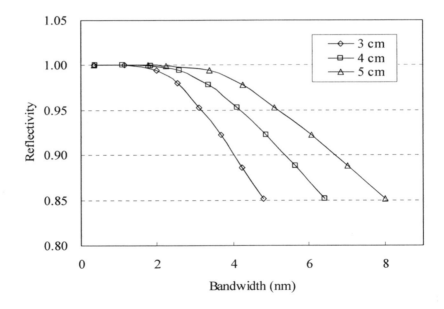

Figure 6. Calculated top-band reflectivity against 3-dB bandwidth for chirp-tuned FBG with various grating lengths of 3, 4, and 5 cm: $n_{eff} = 1.45$, $v\overline{\delta n}_{eff} = 3 \times 10^{-4}$, and $\lambda_B = 1550$ nm.

Figures 6 and 7 show calculated results of the top-band reflectivity and the optical power against 3-dB bandwidth of the chirp-tuned FBG with different grating lengths. Following parameters are used: $n_{eff} = 1.45$, $v\overline{\delta n}_{eff} = 3 \times 10^{-4}$, $\lambda_B = 1550$ nm, and $L_g = 3$, 4, or 5 cm. It is shown that a longer grating can stand up for high reflectivity and linear response of optical power better than a shorter grating when the gratings are chirped. This is because the shorter grating is chirped more seriously than the longer grating for achieving the same bandwidth.

Based on above theoretical studies, we find that high reflectivity and good linear response of reflected optical power of a chirp-tuned FBG to bandwidth can be achieved to a certain extent. The limitation is related to the amplitude of the refractive index change and the length of the grating. A larger refractive index change and/or grating length may lead to a larger linear response range. Therefore, for sensor applications, strong and long gratings are more preferable than weak and short ones. In the following two sections, two optical fiber grating sensors based on the chirp tuning and optical power monitoring technique are proposed.

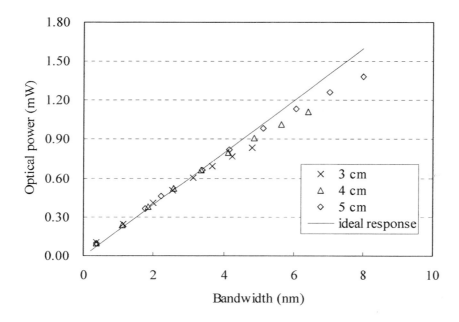

Figure 7. Calculated detectable optical power against 3-dB bandwidth for chirp-tuned FBG with various grating lengths of 3, 4, and 5 cm: $n_{eff} = 1.45$, $v\overline{\delta n}_{eff} = 3 \times 10^{-4}$, $\lambda_B = 1550$ nm (central Bragg wavelength), $\rho = 1$ mW/nm, and $V = 0.2$.

Displacement Sensor with Chirp-tuned FBG

To measure displacement using the optical power detection based technique, the variation of displacement should be related to the bandwidth of the FBG sensing element. To achieve that, we use a specially designed flexible cantilever beam, on which the FBG is attached on the lateral side at a slant orientation so that a varying strain field is generated and transferred to the grating along the length, resulting in a variable chirp in the grating period.

The schematic diagram of the proposed FBG-based displacement sensor as well as the experimental setup is shown in Figure 8. An originally uniform 10-cm long optical fiber grating was fabricated by exposing a deeply hydrogen-loaded single-mode fiber to a 244-nm laser beam through a phase mask. The fabricated grating has a very high reflectivity of ~100%. The original 3-dB bandwidth and the center wavelength of the grating are 0.26 nm and 1550.9 nm, respectively. The cantilever beam is of a right-angle triangle shape, which has been proved able to provide uniform and large chirp to the attached fiber grating [23]. The design parameters of the beam are length $L_b = 18$ cm, thickness $h = 0.8$ cm, and width of the fixed end $b_0 = 3$ cm. The angle between the axis of the grating and the neutral layer of the beam is $\theta = 4.5°$.

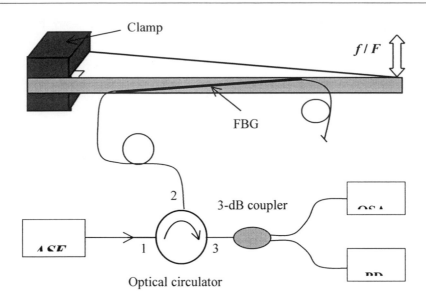

Figure 8. Schematic diagram and experimental setup of the proposed chirp-tuned FBG based displacement sensor: ASE, gain-flattened erbium-doped fiber amplified spontaneous emission source; OSA, optical spectrum analyzer; PD, photodetector.

When the cantilever beam is bent by applying a vertical displacement (or a force) on the free end of the beam, half of the grating is under a varying tension whereas the other half is under a varying compression. The strain on the neutral layer of the beam is zero. If the center of the grating is located well to the neutral layer of the beam, there will be no strain effect at the center of the grating. Therefore, the center wavelength of the chirped grating may be kept fixed since the strain applied to the two halves of the grating is symmetrical. An output power flattened amplified spontaneous emission (ASE) source with erbium-doped fiber was used as the input light to the fiber grating via Port 1 of an optical circulator which directs light to Port 2 and then to the fiber grating. The reflected light from Port 3 of the circulator was measured with both an optical spectrum analyzer (OSA) and a photodetector (PD) through a 3-dB fiber coupler.

If we set a coordinate along the length of the beam with zero point locating at the fixed end, the beam width, the moment of inertia of the cross-section, and the bending moment of the beam at any point x ($0 \leq x \leq L$) can be described respectively as

$$b(x) = \frac{b_0}{L_b}(L_b - x), \tag{20}$$

$$I(x) = \frac{1}{12}h^3 b(x) = \frac{b_0 h^3}{12 L_b}(L_b - x), \tag{21}$$

$$M(x) = F(L_b - x), \tag{22}$$

where F is the vertical force applied at the free end of the cantilever beam. The beam deflection at the free end f, i.e. the displacement we are going to measure, is directly proportional to the applied force. The relationship is given by

$$f = \frac{6FL_b^3}{Eh^3 b_0},$$
(23)

where E is the Young modulus of the beam material. Therefore the deflection-induced curvature for the neutral layer of the beam can be described by

$$\kappa(x) = \frac{M(x)}{EI(x)} = \frac{2f}{L_b^2}.$$
(24)

The above equation shows that the curvature of the neutral layer of the beam is independent of the position x, namely κ is uniform along the beam. This is very important for achieving a linear chirp in the grating period, as shown below.

Bending of the beam may induce various strains along the beam at different layers except for the neutral layer, where no strain is produced. If we set a coordinate t along the thickness direction of the beam and make the point at the neutral layer as zero point, the strain field with regarding to t ($-0.5h \le t \le 0.5h$) can be given by

$$\varepsilon(t) = \kappa t .$$
(25)

This strain field is transferred to the attached grating. Due to the existence of angle θ, the achieved strain of the grating is only part of $\varepsilon(t)$, which can be described by

$$\varepsilon_{ax}(t) = C\varepsilon(t)\cos(\theta),$$
(26)

where C ($0 < C < 1$) is a constant that represents the efficiency of the strain transfer from the beam to the grating. By considering $t = z\sin(\theta)$, the strain field along the grating can be rewritten as

$$\varepsilon_{ax}(z) = \frac{Cf}{L_b^2}\sin(2\theta)z .$$
(27)

Therefore, we get the specific description of ε_0 and K_ε, which are given respectively by

$$\varepsilon_0 = 0 ,$$
(28)

$$K_\varepsilon = \frac{Cf}{L_b^2}\sin(2\theta).$$ (29)

Substituting the descriptions into Eqs (8) and (9) yields

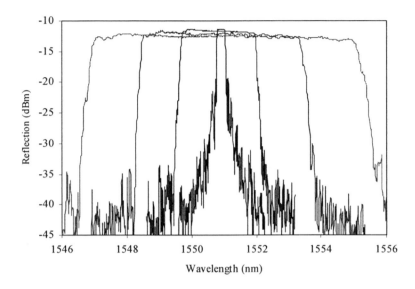

Figure 9. Reflection spectra of the chirp-tuned FBG sensor under different displacements.

$$R_{ch} = \frac{C\lambda_B}{L_b^2}\sin(2\theta)(1 - p_e)f,$$ (30)

$$\Delta\lambda_{bw} = \Delta\lambda_{bw0} + \frac{C\lambda_B L_g}{L_b^2}\sin(2\theta)(1 - p_e)f.$$ (31)

From Eq. (31), it can be seen that the variation in the bandwidth of the grating is directly proportional to the displacement f, on the free end of the cantilever beam. Therefore, within the above discussed linear response range of optical power to bandwidth, the detected optical power from the chirped-FBG is also directly proportional to f. This is the basic sensing principle of this FBG displacement sensor. And the sensor is temperature-insensitive because temperature variation cannot change the bandwidth but only the center wavelength of the reflection band of the grating.

Figure 9 shows the measured spectra of the chirped FBG for different displacements of 0, 6, 13 and 22 mm. It is noted that the original spectrum of the uniform FBG corresponds to zero displacement. In this case, high and uniform reflectivity of the grating was maintained even for the case of the largest 3-dB bandwidth of 8.12 nm. The measured 3-dB bandwidth and the PD output signal for displacements varying from 0 to 15 mm are shown in Figures 10 and 11, respectively. A linear fit to the data of the 3-dB bandwidth gives a high R-squared value of 0.9999, showing a good linearity of the response of bandwidth with displacement.

Figure 11 also shows a linear response of the PD output signal with displacement, in which the *R*-squared value is > 0.999 within the displacement range of 9 mm. For displacements greater than 9 mm, the PD output levels are slightly lower than the expected values because the high reflectivity of the grating cannot be maintained when the bandwidth of the FBG exceeds a large value of 3.46 nm. The responses of the 3-dB bandwidth and the PD output to the displacement are 0.36 nm/mm and 37.9 mV/mm, respectively.

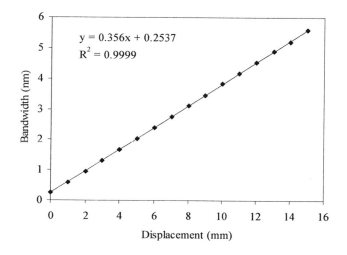

Figure 10. Measured 3-dB bandwidth against displacement.

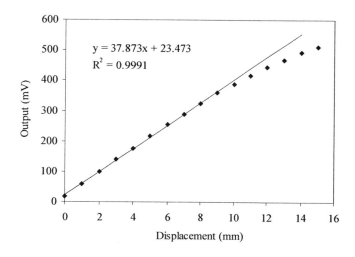

Figure 11. Measured PD output against displacement.

To test the temperature stability of the sensor, we measured the FBG spectrum and PD output for a constant displacement of 5 mm at various temperatures. The center wavelength of the FBG and the PD output as a function of temperature are shown in Figure 12. When the temperature was increased from -15°C to 40°C, the center wavelength of the FBG shifted to the longer wavelength region by 5.3 nm due to thermal effect. The variation in the 3-dB bandwidth is very small (less than 0.1 nm). The largest fluctuation of the PD output signal of about 220 mV was less than 20 mV. We believe that this signal fluctuation is mainly caused by the non-uniform power spectrum of the ASE source, in which the ripple of the power level is about 0.5 dB over a wavelength range of 10 nm around the wavelength of 1550 nm. By using a light source with a flatter optical power spectrum and/or using a beam with material of lower thermal expansion coefficient to reduce the temperature-induced wavelength shift of the FBG, the temperature instability can be greatly minimized.

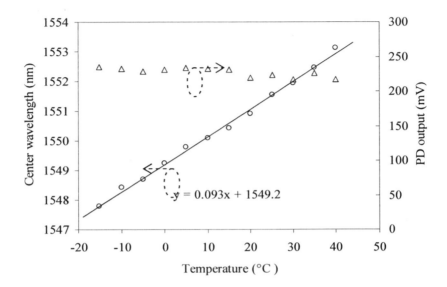

Figure 12. Temperature responses of the displacement sensor for a constant displacement of 5.0 mm.

Tilt Sensor with Chirp-Tuned FBGs

Most recently, a vertical-pendulum-based FBG tilt sensor has been reported [24], which can detect the magnitude as well as the direction of the inclination from the horizontal direction with much higher accuracy and resolution over the conventional tilt sensors. It used two pairs FBGs and the wavelength separation between the FBGs in each pair is used as encode signal for eliminating the thermal effect. However, as a conventional wavelength-coding-based FBG sensor, it always needs a wavelength detection system, which may increase the complexity and the cost of the sensor system. In this part, we propose and demonstrate a novel tilt sensor based on the optical power detection technique. The information of the inclination is encoded by the reflected optical powers from three FBGs whose bandwidths are tuned by nonuniform strain fields from inclinations. It has significant advantages of inherently insensitive to temperature and straightforward in demodulation.

The proposed sensor, as shown in Figure 13(a), is based on a vertical pendulum structure. A free moving plate is hanged at the center of the top plate of a base frame through a free rotational joint. The pendulum, which always points to the earth, is fixed on the moving plate. There are three thin and trapezoidal steel flakes fixed on the top plate and each is separated 120° from the others. They function as cantilevers and the carriers of the three FBGs. Each FBG (originally uniform in period) is glued on the center line of the top surface of a steel flake, as shown in Figure 13(b). When a lateral deflection is applied at the free end of the

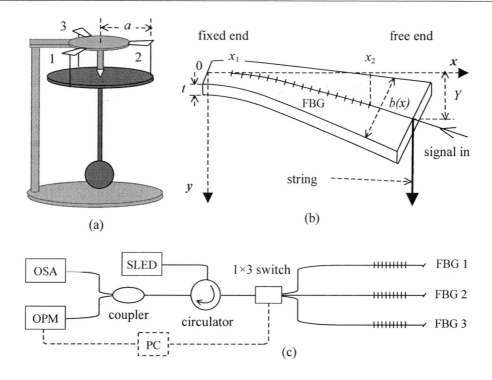

Figure 13. Schematic diagrams of (a) the proposed chirp-tuned FBG based tilt sensor, (b) a steel flake, and (c) the interrogation system.

steel flake, nonuniform strain field will be produced on the surface of the flake, resulting in a chirp in the period of the originally-uniform FBG. The free end of each steel flake is connected to the moving plate through a string. A pre-deflection is applied so that any relative motion of the moving plate (namely the pendulum) can be transferred to the steel flakes and sensed by these three FBGs. By measuring the reflected optical powers from the three strain-chirped FBGs, the deflections of the three steel flakes, hence the inclination (both the tilt angle and direction) of the base frame can be evaluated out.

Based on the analysis performed in [25], the deflection-induced nonuniform strain field in x direction [see Figure 13(b)] for such a linearly tapered cantilever beam can be described as

$$\varepsilon(x) = f(x)Y, \tag{32}$$

where $f(x)$ is a monotonic varying nonlinear function related to the parameters of the steel flake, which include the thickness, length and widths of the both ends; Y is the lateral deflection. The variation of the bandwidth of the FBG due to the nonuniform strain distribution therefore can be written as

$$\Delta\lambda_{bw} - \Delta\lambda_{bw0} = (1 - p_e)C\lambda_B[\varepsilon(x_1) - \varepsilon(x_2)]$$
$$= (1 - p_e)C\lambda_B FY, \tag{33}$$

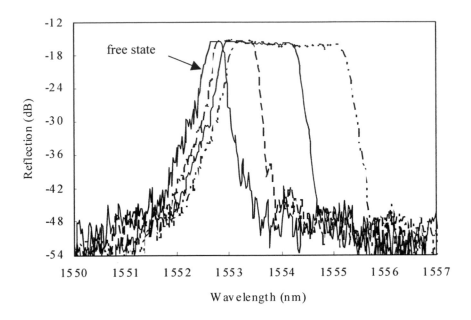

Figure 14. Reflective spectra of one of the FBGs measured under different beam deflections.

where C is the constant describing the strain transfer efficiency, x_1 and x_2 ($x_2 > x_1$) are the positions of the two ends of the FBG [see Figure 13(b)], and $F = f(x_1) - f(x_2)$ has a constant value depending on positions x_1 and x_2. The grating bandwidth and optical power, according to Eq. (2), will be a linear function of beam deflection Y. Therefore, the deflection variation (ΔY_i) over the applied pre-deflection (Y_{i0}) for each of the three steel flakes, in terms of the reflected optical power (P_i), will be expressed as

$$\Delta Y_i = Y_i - Y_{i0} = k_i (P_i - P_{i0}), \quad (i = 1, 2, 3), \tag{34}$$

where the subscript i corresponds to the three steel frakes, k_i is a coefficient that can be found out from experimental measurement, and P_{i0} is the reflected optical power of the i-th FBG in the pre-deflected state.

When an inclination is applied to the sensor device, ΔY_i can be found out from the reflected optical power of the i-th FBG. The direction can be figured out from the values of ΔY_i, and the tilt angle, θ, can be calculated by using the following equation

$$\theta = \sin^{-1}\left(\frac{2}{3a} \sqrt{\Delta Y_1^{\,2} + \Delta Y_2^{\,2} + \Delta Y_3^{\,2} - \Delta Y_1 \Delta Y_2 - \Delta Y_1 \Delta Y_3 - \Delta Y_2 \Delta Y_3} \right) \tag{35}$$

where a is the distance from the center of the top plate to the free end of the steel flake where the string is connected to, as shown in Figure 13(a).

In the experiment, the center wavelengths of the strain-free FBGs are ~1553 nm. The reflectivity, 3-dB bandwidth and the grating length are about 30 dB, 0.3 nm and 27 mm, respectively. The steel flake has a length of 40 mm, a thickness of 0.12 mm, widths of 9 and

24 mm at the fixed and free end, respectively. The value of a is 80 mm. The sensor was of steel flakes

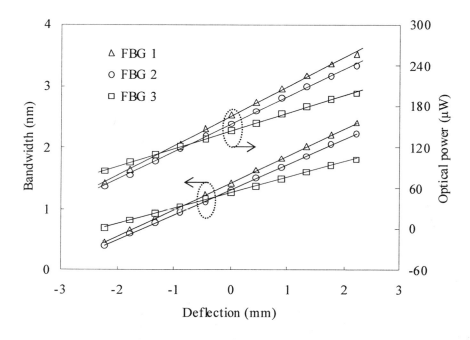

Figure 15. Measured 3-dB bandwidth and optical power of the three FBGs versus deflections.

placed on a tilt adjustable platform. The interrogation system is shown in Figure 13(c). A 1550-nm super-luminescent light emitted diode with a 1-dB flat bandwidth of over 50 nm was used to illuminate the three FBGs through an optical circulator and a 1×3 optical switch. The reflected light from the selected FBG was measured with an OSA and an optical power meter (OPM) at the same time by using a 3-dB coupler.

Figure 14 shows four reflective spectra of one of the FBGs, which were measured under free-strain state and three different deflections. The 3-dB bandwidths are 0.32, 0.8, 1.52, and 2.25 nm, respectively. For tilt measurement, the steel flakes were pre-deflected by adjusting the length of the strings to make each FBG has an initial bandwidth. In a pretest, the broadest bandwidth, by which the good linear response of optical power to bandwidth broadening can be maintained, was found out to be ~2.4 nm, so the initial bandwidth was set to around 1.35 nm, the middle of the linear response range.

Figure 15 shows the measured reflected optical power and bandwidth of each FBG against deflection. Both bandwidth-deflection and optical power-deflection curves show good linearity. From these data, P_{i0} and k_i in Eq. (34) can be evaluated out, which are 165.8, 153.9, 144.3 µW and 0.0232, 0.0246, 0.0386 mm/µW for P_{01}, P_{02}, P_{03} and k_1, k_2, k_3, respectively. The relatively larger value of k_3 was mainly caused by the offset of FBG3 from the designed location on the steel flake. The values of x_1 and x_2 for FBG3 were relatively larger than those of other two FBGs. The Eq. (34) is hereby rewritten as

$$\Delta Y_1 = 0.0232(P_1 - 165.8) \tag{36a}$$

$$\Delta Y_2 = 0.0246(P_2 - 153.9) \tag{36b}$$

$$\Delta Y_3 = 0.0386(P_3 - 144.3) \tag{36c}$$

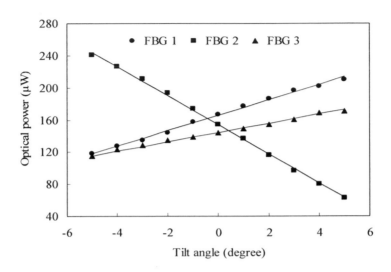

Figure 16. Measured (points) and predicted (lines) optical power of the three FBGs as inclination applied along steel flake 2.

To verify the accuracy of Eq. (36), the range of angles from -5° to 5° with a step of 1° were applied on steel flake 2, such that the steel flake 1 and 3 will experience half of the deflection of steel flake 2 according to the equilateral triangle theory. The measured optical power against tilt angle has a good agreement with the theoretical prediction with Eq. (36), as shown in Figure 16.

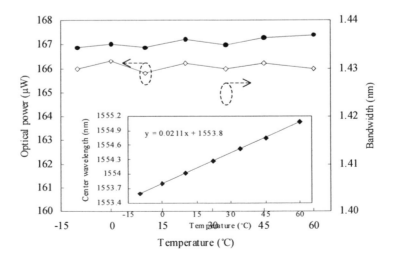

Figure 17. Temperature sensitivity measurement. The inset shows the center wavelength versus temperature.

The accuracy of tilt angle measurement, based on comparison of the applied angle value with the calculated value from the experimental data, is ±0.13°. The resolution of the OPM of 0.01 dB results the resolution of tilt angle measurement of ~0.02° for the cases of ±5° (smaller tilt angle; better resolution). The accuracy and resolution are relatively lower than those, ±0.1° and 0.007°, reported in [24], but can be improved by further optimization of the sensor design such as using thicker steel flakes or increasing a by using a bigger top plate. The measurement range, from -5° to 5°, is limited by the settings of the initial bandwidths of FBGs (or pre-deflections of steel flakes), which cannot be set to arbitrarily large values because the nonlinear response of optical power to deflection arising from a fast reduction in reflectivity of the FBGs with broadening of bandwidth will reduce the accuracy of tilt angle measurement. This problem may be partially overcome by using the originally chirped FBGs instead of the uniform ones, at a cost of FBG fabrication cost increasing.

The experiment on temperature sensitivity was carried out by using only one FBG sensor element (steel flake 1) because the size of the setup (~22 cm × 22 cm × 30 cm) is relatively large and cannot be placed into a temperature-controlled cabinet. The steel flake was kept under pre-deflected state. The reflected optical power, 3-dB bandwidth and center wavelength of the FBG were measured within a temperature range from -10°C to 60°C. As the temperature increased, the center wavelength changed at a rate of 21.1 pm/°C, while there is no relationship between the optical power (or bandwidth) and the temperature, as shown in Figure 17. Here, the thermal sensitivity of wavelength is about twice of the typical value of FBGs due to the influence of the steel flake which has a higher thermal expansion coefficient than normal optical fibers. The variation of bandwidth is smaller than 0.02 nm. The maximum variation in optical power is only 0.57 µW (~ 0.3%). Therefore, the sensor can be regarded as insensitive to temperature.

In practical situation, there is no need for measurement of the optical spectrum, so only OPM is used. For quasi-realtime monitoring, a personal computer (PC) can be included in the system, as shown in Figure 13(c), to control the optical switch and, at the same time, acquire and process data from the OPM. The proposed FBG-based tilt sensor can be used as a high sensitive universal inclinometer, and may have potential applications in detecting of earthquakes.

Conclusion

Wavelength-encoded FBG sensor systems usually need wavelength interrogators and temperature compensators in practical applications, so that the cost may be increased and the system become complex. We have theoretically studied the basic sensing principle of another FBG sensor prototype based on chirp-tuned FBGs and optical power-encoding technique. Based on this study, two novel sensor designs for displacement and tilt measurement have been proposed and demonstrated, respectively. These sensors showed great advantages including simple construction (no need of wavelength interrogator) and inherently insensitive to temperature thus eliminating the need for temperature compensation. Some disadvantages such as the limited measurement range due to the inevitable failure in keeping near 1 reflectivity when the bandwidth of the grating is increased significantly, can be partly overcome by optimizing grating parameters, such as increasing the grating length and/or index modulation depth. The proposed sensor designs may have potential applications in the optical fiber sensor area.

Acknowledgements

The authors would like to thank Dr. Nam Quoc Ngo with Network Technology Research Centre (NTRC), Nanyang Technological University (NTU), Singapore, Dr. Xiufeng Yang with Institute for Infocomm Research, Singapore, Dr. Lei Ding with Institute of Physics, Nankai University, China, and Dr. Chun-Liu Zhao with Department of Electrical Engineering, The Hong Kong Polytechnic University, China, for fruitful discussions, and Mr. Chunlei Zhan and Miss Kun Hu, both with NTRC, NTU, Singapore, for assistances in optical measurements. The author Dr. Xinyong Dong would also like to thank the Singapore Millennium Foundation (SMF) for its support through a postdoctoral fellowship award.

References

[1] K. O. Hill, Y. Fujii, D. C. Johnson, and B. S. Kawasaki, Photosensitivity in optical fiber waveguides: Application to reflection filter fabrication, *App. Phys. Lett.*, **32**, pp. 647-649, 1978.

[2] T. Erdogan, Fiber grating spectra, *J. Lightwave Technol.*, **15**, pp. 1277-1294, 1997.

[3] R. Kashyap, *Fiber Bragg gratings*, Academic Press, New York, 1999.

[4] C. R. Giles, Lightwave applications of fiber Bragg gratings, *J. Lightwave Technol.*, **15**, pp. 1391-1404, 1997.

[5] D. Kersey, M. A. Davis, H. J. Patrick, M. LeBlanc, K. P. Koo, C. G. Askins, M. A. Putnam, and E. J. Friebele, Fiber grating sensors, *J. Lightwave Technol.*, **15**, pp. 1442-1463, 1997.

[6] Y. J. Rao, In-fiber Bragg grating sensors, *Meas. Sci. & Technol.*, **8**, pp. 355-375, 1997.

[7] L. Zhang, W. Zhang, and I. Bennion, In-fiber grating optic sensors, in: F. T. S. Yu, S. Yin (Eds), Fiber optical sensors, Marcel Dekker, New York, 2002, pp. 123-181.

[8] M. G. Xu, L. Reekie, Y. T. Chow, and J. P. Dakin, Optical in-fiber grating high pressure sensor, *IEEE Photon. Technol. Lett.*, **29**, pp. 398-399, 1993.

[9] Y. Liu, Z. Guo, Y. Zhang, K. S. Chiang, and X. Dong, Simultaneous pressure and temperature measurement with polymer-coated fiber Bragg grating, *Electron. Lett.*, **36**, pp. 564-566, 2000.

[10] T. A. Berkoff, and A. D. Kersey, Experimental demonstration of a fiber Bragg grating accelerometer, *IEEE Photon. Technol. Lett.*, **8**, pp. 1677-1679, 1996.

[11] M. D. Todd, G. A. Johnson, B. A. Althouse, and S. T. Vohra, Flexural beam-based fiber Bragg grating accelerometers, *IEEE Photon. Technol. Lett.*, **10**, pp. 1605-1607, 1998.

[12] L. A. Ferreira, A. B. Lobo Ribeiro, J. L. Santos, and F. Farahi, Simultaneous measurement of displacement and temperature using a low finesse cavity and a fiber Bragg grating, *IEEE Photon. Technol. Lett.*, **8**, pp. 1519-1521, 1996.

[13] X. Dong, Y. Liu, Z. Liu, and X. Dong, Simultaneous displacement and temperature measurement with cantilever-based fiber Bragg grating sensor, *Opt. Commun.*, **192**, pp. 213-217, 2001.

[14] M. J. Gander, W. N. MacPherson, R. McBride, J. D. C. Jones, L. Zhang, I. Bennion, P.

M. Blanchard, J. G. Burnett and A. H. Greenaway, Bend measurement using Bragg gratings in multicore fibre, *Electron. Lett.*, **36**, pp. 120-121, 2000.

[15] K. O. Lee, K. S. Chiang, and Z. H. Chen, Temperature-insensitive fiber-Bragg-grating-based vibration sensor, *Opt. Eng.*, **40**, pp. 2582-2585, 2001.

[16] P. M. Cavaleiro, F. M. Araujo, and A. B. Lobo Ribeiri, Metal-coated fiber Bragg grating sensor for electrical current metering, *Electron. Lett.*, **34**, pp. 1133-1135, 1998.

[17] N. E. Fisher, D. J. Webb, C. N. Pannell, D. A. Jackson, L. R. Gavrilov, J. W. Hand, L. Zhang, I. Bennion, Ultrasonic field and temperature sensor based on short in-fibreBragg gratings, *Electron. Lett.*, **34**, pp. 1139-1140, 1998.

[18] W. L. Schulz, E. Udd, J. M. Seim, and G. E. McGill, Advanced fiber grating strain sensor systems for bridges, structures, and highways, *SPIE*, *3325*, pp. 212-222, 1998.

[19] Lee, and Y. Jeong, Interrogation techniques for fiber grating sensors and the theory of fiber gratings, in: F. T. S. Yu, S. Yin (Eds), *Fiber optical sensors*, Marcel Dekker, New York, 2002, pp. 295-381.

[20] X. Dong, H. Meng, G. Kai, Z. Liu and X. Dong, Bend measurement with chirp of fiber Bragg grating, *Smart Mater. Struct.*, **10,** pp. 1111-1113, 2001.

[21] M. G. Xu, L. Dong, L. Reekie, J. A. Tucknott, and J. L. Cruz, Temperature-independent strain sensor using a chirped Bragg grating in a tapered optical fibre, *Electron. Lett.*, **31**, pp. 823-825, 1995.

[22] Y. Zhu, P. Shum, C. Lu, B. M. Lacquet, P. M. Swart, and S. J. Spammer, Fiber Bragg grating accelerometer with temperature insensitivity, *Microwave Opt. Technol. Lett.*, **37**, pp. 151-153, 2003.

[23] X. Dong, P. Shum, N.Q. Ngo, C.C. Chan, J.H. Ng, and C.-L. Zhao, Largely tunable CFBG-based dispersion compensator with fixed center wavelength, *Opt. Express*, **11**, pp. 2970-2974, 2003.

[24] B-O. Guan, H.-Y. Tam, and S.-Y. Liu, Temperature-independent fiber Bragg grating tilt sensor, *IEEE Photon. Technol. Lett.*, **16**, pp. 224-226, 2004.

[25] S. Goh, S. Y. Set, K. Taira, S. K. Khijwania, and K. Kikuchi, Nonlinearly strain-chirped fiber Bragg grating with an adjustable dispersion slope, *IEEE Photon. Technol. Lett.*, **14**, pp. 663-665, 2002.

In: Progress in Smart Materials and Structures
Editor: Peter L. Reece, pp. 227-290

ISBN: 1-60021-106-2
© 2007 Nova Science Publishers, Inc.

Chapter 8

WAVELET-BASED UNSUPERVISED AND SUPERVISED LEARNING ALGORITHMS FOR ULTRASONIC STRUCTURAL MONITORING OF WAVEGUIDES

Piervincenzo Rizzo[*1] *and Francesco Lanza di Scalea*[*2]

[1] Department of Civil and Environmental
University of Pittsburgh, 942 Benedum Hall, Pittsburgh, PA 15261, USA
[2] NDE and Structural Health Monitoring Laboratory
Department of Structural Engineering
University of California, San Diego, 9500 Gilman Drive, M.C. 0085
La Jolla, CA 92093-0085, USA

Abstract

Guided Ultrasonic Waves (GUWs) are a useful tool in those structural health monitoring applications that can benefit from built-in transduction, moderately large inspection ranges and high sensitivity to small flaws. This chapter describes two complementary methods, one based on unsupervised learning algorithms, and one based on supervised learning algorithms, for structural damage detection and classification based on GUWs. Both methods combine the advantages of GUW inspection with the outcomes of the Discrete Wavelet Transform (DWT), that is used for extracting robust defect-sensitive features that can be combined to perform a multivariate diagnosis of damage. In particular, the DWT is exploited to de-noise and compress the ultrasonic signals in real-time and generate a set of relevant wavelet coefficients to construct a uni-dimensional or multi-dimensional damage index. The damage index is then fed to an outlier analysis (unsupervised algorithm) to detect anomalous structural states, or to an artificial neural network (supervised algorithm) that classifies the size and the location of the defects.

The general framework proposed in this chapter is applied to the detection of crack-like and notch-like defects in seven-wire steel strands and in railroad tracks. In the first application, the probing hardware consists of narrowband magnetostrictive transducers used for both ultrasound generation and detection. In the second application, the hardware consists

---
[*1] E-mail address: prizzo@engr.pitt.edu; Phone: (412) 624-9575 ; Fax : (412) 624-0135.
[2] E-mail address: flanza@ucsd.edu, Phone (858) 822-1458 ; Fax (858) 534-6373.

of a hybrid laser/air-coupled system for broadband ultrasound generation and detection. These applications demonstrate the effectiveness of the DWT-aided structural diagnosis of defects that are small compared to the waveguide cross-sectional area.

The proposed signal analysis approaches are general, and are extendable to many other structural monitoring applications using GUWs as the main defect diagnosis tool.

Keywords: Guided ultrasonic waves, damage index, feature extraction, Discrete Wavelet Transform, Artificial Neural Networks, Multi-wire strands, rail inspection

1. Introduction

The progressive aging of engineering systems increases the demand for structural monitoring, which includes damage detection and sizing. The development of an effective and robust monitoring strategy can prevent fatalities and reduce maintenance/repair costs. The use of Guided Ultrasonic Waves (GUWs) is a well-established technique for the non-destructive evaluation (NDE) and/or the structural health monitoring (SHM) of civil, mechanical and aerospace structures (Rose 1999). GUWs are very attractive as they enable the inspection of a large area from a single or few probe positions thanks to the guidance of stress waves by the structure acting as an acoustic waveguide. The application of GUWs in NDE has already become a standard practice among a number of industries, particularly the oil industry (Rose 2002, Alleyne *et al.* 2002). In the last few years the necessity for improvements in the field of signal conditioning, feature extraction and defect classification became evident in order to enhance the monitoring performance in terms of signal-to-noise ratio (SNR), and in terms of defect detection, sizing, location and classification. Improvements in SNR make the technique also attractive to those applications where the structure is embedded as in the case of rebars in concrete, grouted strands, pipes partially or completely buried in soil or when environmental noise is an issue. Generally, these improvements are based on the evaluation of certain damage-sensitive features that are measured on the structure of interest. Traditionally these features are extracted from the time domain or from the frequency domain. In this chapter features extracted in the time, frequency and joint time-frequency domain, particularly through the Discrete Wavelet Transform (DWT), are considered. Compared to the Continuous Wavelet Transform that is not computationally efficient, the DWT can be performed in real-time owing to the existence of a fast orthogonal wavelet transform based on a set of filter banks (Mallat 1999). The two main outcomes of DWT processing are data de-noising and data compression (Abbate *et al.* 1997), which have made such processing attractive in various SHM applications (Staszewski 1998, 2002, Staszewski *et al.* 1997, 2004, Paget *et al.* 2003, McNamara and Lanza di Scalea 2004, Rizzo and Lanza di Scalea 2004a, 2005, Kim and Melhem, 2004).

Once the damage-sensitive features are extracted, they can be coupled to an automatic classification algorithm able to detect the presence of defects (unsupervised learning class), or to determine the presence, the size and the location of defects (supervised learning class). The first class offers the fundamental advantage that the information on the damage conditions do not need to be known a priori, contrarily to the supervised learning class. The unsupervised approach is particularly useful in the case of complex or expensive structures where it is difficult to simulate damage prior to the deployment and the use of the monitoring system. In this chapter both classes are discussed and investigated.

As an unsupervised learning algorithm, the outlier analysis is used. This statistical approach is a novelty detection method that establishes whether a new configuration of the system is discordant or inconsistent from the baseline configuration. The baseline describes the normal operative conditions and it usually consists of an existing set of data or patterns.

The framework of outlier analysis is available in classical statistical textbooks (Barnett 1994). The application of the method to diagnose structural damage was first presented by Worden (1997). In this work the reduced stiffness of a simulated lumped-mass system was detected by using the dynamic transmissibility function as the damage-sensitive feature. A 50-dimensional feature vector was constructed by discretizing the transmissibility function into 50 points. The approach was later extended by Worden *et al.* (2000a) to the detection of three different damage states, simulated by reducing the stiffness between two masses by 1%, 10% and 50% of the original, pristine value. The 50-point transmissibility function between the two masses was, again, considered as the damage-sensitive feature vector. Fifty-dimensional feature vectors were also considered in novelty detection for the structural monitoring of composite plates (Worden *et al.* 2000b) and wings of a Gnat aircraft (Manson and Worden 2003). The feature vectors in these work consisted of discretized ultrasonic Lamb waves and transmissibilities along the stringers, respectively. Another application was the defect detection in running rotors (Guttormsson *et al.* 1999).

In this chapter, the novelty detection technique is applied to GUWs processed through the DWT in order to construct a set of damage-sensitive features. The use of features extracted from the DWT analysis for novelty detection is not entirely new. At least one recent work (Omenzetter *et al.* 2004) used the DWT coefficients of strain time series to formulate a vector dynamic regression model for the identification of outlier events in a bridge. The purpose of using the DWT in the present work is to de-noise and compress GUW measurements taken from structural components with finite cross-sectional dimensions, as opposed to vibration measurements representative of the global behavior of the structure.

When it is possible to simulate the damage *a priori*, the defect detection and classification are achievable from supervised learning algorithms, such as the Artificial Neural Networks (ANNs), that are able to learn from training samples through iterations. In this chapter a wavelet-based damage index vector is computed and used as the input of a feedforward backpropagation (FF BP) artificial neural network. The network provides automatic classification of the damage size and the damage location. It will be shown that the network performance, i.e. the capability to classify the defects properly, can be enhanced by an intelligent selection of the signal features and of the network parameters.

Two applications are presented for the DWT-aided guided-wave SHM using both unsupervised and supervised learning algorithms. The first application is the monitoring of multi-wire steel strands widely used in civil infrastructures as tensioning members in cable-stayed bridges, suspension bridges and prestressed concrete. The proposed technique uses narrowband magnetostrictive sensors resonant at 320 kHz to excite and detect longitudinal cylindrical waves. The damage-sensitive features are extracted from GUW signals that are reflected from an artificial notch, cut at different depths. The second application is the inspection of railroad tracks based on a hybrid laser/air-coupled ultrasonic system sensitive to surface-breaking transverse cracks by means of high-frequency surface waves, ranging from 100 kHz to 900 kHz. A pulsed laser is used to generate the waves in the rail and a pair of air-coupled sensors is used to detect the waves. The rail monitoring system stays beyond the generally-recommended clearance envelope of 64 mm (2.5") from the top of the rail head.

The cracks are detected and quantified by monitoring the energy decay of the transmitted waves.

The outline of the chapter is as follows. Section 2 provides a brief overview of the GUW concepts. Section 3 describes the DWT and the strategy for performing an effective signal de-noising. Section 4 introduces the concept of damage index and it presents the specific monitoring schemes adopted. Section 5 describes the architecture of outlier analysis and artificial neural networks. Sections 6 and 7 show the SHM approach at work, by describing the experimental setups, the test protocols and the results of two specific applications: the monitoring of seven-wire steel strands by permanently-attached sensors (Section 6), and that of railroad tracks by non-contact sensors (Section 7).

2. Guided Waves for Structural Health Monitoring

When an ultrasound propagates into a bounded media, a guided wave is generated. The wave is termed "guided" because it travels along the medium guided by its geometric boundaries. Different types of guided waves exist, including Rayleigh (surface) waves, Lamb waves, and cylindrical waves. Rayleigh waves are waves propagating along the surface of a semi-infinite space. The particle motion is elliptical and the amplitude decays with the space depth. Lamb waves propagate in plate-like structures. They occur in two different basic modes, the symmetrical or dilatational mode, and the asymmetrical or bending mode. The particles of the neutral fibre (the middle zone in an unstressed plate) perform pure longitudinal displacements in the case of symmetric modes and pure shear oscillations in the case of anti-symmetric modes.

Generally, all structures having one dimension much smaller than the other two can sustain guided waves. Examples of these structures are: plates, solid cylinders, multi-wire strands, hollow cylinders (pipes), railroad tracks; they are all good candidates for monitoring by means of GUWs. For interested readers, a wide selection of excellent references covering guided stress waves includes the books by Kolsky (1963), Achenbach (1973), Auld (1990), Krautkramer (1990), Graff (1991) and Rose (1999).

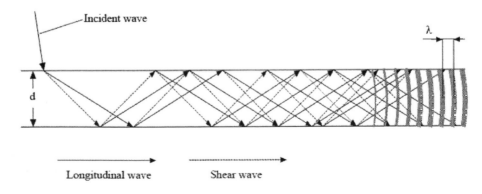

Figure 2.1. Lamb wave generated by an oblique incidence of a longitudinal wave from longitudinal and shear partial waves.

A guided wave can be thought of as a superposition of partial plane waves, which are reflected inside the bounded structure; this is the partial wave representation schematized in Fig. 2.1.

In a free waveguide, i.e. surrounded by vacuum, the guided wave is attenuated by material damping only; when the waveguide is surrounded by another material the attenuation is also due to the leakage of waves into the surrounding medium.

The study and the application of GUWs are challenging since they are multimode (many vibrating modes can propagate simultaneously) and dispersive (the propagation velocity and the attenuation depend on the wave frequency f). The dispersive behavior is represented by the dispersion curves that describe the relation between the wave velocities and the frequency. Sometimes the frequency is replaced by the wavenumber k, that is related to wavelength λ by the equation $k = 2\pi/\lambda$.

The phase velocity c_p of a guided wave is the speed at which an individual crest of a wave moves; it is related to wavelength λ and the circular frequency $\omega=2\pi f$ by the equation $c_p = \omega\lambda/2\pi = \omega/k$. The group velocity c_g is the speed at which a wave packet travels in the element. This parameter tells how quickly the energy propagates. Group velocity is related to the circular frequency and to the phase velocity by the following equations (Rose 1999):

$$c_g = \frac{d\omega}{dk} = c_p - \lambda\frac{dc_p}{d\lambda} \qquad (2.1)$$

The dispersion (velocity vs. frequency) curves can be calculated analytically or they can be computed by approximate solutions derived from numerical methods. Exact solutions for cylinders and plates are well established (Rose 1999). When the waveguide cross-sectional geometry is complex, a closed form solution is not available, and a numerical approach is necessary.

An example of closed form solution for a solid cylinder is given below. In a slender, isotropic, traction-free cylindrical waveguide three types of vibrational modes can exist: longitudinal, flexural and torsional waves. The first two waves are analogous to symmetric and anti-symmetric Lamb waves in plates, respectively. The family of the torsional mode results when only the angular displacement exists along the cylinder cross-sectional surface.

The study of the guided wave propagation in a free bar began at the end of the 19th century when Pochhammer (1876) and Chree (1889) proposed the frequency equation for the longitudinal modes. Due to the complexity of the equations, several decades elapsed before numerical solutions were proposed. In the middle of the past century different authors contributed to the theoretical understanding of ultrasonic wave propagation in isotropic solid cylinders (Davies 1948, Onoe et al. 1962, Pao and Mindlin 1960, Pao 1962, Meitzler 1961, Zemanek 1971). Others analyzed the behavior of waves in transversely isotropic or anisotropic bars (Morse 1954, Mirsky 1964a and 1964b, Xu and Datta 1991).

Wave propagation in solid cylinders immersed in a fluid or embedded in solid material differs from wave propagation in cylinders in vacuum. Theoretical and experimental advancements were possible thanks to Thurston (1978), Nagy (1995), and by researchers at the Imperial College as Pavlakovic (1998, 1999 and 2001) and Beard (2002 and 2003). The group at Imperial College was responsible for the creation of a software, DISPERSE

(Pavlakovic *et al.* 1997), that calculates the dispersion curves of structures such as plates and rods in vacuum or embedded.

According to the formulation by Meitzler (1961) the vibrational modes of a cylindrical waveguide are the solutions of the following general frequency equation:

$$\begin{vmatrix} \beta^2 - 1 - (qr)^2 \dfrac{v-1}{2v-1} & \beta^2 - 1 - (qr)^2 & 2\left(\beta^2 - 1\right)\varphi_\beta(qr) - (qr)^2 \\ \\ -1 & v\varphi_\beta(qr) - 1 & \beta^2 - 2\varphi_\beta(qr) - (qr)^2 \\ \\ \varphi_\beta(pr) & (1-v)\varphi_\beta(qr) & \beta^2 \end{vmatrix} = 0 \qquad (2.2)$$

where r is the waveguide radius; $p^2 = (\omega/c_L)^2 - k^2$; $q^2 = (\omega/c_T)^2 - k^2$; $v = 0.5\,(\omega/kc_T)^2$; $\varphi_\beta(pr) = (pr)\,J_\beta{}'(pr)\,/\,J_\beta(pr)$; $\varphi_\beta(qr) = (qr)\,J_\beta{}'(qr)\,/\,J_\beta(qr)$; J_β is a Bessel function of order β and $J_\beta{}'(x) = 0.5\,[J_{\beta-1}(x) - J_{\beta+1}(x)]$.

The corresponding displacement components (cross-sectional mode shapes) expressed in a cylindrical coordinate system (r, θ, z), with z coincident with the rod axis, can be written as

$$\begin{cases} u_r = U(r)\cos\beta\theta\, e^{i(kz-\omega t)} \\ u_\theta = V(r)\sin\beta\theta\, e^{i(kz-\omega t)} \\ u_z = W(r)\cos\beta\theta\, e^{i(kz-\omega t)} \end{cases} \qquad (2.3)$$

where the amplitudes $U(r)$, $V(r)$, and $W(r)$ are found by imposing the appropriate boundary conditions at the stress-free surface of the rod and field equations.

Longitudinal modes are found for $\beta = 0$ and $u_\theta = 0$. The flexural modes correspond to $\beta \geq 1$ with all three displacement components different from zero. Finally, the torsional modes are given for $\beta = 0$ and u_θ as the only non-zero displacement component. Explicit expressions for the displacement mode shapes associated to the longitudinal and to the flexural modes can be found in Love (1944).

The phase velocity $c_p(f)$, and group velocity curves $c_g(f)$ dispersion curves are directly obtained from Eq. (2.2) using the expressions $c_p = \omega/k$ and $c_g = d\omega/dk$. A possible search algorithm fixes a frequency f (related to the angular frequency $f = \omega/2\pi$) and sweeps along values of c_p (linked to the wavenumber k by the relation $c_p = \omega/k$) to find those values of c_p that satisfy the characteristic equation Eq. (2.2).

The dispersion curves for a steel rod in vacuum are plotted in Figs. 2.2 (a-b). The following material parameters are used for steel: $c_L = 5.890$ mm/μsec, $c_T = 3.233$ mm/μsec and $\rho = 7843$ kg/m^3. The notation of Meitzler (1961) is employed for mode identification. Letters L, F and T indicate the longitudinal, flexural and torsional waves, respectively. The letter is then followed by two integer numbers in parenthesis. The first number is associated to the β value in Eq. (2.2) and describes the number of wavelengths around the circumference of the solid cylinder. The second number is a counter variable that represents the order modes of the branches. The numbering starts with 1 for those modes (the fundamental modes) that can propagate at zero frequency.

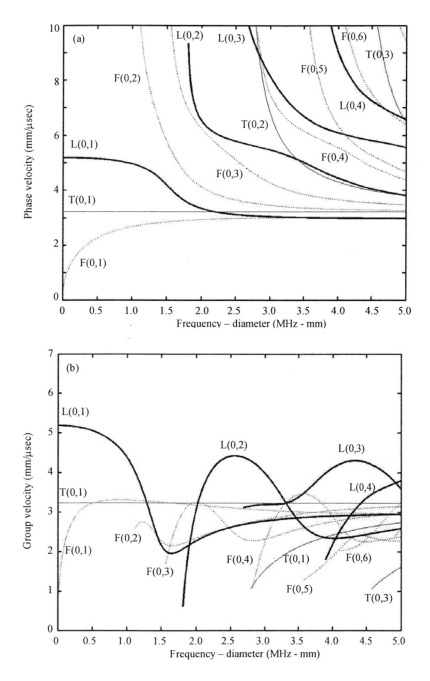

Figure 2.2. Dispersion curves for a steel bar in vacuum. (a) Phase velocity; (b) Group velocity.

An alternative approach to trace the dispersion curves is through the numerical methods; one of these methods is the 2-D Finite Element Method, or Semi-analytical Finite Element (SAFE) Method, that can model waveguides of arbitrary cross-section by simply discretizing a bi-dimensional domain. This method goes back to the early work of Aalami (1973).

The SAFE method can be implemented in a Matlab software and using the PDE toolbox for creating the finite element mesh. A brief description of the formulation adopted for dispersion curves of hollow cylinders is described below and it is summarized from Rizzo *et al.* (2005). We maintained here the same symbolism adopted in that paper.

At a frequency ω, each mode is represented by the wavenumber, k, and the displacement components u_1, u_2 and u_3, where (x_1, x_2) is the cross-sectional plane and x_3 is the axial direction of the pipe. Since axially-propagating waves were of interest in this study, the wave propagation direction coincides with direction x_3. At every point of the pipe, the displacement vector u can be approximated by the relations

$$\mathbf{u}(x_1,x_2,x_3,t)=\begin{bmatrix} u_1(x_1,x_2,x_3,t) \\ u_2(x_1,x_2,x_3,t) \\ u_3(x_1,x_2,x_3,t) \end{bmatrix}=\begin{bmatrix} \sum_{k=1}^{n} N_k(x_1,x_2)U_k\exp\left[\sqrt{-1}\left(k\,x_3-\omega t\right)\right] \\ \sum_{k=1}^{n} N_k(x_1,x_2)V_k\exp\left[\sqrt{-1}\left(k\,x_3-\omega t\right)\right] \\ \sum_{k=1}^{n} N_k(x_1,x_2)W_k\exp\left[\sqrt{-1}\left(k\,x_3-\omega t\right)\right] \end{bmatrix} \tag{2.4}$$

In the above expressions, N_k are the shape functions, U_k, V_k and W_k are the amplitudes of the nodal displacement components, and t is the time variable. It can be noted that the displacement components u_i are interpolated by the finite element approach only in the cross-sectional plane (x_1, x_2). No interpolation is performed in the direction of propagation, where the harmonic motion is simply described by the term $exp[\sqrt{-1}(k\,x_3-\omega t)]$.

The known strain displacement relations, $\varepsilon_{ij}=1/2(u_{i,j}+u_{j,i})$, and the constitutive laws, $\sigma_{ij}=C_{ijkl}\varepsilon_{kl}$, are employed. In the previous expressions the derivation convention $(..)_{,j} = \partial(..)/\partial x_j$ is assumed and repeated indices imply the use of the summation convention. σ_{ij} and ε_{kl} represent the components of the stress and strain tensors, σ and ε, respectively. C_{ijkl} is the generic element of the constitutive tensor that depends only on the two Lamé constants λ and μ for an isotropic material. The last two quantities can be expressed as a function of the bulk longitudinal and shear velocities, $*c_L$ and $*c_S$. For a viscoelastic material such velocities become complex and can be evaluated as follows (Bernard *et al.* 2001)

$$*c_{L,S} = c_{L,S}\left(1-\sqrt{-1}\frac{\alpha_{L,S}}{2\pi}\right)^{-1} \tag{2.5}$$

where $\alpha_{L,T}$ are the longitudinal and shear attenuation in the material, expressed in Nepers per wavelength. As a result, the material constants λ, μ are complex and, consequently, each component of the constitutive tensor C_{ijkl} has both a real and an imaginary part.

In order to solve the problem the following expression of the principle of virtual works can be used

$$\int_V \delta u_i\left(\rho\ddot{u}_i\right)dV + \int_V \delta\varepsilon_{ij}\sigma_{ij}dV = 0 \tag{2.6}$$

where V represents the volume of the waveguide. By substituting in Eq. (2.6) the constitutive laws, the strain-displacement expressions and introducing the interpolation described of Eq. (2.4), the following eigenvalue problem can be obtained:

$$(A - \xi B)Q = 0 \tag{2.7}$$

Details on the calculation of matrices A and B can be found in (Gavric 1995, Hayashi *et al.* 2003). At each circular frequency ω, *2M* eigenvalues k_m and, consequently, *2M* eigenvectors are obtained where M is the number of total degrees of freedom in the system. The eigenvectors are the M forward and the corresponding M backforward modes. The *m-th* eigenvalue k_m represents the *m-th* wavenumber, and Q_m is the corresponding eigenvector in which the first M components describe the mode shapes (Hayashi *et al.* 2003). Once the wavenumbers are known as a function of frequency, the dispersion curves can be easily computed.

However, when considering material attenuation the matrices A and B become complex as a result of the complex constitutive tensor C. Consequently all of the wavenumbers and the eigenvectors become complex. The phase velocity can be evaluated by the expression $c_{ph} = \omega / k_{real}$ where k_{real} is the real part of the wavenumber. The imaginary part of the wavenumber is instead the attenuation, in Nepers per meter, of the *m-th* mode at the circular frequency ω. As reported in (Bernard *et al.* 2001), the conventional approach used to evaluate the group velocity is not valid for attenuative waves. In this case the energy velocity, V_e, is the physically meaningful parameter. Energy velocity coincides with group velocity only when material attenuation is neglected and it can be expressed as (Achenbach 1973)

$$V_e = \frac{\frac{1}{T}\int\left(\frac{1}{S}\int P \bullet \hat{x}_3 dS\right)dt}{\frac{1}{T}\int\left(\frac{1}{S}\int E dS\right)dt} \tag{2.8}$$

where $1/T\int_T (..)dt$ denotes the time average over one period and $1/S\int_S (..)dS$ allows to evaluate the flow of the Poynting vector P (real part only) and the average total energy (kinetic and potential) E over the section surface S. The dispersion curves for a steel pipe in vacuum, with an outer external diameter of 60.2 mm (2 3/8 in) and a wall thickness of 5.08 mm (0.2 in), are plotted in Fig. 2.3. The corresponding SAFE mesh on the cross-section is shown in Fig. 2.3a. The analysis used three-node, triangular elements with linear shape functions. A total of 144 elements and 104 nodes were used. The following properties are assumed in the model: c_L= *5,960 km/sec*, c_S= *3,260 km/sec*, α_L = *0.003* Np/λ, α_T = *0.043* Np/λ and ρ=*7700 kg/m³* where ρ is the material density. The resulting phase velocity, energy velocity and attenuation curves are shown in Figs. 2.3b, 2.3c, and 2.3d, respectively.

(a)

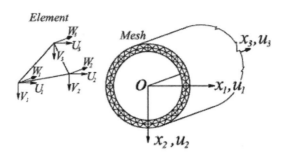

(b)

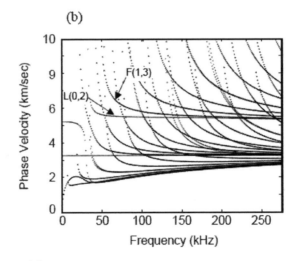

(c)

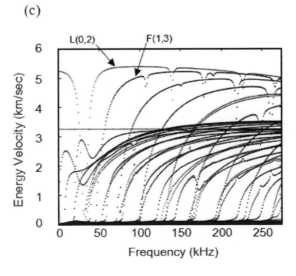

Figure 2.3. continued on next page.

(d)

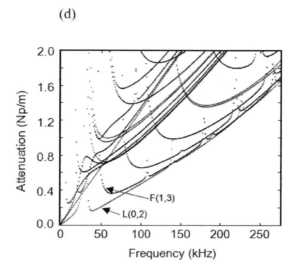

Figure 2.3. (a) Notation and mesh for the SAFE model of the pipe; (b) phase velocity dispersion curves; (c) energy velocity dispersion curves; (d) attenuation curves.

3. Joint Time-Frequency Analysis: The Discrete Wavelet Transform

The DWT is a multi-resolution analysis technique that can be used to obtain the joint time-frequency information of a signal. When compared to traditional Fourier transforms that lose the time resolution of non-stationary signals, wavelet transforms retain both the time and the frequency resolution. The wavelet transforms decompose the original signal by computing its correlation with a short-duration wave called the mother wavelet, that is flexible in time and in frequency. For high-speed applications, the DWT implemented with parallel filter banks is a good choice because of its computational efficiency. The efficiency results from the existence of a fast orthogonal wavelet transform algorithm based on a set of filter banks (Mallat 1989, 1999). The DWT may be intuitively considered as a decomposition of a function (signal) following hierarchical steps (levels) of different resolutions (Fig. 3.1a). At the first step the function is decomposed into wavelet coefficients; low-frequency components (low-pass filtering) and high-frequency components (high-pass filtering) of the function are retained. The signal is therefore decomposed into separate frequency bands (scales). The filtering outputs are then downsampled. The number of wavelet coefficients for each branch is thus reduced by a factor of 2 such that the total number of points at a given level is that of the original signal. Each level j corresponds to a dyadic scale 2^j at the resolution 2^{-j}. According to the casualty property, an approximation at resolution 2^{-j} contains all necessary information to compute the next approximation at resolution 2^{-j-1} (scale 2^{j+1}). Furthering the decomposition means increasing the scale that corresponds to zooming into the low-frequency portions of the spectrum. Because of the downsampling, a signal of 2^u points can be decomposed into u levels, which will produce a total of 2^{u+1} sets of coefficients. The final level u has 2^u coefficients and each branch has only one coefficient.

A perfect dyadic scale decomposition is only achieved when the Haar mother wavelet is employed. The Haar wavelet transforms a signal of even length into two equal parts, each having half of the original signal length.

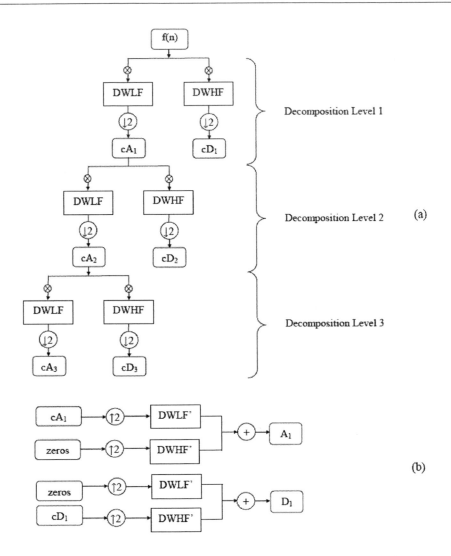

Figure 3.1. (a) Wavelet decomposition by filter bank tree; (b) signal reconstruction from wavelet coefficients; (c) reconstruction of original signal.

As will be discussed shortly, an efficient wavelet decomposition is one where the mother wavelet resembles the signal of interest. For this reason the Haar wavelet is not suitable, for example, for processing a narrowband ultrasonic test, where a function among the Daubechies wavelet family is more appropriate. In these conditions, a problem arises at the edges of the time domain where the wavelet function spills over the time window containing the signal. At these boundaries the signal is convoluted with extreme values of the wavelet and the result is the presence of spurious wavelet coefficients. It can be proven (Jensen and La Cour-Harbo 2001) that the total number of wavelet coefficients at a detail (or approximation) level j is equal to:

$$N_j = \frac{N_{j-1}}{2} + \frac{L}{2} - 1 \; (j = 1, 2..., u) \tag{3.1}$$

where L is the mother wavelet support length and N_0 is the length of the original signal. Each decomposition detail thus possesses $L/2-1$ extra coefficients. The support length of the Haar wavelet is $L = 2$ corresponding to perfect downsampling.

The presence of the spurious wavelet coefficients at the boundary of the time signal constitutes a problem for any feature extraction in the wavelet coefficient domain. In structural monitoring applications, this problem may lead to an overestimation of the defect-related features that is created by a numerical artifact. To overcome the boundary problem several methods have been proposed in the past as reviewed in (Ogden 1997, Karlsson and Vetterli 1989). These include the zero-padding method that is employed in the results presented here for its simplicity.

Analytically, the wavelet transform of a time signal f(t) with N number of points can be described by the following equation

$$Wf(n,s) = N^{-1/2}Wf(Nn,Ns)\qquad(3.2)$$

where n is known as the translation parameter and s as the scaling parameter. The parameter n shifts the wavelet in time and s controls the wavelet frequency bandwidth hence the time-frequency resolution of the analysis. The DWT of the discrete signal f(t) is computed at scales s = 2 j. Let ψ(t) be a wavelet such that the dilated and the translated family

$$\psi_{j,n}(t) = \frac{1}{\sqrt{2^j}} \cdot \psi\left(\frac{t}{2^j} - n\right)\qquad(3.3)$$

is an orthonormal basis of $\mathbf{L}^2(R)$, the space of the finite energy functions $g(t)$

$$\int_{-\infty}^{+\infty} |g(t)|^2 \, dt < +\infty\qquad(3.4)$$

Given a wavelet basis as in Eq. (3.3), the DWT of the function f(t) is calculated by the following inner product

$$W_{j,n} = \int_{-\infty}^{+\infty} f(t)\, \psi(t)^*_{j,n}\, dt\qquad(3.5)$$

where $\psi(t)^*$ is the conjugate of $\psi(t)$, and $W_{j,n}$ are the detail coefficients. From the wavelet coefficients the initial function is reconstructed by the equation

$$f(t) = \sum_j \sum_n W_{j,n}\psi_{j,n}\qquad(3.6)$$

The filter bank tree used for the wavelet decomposition of the signals presented in this chapter can be seen in Fig. 3.1a where DWLF is the discrete wavelet lowpass filter and DWHF is the discrete wavelet highpass filter. The low-pass filter is given by

$$DWLF[n] = \left\langle \frac{1}{\sqrt{2}} \phi(\frac{t}{2}), \phi(t-n) \right\rangle \tag{3.7}$$

where the symbol $\langle \rangle$ indicates the inner product operator, $\phi(t)$ is the mother scaling function associated with the mother wavelet $\psi(t)$ and n is the coefficient number of the lowpass and highpass filters. In a similar fashion the high-pass filter is given by

$$DWHF[n] = \left\langle \frac{1}{\sqrt{2}} \psi(\frac{t}{2}), \phi(t-n) \right\rangle \tag{3.8}$$

Low- and high-pass filters are related by the formula:

$$DWHF[n] = (-1)^{1-n} DWLF[1-n] \tag{3.9}$$

At the first level, Fig. 3.1a, the output of filtering and downsampling steps is a set of lowpass, cA1, and highpass, cD1, filter coefficients given by

$$cA_1 = [f(n) \otimes DWLF] (\downarrow 2) \tag{3.10a}$$

$$cD_1 = [f(n) \otimes DWHF] (\downarrow 2) \tag{3.10b}$$

where $\otimes$ is the convolution operator and $(\downarrow 2)$ is the downsampling operator. The decomposition then proceeds to the higher levels as shown in Fig. 3.1a.

De-nosing and compression of the original signal can be achieved if only a few wavelet coefficients representative of the signal are retained and the remaining coefficients, related to noise, are discarded. When all the coefficients belonging to one or more decomposition levels are retained, the procedure is referred to as pruning (Van Nevel et al. 1996, Abbate et al. 1997). The process of reconstructing the time signal from a set of wavelet coefficients is illustrated in Fig. 3.1b. The coefficients are upsampled to regain their original number of points and then passed through a reconstruction lowpass filter, DWLF', and reconstruction highpass filter, DWHF'. The reconstruction filters are closely related but not equal to those of the decomposition tree and are given by (Strang and Nguyen 1996)

$$DWLF'[n] = DWHF[-n] \tag{3.11a}$$

$$DWHF'[n] = - DWLF[-n] \tag{3.11b}$$

Reconstruction by using the decomposition level $\square$ (scale $2\square$), for example, is achieved by setting the wavelet coefficients from other scales equal to zero:

$$W_{j,n} = 0 \text{ for } j = 1, 2,..., \alpha -1, \alpha +1,..., u. \tag{3.12}$$

Finally, the linear combination of the reconstructions from various decomposition levels results in the reconstruction of the original time signal.

The selection of the wavelet levels where the signal is thought to live is driven by the following formula

$$f_j = \Delta \times F / 2^j \qquad (3.13)$$

relating the reconstructed frequency f_i at level i to the center frequency F of the mother wavelet, the scale 2^j, and the signal sampling frequency Δ. By using, for example, the Daubechies mother wavelet of order 40 (db40), whose center frequency is $F=0.671$ rad, and a sampling frequency $\Delta = 33 \times 10^6$ Hz, the center frequency of the sixth detail level is $f_6 = 346$ kHz.

A thresholding step can be used after the pruning process to further increase the SNR or can be used as an independent de-noising and compression procedure. In this case, a threshold is applied to the magnitude of the coefficients that are retained. This step assumes that the smaller coefficients represent noise, and can be safely omitted. Hence the data compression outcome of the DWT processing. Previous studies by the authors on DWT-processing of GUWs demonstrated signal compression performances on the order of 2% with pruning only, and on the order of 0.3% with pruning and thresholding (Rizzo and Lanza di Scalea 2004a). In that study the defect-sensitive information of a 320-kHz toneburst GUW signal, originally consisting of 9,000 points in the time domain, was reduced down to only 27 wavelet coefficients. Different strategies may be pursued for the selection of the threshold and are reviewed in Staszewski (1998).

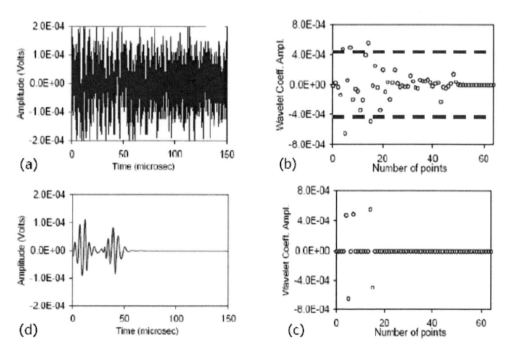

Figure 3.2. DWT de-noising and compression of guided-wave health monitoring signals. (a) Raw time signal propagating through simulated UAV wing skin-to-spar joint; (b) DWT decomposition at level 6; (c) 70% max amplitude thresholding of DWT coefficients; (d) de-noised time signal reconstructed from thresholded DWT coefficients.

An example of DWT processing of GUW signals is shown in Fig. 3.2. This result refers to signals generated and detected across a composite joint similar to what fond in wing skin-to-spar joints of Unmanned Aerial Vehicles. To represent a high-noise scenario, the excitation voltage to the wave actuator was lowered resulting in a very poor SNR at the wave receiver.

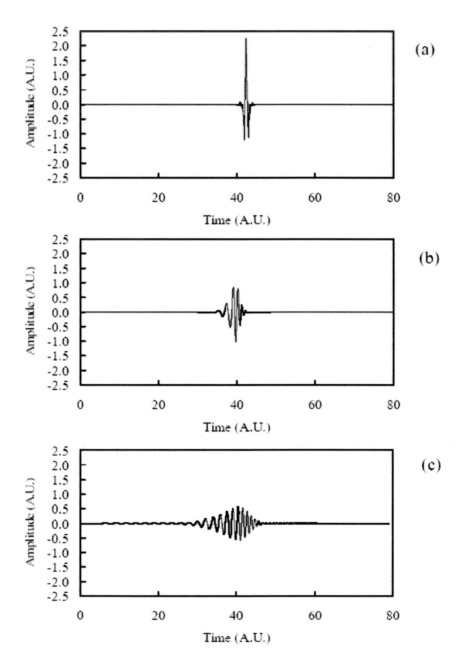

Figure 3.3. (a) the Coiflet wavelet of order 1; (b) the Daubechies wavelet of order 10; (c) the Daubechies wavelet of order 40.

In fact, the wave signals are not even discernable from the incoherent noise in the raw measurements of Fig. 3.2a. The DWT coefficients corresponding to the 6th decomposition level of the raw measurement are shown in Fig. 3.2b. In Fig. 3.2c only the DWT coefficients exceeding 70% of the maximum amplitude value are retained, and the remaining ones are discarded (thresholding). The retained coefficients are only five in this case. Fig. 3.2d shows the time signal that is reconstructed through an inverse DWT from the five coefficients of the previous plot. Compared to the raw measurement of Fig. 3.2a, the GUW signals in Fig. 3.2d are now clearly visible with the noise being dramatically reduced.

The success of a proper DWT decomposition is dependent on choosing a mother wavelet that best matches the shape of the signal that is being analyzed. In Fig. 3.3 three examples of mother wavelets are shown, namely the coiflet of order 1 (coif1), the db 10 and the db 40. It can be seen that the coif1 has a pulse-like shape, effective to de-noise impulsive, broadband signals; on the contrary, the db40 has a more narrowband character that matches the toneburst trace of narrowband signals.

In this chapter the db40 and the db10 are employed to process the ultrasonic signals measured from the strand monitoring system and the rail monitoring system, respectively.

4. Wavelet-Based Feature Extraction

Fig. 4.1 represents a general SHM scenario by means of GUWs: a stress wave (incoming GUW) propagates along the waveguide (plate, bar, rail, etc...), hits a crack and it is partially reflected back (reflected GUW) and partially transmitted (transmitted GUW).

Waves may be generated and detected by contact ultrasonic transducers, non-contact ultrasonic transducers (magnetostrictive sensors, EMAT), or by optical methods (laser ultrasonics). In this chapter two possible SHM schemes are proposed based on the position of the probing devices. In both schemes, a Damage Index (D.I.) is calculated. An efficient D.I. must be robust against noise and must allow for the detection, as well as for the sizing of the defect.

The first scheme is illustrated in Fig. 4.1b where a source (transmitter) generates the stress waves and a single receiver detects the echo signal (reflected GUW). This defect detection scheme is based on reflection measurements ("reflection mode"). The proposed "reflection" D.I. uses the ratio between certain features of the reflected signal, $F_{reflection}$, and the same features of the direct signal, F_{direct}

$$\text{D.I.} = \frac{F_{reflection}}{F_{direct}} \tag{4.1}$$

The "reflection" D.I. is expected to increase with increasing defect size.

The second scheme (Fig. 4.1c) is based on transmission measurements ("transmission mode") and it employs two receivers located on either side of the defect; in this case the proposed "transmission" D.I. uses the ratio between certain features of the signal detected by the further sensor (receiver #2), F_{sens2}, over the same features from the closer sensor (receiver #1), F_{sens1}

$$D.I. = \frac{F_{sens2}}{F_{sens1}} \qquad (4.2)$$

Contrarily to the reflection mode, the "transmission" D.I. is expected to decrease with increasing defect size.

The "reflection" scheme has the advantage of exploiting one receiver only and probing the portion of the waveguide ahead of the receiver at once; the length of this portion being limited only by wave attenuation. The second scheme necessitates two receivers and it gauges the portion between sensors #1 and #2. The transmission mode, however, is the most robust since the SNR is generally higher.

Several GUW features can be considered to construct a multi-dimensional D.I. vectors for the "reflection" mode and the "transmission" mode. These features include the variance (Var), root mean square (RMS), kurtosis, peak amplitude and peak-to-peak (Ppk) amplitude of pruned and thresholded wavelet coefficient vectors. Other features may be then obtained by further processing the DWT reconstructed signals through the Fast-Fourier Transform (FFT) and the Hilbert Transform (HT); the additional features include the peak amplitude and the area below the FFT frequency spectrum within the frequency range of interest, the area below the HT, the peak amplitude and the corresponding arrival time of the HT.

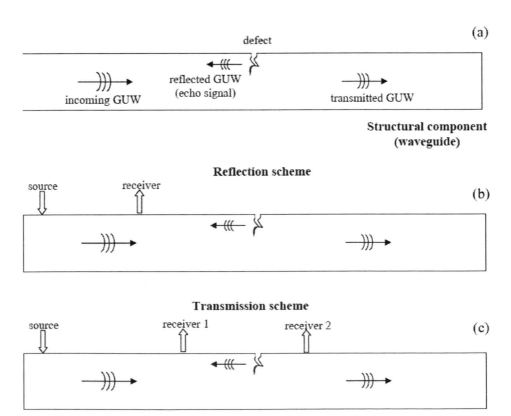

Figure 4.1. (a) Basic interaction of guided wave with damage; (b) defect detection scheme in the "reflection" mode; (c) defect detection scheme in the "transmission" mode.

The overall SHM strategy proposed in this chapter is illustrated in the flowchart of Fig. 4.2. It essentially consists of three main steps. GUW signals are generated and detected, according to one of the two schemes illustrated in Figs. 4.1b and 4.1c. Digitized signals are processed in real time through the DWT algorithm in order to calculate the associated wavelet coefficients and to reconstruct de-noised signals. In the second step, the damage-sensitive features are extracted from the time domain (HT), the frequency domain (FFT), and the joint time-frequency domain (DWT). These features are then assembled in a multi-dimensional D.I. vector. The D.I. vector, in turn, represents the input to the unsupervised or the supervised pattern recognition algorithms that will be discussed in the next section.

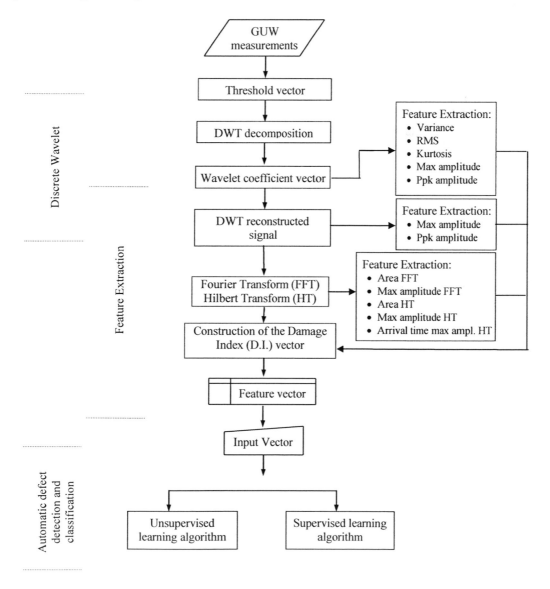

Figure 4.2. Flowchart of the defect detection and classification procedure.

5. Unsupervised and Supervised Methods for Automatic Defect Detection and Classification

5.1. Introduction

The third step of the overall strategy for the SHM of waveguides consists on the automatic defect detection and classification. The unsupervised learning algorithm discussed here is used for defect detection and it is based on outlier analysis, a novelty detection method. The supervised learning algorithm is, instead, required for damage detection, location and sizing; the ANN algorithm is the supervised learning method discussed here. The general procedure illustrated in the flowchart of Fig. 4.2 remains unchanged except for the last step which is interchangeable depending on the type of pattern recognition class being used.

5.2. Outlier Analysis

The outlier analysis is a novelty detection method that establishes whether a new configuration of the system is discordant or inconsistent from the baseline configuration which consists of an existing set of data (or patterns) that describe the normal operative conditions.

Ideally, if outlier analysis is used for detecting damaged states, the baseline should include normal variations in environmental or operative conditions of the structure (e.g. temperature, humidity, loads) (Worden *et al.* 2002). However, it is generally difficult to account for all of the environmental variables that may affect a damage-sensitive set of features.

The outlier analysis will vary if one or more parameters are considered simultaneously. In the approach presented here the number of parameters is equal to the dimension of the D.I.s feature vector which in turn depends on the number of parameters extracted from the DWT processing.

In the analysis of one-dimensional elements, the detection of outliers is a straightforward process based on the determination of the discordancy between the one-dimensional datum and the baseline. The most common discordancy tests is based on the deviation statistics, z_ζ, defined as:

$$z_\zeta = \frac{|x_\zeta - \bar{x}|}{\sigma} \tag{5.1}$$

where x_ζ is the potential outlier, and $\bar{x}$ and σ are the mean and the standard deviation of the baseline, respectively. The mean and the standard deviation can be calculated with or without the potential outlier depending upon whether inclusive or exclusive measures are preferred. The value of z_ζ is then compared to a threshold value in order to determine whether the datum x_ζ is an outlier (above the threshold) or not. By simple algebra manipulations, Eq. (5.1) can be rewritten as:

$$z_\zeta = \left| \frac{x_\zeta}{\sigma} - \frac{\overline{x}}{\sigma} \right| = |m \cdot x_\zeta + q| \tag{5.2}$$

where $m = 1/\sigma$ and $q = -\overline{x}/\sigma$. The deviation statistics thus belongs to a straight line whose slope, m, is inversely proportional to the standard deviation of the baseline. The parameter m can be viewed as the sensitivity of the damage detection procedure that increases with decreasing variability of the baseline conditions.

In the analysis of multi-dimensional elements (multivariate data), the discordancy test of Eq. (5.1) is replaced by the Mahalanobis Squared Distance (MSD), D_ζ, which is a scalar defined as:

$$D_\zeta = \left(\{x_\zeta\} - \{\overline{x}\} \right)^T \cdot [K]^{-1} \cdot \left(\{x_\zeta\} - \{\overline{x}\} \right) \tag{5.3}$$

where $\{x_\zeta\}$ is the potential outlier vector, $\{\overline{x}\}$ is the mean vector of the baseline, $[K]$ is the covariance matrix of the baseline and T represents a transpose matrix.

As in the univariate case, the baseline mean vector and covariance matrix can be inclusive or exclusive. In the present study, since the potential outliers are always known a priori, both z_ζ and D_ζ are calculated exclusively without contaminating the statistics of the baseline data.

In order to determine whether a new datum (uni- or multi-dimensional) is an outlier, the corresponding value of z_ζ or D_ζ has to be compared to a threshold computed from the baseline. Usually, the baseline data set consists of a large number of samples that can be obtained from a large number of experimental acquisitions of more realistically can be constructed by corrupting a certain number of experimental ultrasonic measurements with digital random noise. Once the values of z_ζ and D_ζ of the baseline distribution are determined, the threshold value is taken as the upper value of 3σ, equal to 99.73% of the Gaussian confidence limit. Thus if a new datum was classified as an outlier, theoretically there is only a 0.27% chance of a "false positive" reading. This approach assumes that samples sufficiently represent the statistical distribution as a Gaussian curve.

The general procedure of defect detection by means of outlier analysis is illustrated in the flowchart of Fig. 5.1. The input vector, made of one- or (multi-) dimensional D.I. feature vector and calculated from a new datum, is fed into the classifier in order to compute the value in Eq. 5.1. (Eq. 5.3). If the discordancy value is below the threshold, the datum is representative of a pristine structure; otherwise, the new datum is an outlier and it becomes an indicator of a damaged structure.

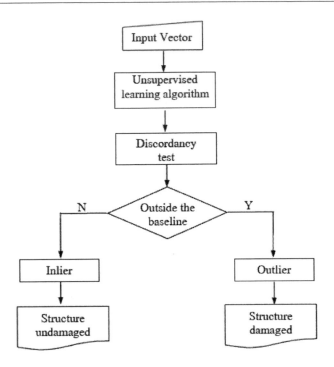

Figure 5.1. Flowchart of the defect detection procedure based on unsupervised learning algorithms (outlier analysis).

5.3. Artificial Neural Network

The supervised approach proposed in the present study couples the damage-sensitive features to an ANN able to learn from training samples through iterations in order to determine the type, size and location of the defects. ANNs have been used successfully for classifying crack damage in honeycomb sandwich plates (Yam *et al.* 2003), railway wheels (Zang and Imregun 2001) and acoustic emission signals from composite structures (Godin *et al.* 2004, Kim *et al.* 2004).

As discussed in section 4, the values of damage-sensitive features, as computed in Eqs. (4.1) and (4.2), are dependent on damage size or on damage location. The aim of the supervised learning algorithms is to learn the pattern, i.e. the relation between the features and the size of damage, and to recognize the type of damage whenever a new feature value is presented to the classifier.

A feed-forward, backpropagation (FF-BP) ANN with three layers was used as neural network architecture in the results presented here. The input layer receives as input data the multi-dimensional D.I. vectors along with the codifications of their classes (targets). The hidden layer processes the data by multiplying the input vectors by weights and adding biases. The results constitute the argument of a transfer function that squashes the output values into a certain range. Since the target classes were coded with binary numbers, the following log-sigmoid transfer function can be employed

$$f(x) = \frac{1}{1 + e^{-x}} \qquad (5.4)$$

This function squashes the output values between zero and one for the binary representation. The output layer provides the network outputs and compares the outputs with the targets. The error E is calculated as

$$E = \frac{1}{N} \sum_{k=1}^{N} \sum_{j=1}^{m} (y_{kj} - \hat{y}_{kj})^2 \qquad (5.5)$$

where N is the number of training samples, m is the number of output nodes, y_{kj} is the desired target, and $\hat{y}_{kj}$ is the network output. If the error is above a certain value, the training process is continued by transmitting the errors backwards from the output layers, and adjusting the weight and biases. If the error is below an established value, the learning process is stopped. The training process is also stopped when a minimum on the error gradient is reached. Each individual weight change is in the direction of a negative gradient and at the iteration step n, the new weight vector

$$\underline{w}(n+1) = \underline{w}(n) - \eta \frac{\partial E}{\partial w_{jp}} \quad (j = 1, ..., m; \, p = 1, ..., N\_features) \qquad (5.6)$$

where $0 < \eta < 1$ is the learning rate. The learning rate determines the magnitudes of the weight change. The smaller η is the smoother is the convergence of the search but at the price of higher number of iteration steps. For higher η, the algorithm may become unstable possibly leading to oscillation and preventing the error to fall below a certain range. In order to control the network oscillations during the training process, an additional coefficient η_m (momentum coefficient or additional momentum) comprised between 0 and 1 is added to the definition in Eq. (5.6); the weight change now becomes:

$$\underline{w}(n+1) = \underline{w}(n) - \eta \frac{\partial E}{\partial w_{jp}} - \eta_m [\underline{w}(n) - \underline{w}(n-1)] \qquad (5.7)$$

The additional coefficient scales the influences of the previous step on the current one.

Fig. 5.2 schematizes the FF-BP architecture. The selection of the network parameters $(\eta, \eta_m,$ number of hidden layers, dimension of the input vector, the type of features) may affect the performance of the network, i.e. the percentage of data correctly classified. For this reason, it is advisable to pursue a parametric analysis such as that described in Fig. 5.3: the network parameters are swept in a certain range and the network performance is tested at each run. The importance of the proper choice of the network parameters on the defect classification performance will be shown in the next two sections.

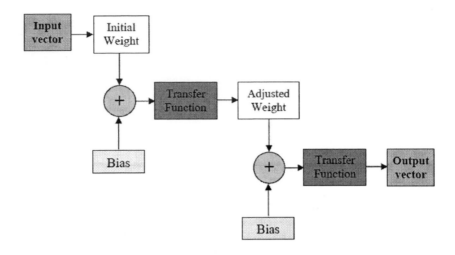

Figure 5.2. Artificial Neural Network for pattern recognition.

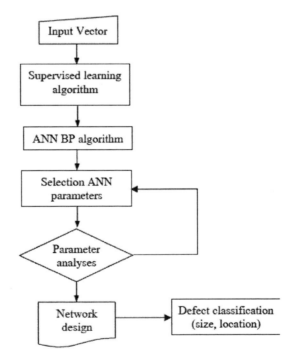

Figure 5.3. Flowchart of the defect detection procedure based on supervised learning algorithm (Artificial Neural Networks).

6. Strand Monitoring

The remaining part of the chapter illustrates two applications. The first one is the monitoring of seven-wire steel strands and it is described in this section. The second application is the monitoring of railroad tracks and it will be described in section 7.

6.1. Introduction

High-strength, multi-wire steel or CFRP strands are widely used in civil engineering such as in prestressed concrete structures, in cable-stayed and suspension bridges, and in external prestressing. Material degradation of the strands, usually consisting of indentations, corrosion or even fractured wires, may result in a reduced load-carrying capacity of the structure that can lead to collapse. In a survey involving the study of more than one hundred stay-cable bridges Watson and Stafford (1988) pessimistically reported that most of them were in danger mainly because of cable defects. Strand failures that caused bridge collapses were documented in Wales (Woodward 1988), Palau (Parker 1996a, 1996b), and North Carolina (Chase 2001). Hence the need for developing monitoring systems for strands that can detect, and possibly quantify, structural defects. Structural monitoring methods based on GUWs have the potential for both defect detection and stress monitoring. GUWs were used for the detection of defects in multi-wire strands and reinforcing rods (Kwun and Teller 1994, 1995, Pavlakovic *et al.* 1999, 2001, Beard *et al.* 2003, Reis *et al.* 2005) and for the evaluation of stress levels in post-tensioning rods and multi-wire strands (Kwun *et al.* 1998, Chen and Wissawapaisal 2001, 2002, Washer *et al.* 2002). The authors have used GUWs for defect detection and stress monitoring in seven-wire steel and composite strands (Rizzo and Lanza di Scalea 2001, 2004a, 2004b, 2005 , 2006, Lanza di Scalea *et al.* 2003).

6.2. The Magnetostrictive Sensors

In this application non-contact magnetostrictive ultrasonic sensors (MsSs) were used as ultrasonic source and receiver. These sensors exploit the phenomenon called magnetostriction which consist on the deformation of a ferromagnetic materials when placed into a magnetic field. A material possesses magnetostrictive constants that are dependent on the temperature, the magnetic state, and the manufacturing processes of the material. When an alternating magnetic field is applied, a stress wave of twice the frequency is generated.

Among different types of MsSs, a solenoid type was used. The solenoid is suitable for stress-wave generation and detection in bar like structures such as strands.

According to Faraday's law an electrical current induces a magnetic field that is perpendicular to the current direction. When an alternating voltage passes through a wire, an alternating current is generated which, in turn, creates an oscillating magnetic field within the coil, that produces a change of magnetostriction of the ferromagnetic test material. The subsequent deformation known as Joule's effect (Joule 1847) generates a stress wave. The amplitude of the wave depends on the existing magnetic field, as will be discussed later. The inverse mechanism is used in wave detection: the wave propagating in the ferromagnetic material modulates an existing magnetic field, by Villari's effect (Villari 1865), thereby exciting a voltage pulse in the receiver coil. A constant magnetic field (bias) is usually superimposed; the bias enhances the transduction, i.e. increases the stress wave amplitude generated and detected, and equals the electrical excitation frequency and the stress wave frequency (Kwun *et al.* 2003).

In recent years MsSs have been used for studying GUW propagation in rods (Kwun and Teller 1994), strands (Kwun *et al.* 1998, Washer 2001, Lanza di Scalea *et al.* 2003) and pipes (Kwun *et al.* 2003, Rizzo *et al.* 2005), for monitoring combustion engines and crash events (Kwun and Bartels 1998), for observing epoxy cure (Vogt *et al.* 2003) and for determining torques in steering shafts (Wakiwaka and Mitamura 2001).

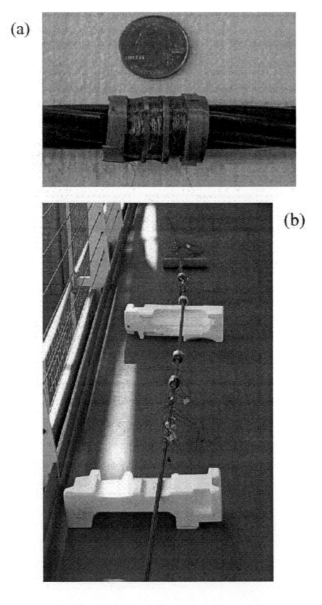

Figure 6.1. Magnetostrictive sensors installed in 0.6-in seven-wire strands: (a) a coil; (b) assembled sensors with permanent magnets to provide bias magnetic field.

For solid cylinder applications, the orientation of the bias field controls the generation and detection of longitudinal, torsional, or flexural waves. When the coil encircles the bar, a uniform alternate field is generated across the section. When the bias is along the bar axis, by applying a U-shape permanent magnet (Kwun and Teller 1994, 1995) or a solenoid electromagnet (Laguerre *et al.* 2002), a longitudinal wave is launched (or detected). When the bias is perpendicular to the bar axis, a "twisting" motion is induced that excites (or receives) a torsional wave (Kim *et al.* 2005). When the static field guarantees an antisymmetric magnetic induction with respect to the radial direction of the bar axis, the alternate axial field generates (or receives) flexural waves only (Lee and Kim 2002, Cho *et al.* 2003).

The dynamic magnetic field H generated by the active sensor is proportional to the number of coil turns per unit length N and to the current i, according to the formula $H = Ni$. Because the current is inversely proportional to the impedance, it is advantageous to employ low impedance wires to build the generator. Conversely, when the coil is employed as a receiver, it is important to have as high a voltage as possible. Because the voltage is proportional to the impedance, high-resistance wires should be used to manufacture receiver coils.

The magnetostrictive devices designed and constructed at UCSD have the wire wounded in an opposite direction for one third of the coil length (Fig. 6.1a) so as to enhance narrow-band stress wave transduction while limiting electrical noise (Kwun and Bartels 1998). The magnetic bias was provided by U-shaped permanent magnets or cylindrical magnets as needed to improve the sensor efficiency. This setup preferentially excited longitudinal waves in the strands. Copper adhesive tape was used to shield the sensors from electromagnetic interference.

A typical installation of MsSs in a seven-wire, 15.2 mm (0.6 in) strand is illustrated in Fig. 6.1b.

6.3. Experimental Setup

The test component under investigation was a high-grade steel 270, seven-wire twisted strand with a total diameter of 15.24 mm (0.6 in). This is a typical strand for stay cables and for prestressed concrete structures. The nominal diameter of each of the wires was 5.08 mm (0.2 in). A notch was machined, perpendicular to the strand axis, in one of the six peripheral wires by saw-cutting with depths increasing by 0.5-mm steps to a maximum depth of 3 mm. A final cut resulted in the complete fracture of the helical wire (broken wire, b.w.), which was the largest defect examined. It should be pointed out that due to geometrical restraints two adjacent wires were damaged when notch depth larger than 2 mm were machined. Fig. 6.2a illustrates the eight defect sizes machined. Fig. 6.2b shows the cross-sectional area reductions corresponding to the various notch depths.

The strand was subjected to a 120 kN tensile load, corresponding to 45% of the material's ultimate tensile strength, that is a typical operating load for stay cables. The load was applied in the laboratory by a hydraulic jack operating in load control.

MsSs, resonant at 320 kHz, were used to excite and detect GUWs. This frequency was chosen since it is known to propagate with little losses in loaded strands (Rizzo and Lanza di Scalea 2004b). The sensors were deployed in the "reflection" mode scheme described in Fig. 4.1b. The distance between the transmitting and the receiving transducers, d_1 in Fig. 6.3a, was fixed at 203 mm (8 in) in all tests. By sliding the transmitter/receiver pair along the strand, tests were conducted at the five different notch-receiver distances, d in Fig. 6.3a, of 203 mm (8 in), 406 mm (16 in), 812 mm (32 in), 1016 mm (40 in), and 1118 mm (44 in). The latter was the largest distance allowed by the rigid frame of the hydraulic loading.

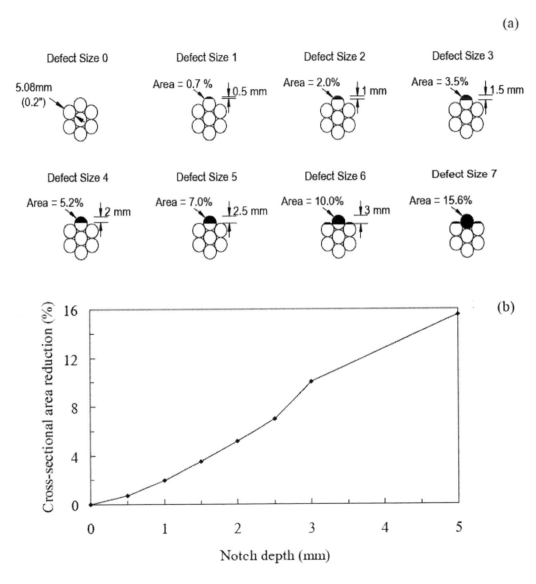

Figure 6.2. (a) The different notch depths examined; (b) reduction of the cross-sectional area of the strand as a function of notch depth.

A National Instruments PXI© unit running under LabVIEW© was employed for signal excitation, detection and acquisition, Fig. 6.3b. Five-cycle tonebursts centered at 320 kHz, modulated with a triangular window, were used as generation signals. Signals were acquired at sampling rate equal to 33 MHz and stored after different number of digital averages, namely 500, 50, 10, 5, 2 and 1 (single generation). The higher the number of the digital averages, the lower is the presence of incoherent noise. Table 1 summarizes the various testing configurations adopted.

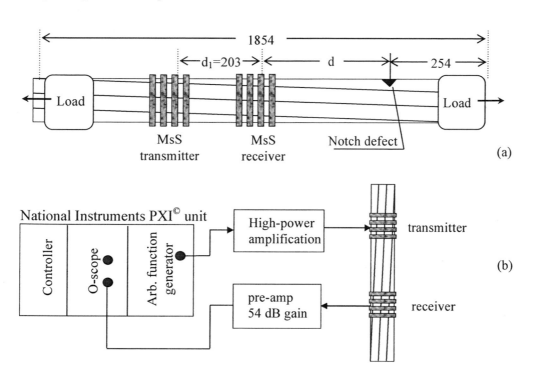

Figure 6.3. (a) Experimental setup for the detection of notch defects in the strand (dimensions in mm); (b) overall schematic of the monitoring system.

Table 6.1. Strand monitoring: the different testing configurations considered.

Defect Size (mm)	Strand cross-sectional area reduction (%)	Notch-receiver distance, d (mm)	Number of averages on GUW signals
0	0.00	203	1
0.5	0.71	406	2
1	1.98	812	5
1.5	3.50	1016	10
2	5.18	1118	50
2.5	6.97		500
3	10.0		
Broken wire	15.6		

6.4. DWT-Based Feature Extraction

Two time windows were selected for the direct signal and the defect reflection. Figs. 6.4a and 6.4b show a typical ultrasonic signal detected after 500 averages when probing the 2-mm defect size, for receiver-notch distance d = 203 mm and d = 1016 mm, respectively. The direct signal window was fixed between 35 μsec and 97 μsec for the given constant transmitter-receiver distance and it consisted of 2048 points. The defect reflection window varied for the different notch-receiver distances. Such window consisted of 4096 points and covered the expected arrival of the reflection. The other ultrasonic trace present in both figures is the reflection from the strand end.

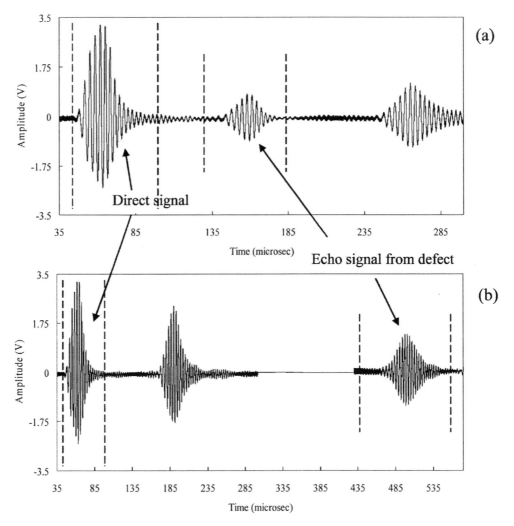

Figure 6.4. Typical ultrasonic signals detected after 500 averages for notch-receiver distance (a) equal to 208 mm and (b) equal to 1016 mm.

Once gated, the ultrasonic signals of interest were processed through the DWT. The bandwidths of the wavelet decomposition levels from eq. (3.13) are illustrated in Fig. 6.5, using the db40 mother wavelet and a 33 MHz sampling frequency. The probing frequency of interest, 320 kHz, was located in the sixth level of the wavelet decomposition. Thus the ultrasonic signals were pruned by considering the sixth DWT level only.

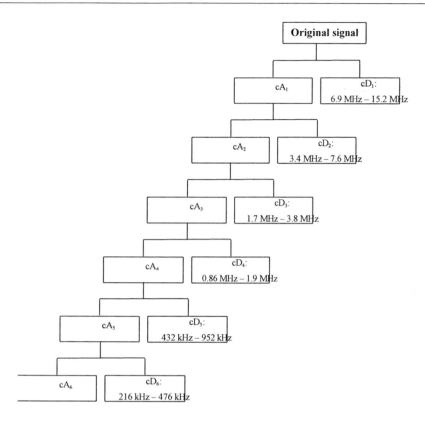

Figure 6.5. Bandwidth of each wavelet level by using the db 40 and a 33 MHz sampling frequency.

Representative results of the pruning process are shown in Fig. 6.6. Fig. 6.6a shows a typical ultrasonic signal detected after 500 averages. The signal detected at around 140 microseconds is the reflection from a 2.5mm-deep indentation in one of the helical wires of the strand. A single signal (no averages) is shown in Fig. 6.6b where the defect reflection is completely buried in noise. The no-average signals reconstructed, by using the db40, from the first six DWT decomposition levels are indicated in this figure as D_1, D_2,...,D_6. Reconstruction D_1 corresponds to what the time signal would look like if only level 1 highpass filter coefficient, cD_1, were used for the reconstruction. D_2 corresponds to what the reconstructed time signal would look like if only level 2 highpass filter coefficient, cD_2, were used to reconstruct the signal, and so on. The result of averaging the raw ultrasonic measurements 500 times was the comparison signal. As expected from the flowchart in Fig. 6.5, the filter associated with D_6 in Fig. 6.6c yields the best reconstruction given the close resemblance with the averaged result in Fig. 6.6a. Since levels 1 to 5 will merely reconstruct noise, they are eliminated in the pruning process.

Subsequently to pruning, the sixth decomposition level was subjected to the thresholding process. The threshold chosen to select the relevant wavelet coefficients for the D.I. computation is an important variable that affects the sensitivity of the defect sizing. An optimum threshold combination for the direct signal and the defect reflection was searched based on obtaining the largest sensitivity to defect size through the variance-based D.I. It was found that the larger sensitivities were obtained when setting more severe thresholds on the defect-reflected signals, with little effect of the thresholds imposed on the direct signal. Based on the findings in Rizzo and Lanza di Scalea (2006), optimum thresholds were fixed at 20%

of the maximum wavelet coefficient amplitude for the direct signal and at 70% of the same quantity for the defect reflection.

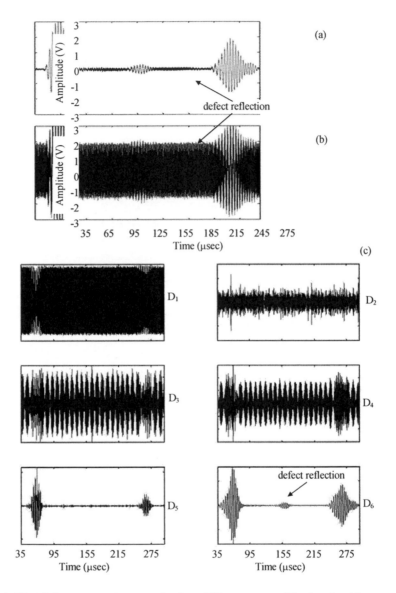

Figure 6.6. (a) Signal in seven-wire strand after 500 averages; (b) signal with no averages; (c) reconstructed signal after pruning the DWT coefficients at the first six decomposition levels.

Due to the "reflection" mode scheme adopted, the D.I.s were computed by using Eq. 4.1 and they were expected to increase with increasing damage size. Some of the individual D.I. features are shown in Fig. 6.7 as a function of the notch depth for all notch-receiver distances. The signals processed are those acquired after 10 averages. The D.I.s shown are based on the Var, RMS, peak amplitude and Ppk amplitude of the wavelet coefficients (Figs. 6.7a-d), on the area below and the peak amplitude of the HT of the reconstructions (Fig. 6.7e and f), and on the peak amplitude and on the area below the FFT spectrum of the reconstructions (Fig. 6.7g and h). All D.I.s show a quite linear dependence in a semi-logarithmic scale on the notch depth, and a relatively negligible dependence on the defect position for notches between 1.5 mm and 3 mm in depth. The results for very small notches, below 1 mm in depth, are less stable against varying distances due to the poorer SNRs of the defect reflections. The results for the broken wire case (5mm-deep notch) also show an increased dependence on the notch-receiver distance, with D.I.s generally increasing for defects located further away from the receiver. This trend is opposite to what would be expected considering wave attenuation effects, and its origin is probably associated with the interference of multiple propagating modes that is distance dependent. One of the most notable results in Fig. 6.7 is that the D.I. based on the Var of the wavelet coefficient vector, Fig. 6.7a, has the largest sensitivity to notch depth compared to all other features presented.

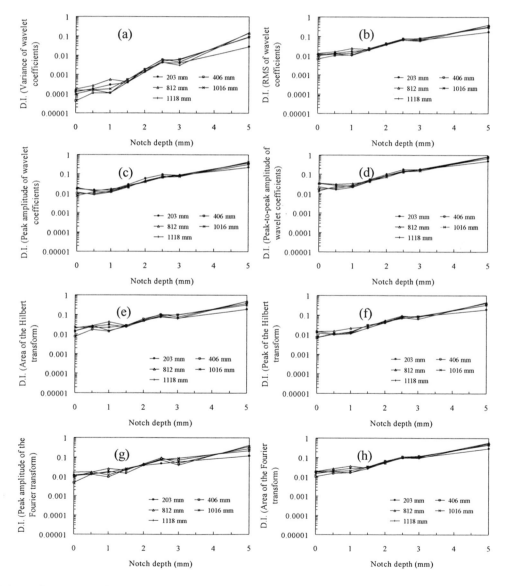

Figure 6.7. Damage index computed from (a) the variance, (b) the RMS, (c) the peak amplitude, (d) the peak-to-peak amplitude of the wavelet coefficient vector cD_6, (e) the area and (f) the peak of the Hilbert transform, (g) the peak amplitude and (h) the area of the Fourier transform.

6.5. Unsupervised Learning Algorithm

In this application the baseline distribution was obtained based on the ultrasonic signals stored after averaging over ten acquisitions and corrupted by two different levels of white Gaussian noise. The noise signals were created by the MATLAB *randn* function. The random noise increased the sample population and simulated possible variations in SNR of the measurements that can be originated, in practice, by a number of factors including changing sensor/structure ultrasonic transduction efficiency, changing environmental temperature affecting ultrasonic damping losses. The *randn* function generates arrays of random numbers whose elements are normally distributed with zero mean and standard deviation equal to 1. The function was pre-multiplied by a factor that determines the noise level. Factors equal to 0.01 and 0.1 were considered as "low noise" and "high noise", respectively. For each noise level, 300 baseline samples were created.

The same approach was taken to generate a large number of data for the damaged conditions. Six of the seven total defect sizes listed in Table 6.1 were considered in this portion of the study. The ten-average signals acquired for each of the six defect sizes were corrupted by the low noise level and the high noise level, generating a total of 300 samples for each damage size. These samples represented the testing data of the algorithm. A total 2100 samples data were thus collected for each noise level.

The added noise can be quantified in terms of SNR by the following expression:

$$\text{SNR [dB]} = 10 \log\left(\sum_{i=1}^{N} s_i^2 / N \middle/ \sum_{i=1}^{N} u_i^2 / N \right) \tag{6.1}$$

where s_i and u_i are the amplitudes of the ultrasonic signal and of the noise signal, respectively, and N is the number of points. The SNR between the direct signal and the two 0.01 and 0.1 noise levels was about 43 dB and 23 dB, respectively. The SNR between the reflection from the 3 mm-deep notch and the two 0.01 and 0.1 noise levels was about 32 dB and 12 dB, respectively. Clearly, the latter two values decreased with decreasing notch depth.

Only the features extracted from the wavelet coefficient vectors were considered for the unsupervised learning algorithm.

6.5.1. Univariate Analysis for Low Noise

This section presents the outlier analysis results for the strand damage detection when the four features of Var, RMS, peak amplitude and Ppk amplitude of the wavelet coefficients were considered separately.

Fig. 6.8 illustrates the deviation statistics of the D.I. (Var) for the baseline and for all damage conditions of the strand, in the low noise level case. The logarithmic scale is used in the graph. Each value was obtained from Eq. (5.1) where the mean and the standard deviation are those of the baseline. Thus the exclusive method was adopted, simulating an unsupervised learning approach where the damage conditions are unknown a priori. The horizontal line is the value of 2.655 representing the 99.73% confidence threshold. Clear steps can be seen in the discordancy plot at sample numbers 301, 601, 901, 1201, 1501 and 1801, corresponding to the progressive reduction of the strand's cross-sectional area as described in Fig. 6.2 and

Table 6.1. For the 0.5mm-deep notch, equivalent to less than 1% reduction in cross-sectional area, there are only four inliers, i.e. false negative indications, with the remaining 296 samples properly indicating damaged conditions. All other damage sizes, between sample numbers 601 and 2100 and corresponding to more than 2% cross-sectional area reduction, are properly classified as damaged conditions. Overall the results are very satisfactory confirming the effectiveness of the wavelet coefficient variance feature for computing the D.I.

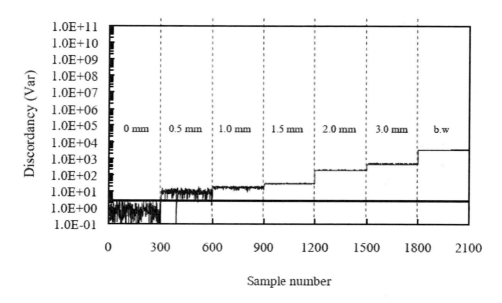

Figure 6.8. Discordancy test for the feature D.I. (Var). Baseline (undamaged) and damaged strand data corrupted with the low level noise.

By applying the same routine to the other three features, and thus computing the D.I. (RMS), D.I. (Max) and D.I. (Ppk), the results were similar. The qualitative trend of increasing discordancy with increasing defect size was the same although the sensitivity to damage (represented by the rate of change of the discordancy plot) was larger for the D.I. (Var) compared to any of the other three features as suggested previously in Fig. 6.7.

6.5.2. Multivariate Analysis for Low Noise

The four wavelet-based features considered separately in the univariate analysis were subsequently used simultaneously to construct a four-dimensional D.I. vector for the outlier analysis. The deviation statistics is now replaced by the MSD expressed in Eq. (5.3).

The MSD of all samples, including the baseline data and the damage data, calculated for the low noise level of 0.01 are summarized in Fig. 6.9a. The mean vector and the covariance matrix were determined from the 300 D.I. vectors associated with the undamaged condition of the strand. As for the univariate analyses, the discordancy values of the damaged conditions were calculated in an exclusive manner. The horizontal line in this figure represents the 99.73% confidence threshold value of 21.579, which is one order of magnitude larger than the corresponding thresholds of the univariate cases. Eight baseline samples are outliers, thus false positive indications. The same stepwise trend observed in the univariate cases is seen in this figure. However, the multivariate analysis outperforms any of the

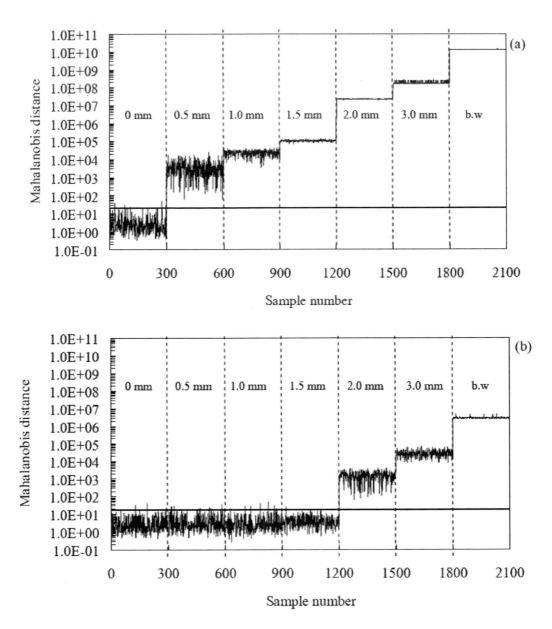

Figure 6.9. (a) Mahalanobis squared distance for the baseline (undamaged) and damaged strand data corrupted with the low level noise; (b) Mahalanobis squared distance for the baseline (undamaged) and damaged strand data corrupted with the high level noise.

univariate results at the same level of noise corruption. First, all damaged conditions are properly classified as outliers, thus there are no false negative indications. Second, the values of the damaged MSDs are several orders of magnitude larger than the corresponding values of the univariate deviation statistics. For example, the MSDs for the 2 mm-deep notch in Fig. 6.9a are five orders of magnitude larger than the deviation statistics in Fig. 6.8. As a consequence, the sensitivity to damage is considerably improved. Finally, the MSD values show good discrimination between all defect sizes, including the smallest notch depths confirming that it is advantageous to combine multiple GUW features to provide a large sensitivity to the defects. Nevertheless, compared to previous multivariate outlier analyses in structural monitoring applications, the dimension of the D.I. vector is still kept at a very low value by selecting only four features of the GUW signals containing the essential information of interest owing to the effectiveness of the DWT decomposition.

6.5.3. Multivariate Analysis for High Noise

Following the same approach, the MSD results of the 4-dimensional D.I. vectors corrupted with the high noise level of 0.1 are shown in Fig. 6.9b. The 99.73% confidence threshold was now computed as 18.137. Compared to the low noise results of Fig. 6.9a, it is clear in Fig. 6.9b that the heavier noise corruption compromises the ability to detect the notch depths below 2.0 mm, corresponding to a 5% reduction in strand's cross-sectional area. The same result was found from the prior univariate analyses, which is not shown here. The ratios of correctly classified outliers below 5% area reduction are only 12/300, 7/300 and 1/300 for notch depths of 0.5 mm, 1.0 mm and 1.5 mm, respectively. Above the 5% area reduction, the sensitivity to defect detection was also degraded with the increasing noise level; for example, the MSD values for the 2 mm notch depth in Fig. 6.9b are four orders of magnitude smaller than the corresponding values in Fig. 6.9a. The reduced number of false positive indications (three against eight) is the only improvement over the low noise level.

Table 6.2 summarizes the number of outliers detected in the multivariate analyses for both levels of noise considered; the outliers are false positive indications for the baseline data (Damage Size 0) and, instead, correct indications of anomalies for the defect data.

Table 6.2. Strand monitoring: multivariate analysis (four features): number of outliers n/300 for the various damage sizes and the two levels of added noise

Corruption Level	Damage Size (notch depth – mm)						
	0	0.5	1.0	1.5	2.0	3.0	5.0 (b.w.)
0.01	8/300	300/300	300/300	300/300	300/300	300/300	300/300
0.1	3/300	12/300	7/300	1/300	300/300	300/300	300/300

6.6. Supervised Learning Algorithm

Table 6.3a summarizes the definition of the classes considered in the NN algorithm. The defect sizes were subdivided into four classes (Classes 1, 2, 3 and 4), corresponding to strand's area reductions in the ranges 0% - 1%, 2% - 5%, 7% - 10%, and 16%. Class 1 can be considered the "no defect" case. Similarly, the notch-receiver distances were subdivided into four classes (Classes A, B, C and D) corresponding to the values $d = 203$ mm, 406 mm, 812 mm and 1016-1118 mm. Each class of defect size and notch-receiver distance was coded with a 2-digit binary number. In total, the classification problem consisted of $4^2=16$ combinations of defect sizes and notch-receiver distances, represented by the four digit codes shown in Table 6.3b.

**Table 6.3. Strand monitoring. (a) Defect classification problem definition;
(b) Coding for the classification problem.**

Class	Defect size (mm)	Strand area reduction (%)	Binary code	Notch-receiver distance (mm)	Binary code
1	0 – 0.5	0 – 1	0 0		
2	1 – 2	2 – 5	0 1		
3	2.5 – 3	7 – 10	1 0		
4	Broken wire	16	1 1		
A				203	0 0
B				406	0 1
C				812	1 0
D				1018-1116	1 1

a.

Combination	Binary code	Combination	Binary code
1A	0 0 0 0	3A	1 0 0 0
1B	0 0 0 1	3B	1 0 0 1
1C	0 0 1 0	3C	1 0 1 0
1D	0 0 1 1	3D	1 0 1 1
2A	0 1 0 0	4A	1 1 0 0
2B	0 1 0 1	4B	1 1 0 1
2C	0 1 1 0	4C	1 1 1 0
2D	0 1 1 1	4D	1 1 1 1

b.

As described in section 5.3, a FF-BP ANN with three layers was used. The input layer receives as input data the multi-dimensional D.I. vectors along with the codifications of their classes (targets). The hidden layer processes the data by multiplying the input vectors by weights and adding biases. The results constitute the argument of the log-sigmoid transfer function that squashes the output values between zero and one. The outputs are compared with the targets in order to estimate the error E as calculated in Eq. 5.5. In this application, the number of output nodes is equal to 4.

Eight (defect sizes) x five (notch-receiver distances) = 40 D.I. vectors used as training data in the present analysis corresponded to all acquisitions after 500 averages. This was considered to be a representative configuration of the structure. Forty x six (all different signal averages) = 240 D.I. vectors were used to test the network, i.e. they represented the testing data. It is evident that acquisitions after 500 averages, which represented the 16.7% (1/6) of all the data measured, were used for both the network training and the network testing.

Eight features, listed in Table 6.4, were used to compute the D.I.s. for the supervised learning algorithm.

Table 6.4. Strand monitoring. ANN parameters considered

Feature #	D.I. feature	η = learning rate	η_m = additional momentum	# of features	# of hidden neurons
1	Variance (WCV)	0.05	0.05	3 to 8	6 to 20
2	Peak position (HT)	0.20	0.20	step 1	step 2
3	Peak-to-peak (WCV)	0.35	0.35		
4	Peak amplitude (HT)	0.50	0.50		
5	Peak amplitude (FFT)	0.65	0.65		
6	RMS (WCV)	0.80	0.80		
7	Peak amplitude (WCV)	0.95	0.95		
8	Area (FFT)				

WCV: wavelet coefficient vector
HT: Hilbert Transform
FFT: Fast-Fourier Transform

The parameter analyses recommended at the end of section 5.3 and schematized in the flowchart of Fig. 5.3 were conducted in order to find the design that provided the best network performance, i.e. the largest percentage of testing data correctly classified. In order to ensure the same network initial conditions, the initial biases and weights of the network were constrained to be constant.

The first analysis considered the values of the learning rate, η, and the additional momentum, η_m, as parameters, columns 3 and 4 in Table 6. The network consisted of eight hidden neurons with eight-dimensional input vectors. The input vector consisted of the D.I. based upon the eight features listed and ordered in columns 1 and 2 of Table 6.4. The analysis demonstrated that for this application the two parameters η and η_m did not substantially affect the performance of the network.

The second analysis considered fixed values for η (= 0.2) and η_m (= 0.5). The goal now was to find the *best* network configuration by selecting the parameters resulting in the best classification performance in terms of (a) the number of features (dimension) of the input D.I. vector, (b) the type of these features, and (c) the number of hidden neurons. The first two features of the D.I. vector were kept constant in the optimization process. These were the Var of the wavelet coefficients (providing the highest sensitivity to the notch size) and the position of the HT peak of the reconstructions (providing the sensitivity to the notch location). The analysis then proceeded by adding the remaining six features of Table 6.4 to the D.I. vector, one at a time and in all possible combinations. Thus D.I. vectors were tested from a minimum of three dimensions to a maximum of eight dimensions. For each input vector, the number of hidden neurons was changed from 6 to 20. This extensive set of trials resulted in a network performance varying between 41.2% for the *worst* network (5-dimensional D.I. input vector, features 1, 2, 3, 6, 8 in Table 6.4, with 6 hidden neurons) and 90.8% for the *best* network (5-dimensional D.I. input vector, features 1, 2, 4, 5, 6 in Table 6.4, and 12 hidden neurons).

The histograms in Fig. 6.10 summarize the performance of the *best* network design. The percentage of correct classification is plotted as a function of various testing parameters. Fig. 6.10a shows the classification performance for varying notch sizes. The best performance is obtained for the two largest defects, above 7% of the strand area reduction. These notches were all properly identified as a result of the large SNR of the reflections. The poorest performance is obtained for the "no defect" case, but still with a success rate as high as 82%. Properly classifying the "no defect" case means avoiding false positives.

The histograms in Figs. 6.10b and 6.10c can be read in a similar way. They are plotted by clustering the number of averages (Fig. 6.10b) and the notch-receiver distances (Fig. 6.10c). Fig. 6.10b shows that the data used to train the network were all properly classified; more interesting is the outcome that the classification performance is rather independent of the number of averages. This confirms and remarks the fact that the DWT-based features chosen for the D.I. vector computation are robust against noise. The classification success is naturally poorer (75%) for the single-event case due to the degraded SNR. Similarly, the classification performance appears substantially independent of the notch-receiver distance, Fig. 6.10c.

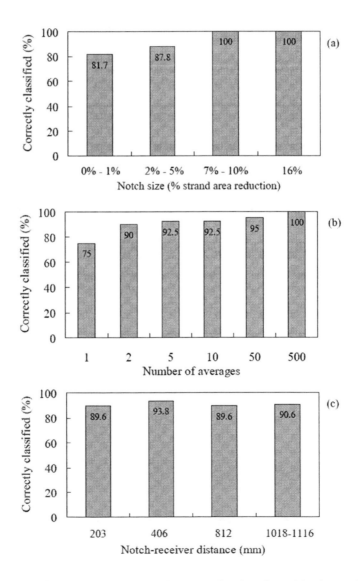

Figure 6.10. Best strand defect classification performance for the 16-combination problem as function of (a) the notch size; (b) the number of signal averages; (c) the notch-receiver distance.

In the *best* network only 9.2% of the cases (22 testing data) were not properly classified. These included false positives (2.9%), false negatives (3.7%), misclassifications (0.8%) and undetermined outputs (1.7%). The 3.7% false negative indications, the most critical in structural monitoring, were nine cases of Class 2 notch sizes incorrectly classified in the "no defect" Class 1. These "missed" defects corresponded to a 2% reduction of the strand's cross-sectional area, thus very small.

7. Rail Monitoring

7.1. Introduction

Safety statistics data from the US Federal Railroad Administration (FRA 2005) indicate that train accidents caused by track failures including rail, joint bars and anchoring resulted in 2,700 derailments and $441M in direct costs during the decade 1992-2002. The associated indirect costs due to disruptions of service are considered to be at least equally significant. The first leading cause of these accidents is the "transverse defect" type that was found responsible for 541 derailments and $91M in cost during the same time period ($17M in year 2001 alone). Transverse defects are cracks developing in a direction perpendicular to the rail running direction, and include transverse fissures (initiating in a location internal to the rail head) and detail fractures (initiating at the head surface as Rolling Contact Fatigue defects). In year 2004 the transverse/compound fissure caused over $14M reportable damage costs.

The most common methods of rail inspection are magnetic induction and contact ultrasonics (Clark 2004). The first method requires a small lift-off distance for the sensors in order to produce suitable sensitivity and it may be affected by environmental magnetic noise. Successful applications were recently reported (Oukheloou *et al.* 1999, Pohl *et al.* 2004). Ultrasonic testing is conventionally performed from the top of the rail head in a pulse-echo configuration. In this system, ultrasonic transducers are located inside a water-filled wheel and are oriented at 0° from the surface of the rail head to detect horizontal cracks, and at 70° to detect transverse cracks. Such an approach suffers from a limited inspection speed and from other drawbacks associated with the requirement for contact between the rail and the inspection wheel. More importantly, horizontal surface cracks such as shelling and head checks can prevent the ultrasonic beams from reaching the internal defects resulting in false negative readings. The problem of surface shelling was highlighted in the June 1992 train derailment in Superior, Wisconsin; the accident was caused by the presence of a transverse crack missed during a previous inspection.

The need to develop more reliable defect detection systems for rails has produced promising results in recent years based on the use of GUWs (Wilcox *et al.* 2003, Cawley *et al.* 2003, Rose *et al.* 2004, Hayashi *et al.* 2003, Lanza di Scalea and McNamara 2003, McNamara *et al.* 2004, Bartoli *et al.* 2005, Kenderian *et al.* 2003, 2004). Since guided waves propagate along, rather than across the rail, they are ideal for detecting the critical transverse defects. The guided waves are also potentially not sensitive to surface shelling since they can run underneath this discontinuity.

The rail monitoring system presented in this chapter is based on non-contact laser ultrasonics and air-coupled transducers, which have been demonstrated effective for non-contact rail testing (Rose et al. 2004, Lanza di Scalea et al. 2005, Kenderian et al. 2003, 2004).

The present study refines the non-contact rail monitoring technique by adding the wavelet processing and the pattern recognition algorithms. The refinements are demonstrated for the quantitative detection of surface-breaking transverse cracks by high-frequency waves, between 100 kHz and 900 kHz. The cracks are detected by the "transmission" mode.

7.2. Non-contact Ultrasonic Transduction

Laser ultrasonic techniques make use of laser beams to generate and detect mechanical waves. By heating the surface of a body suddenly ("heat shock") the thermal expansion of the material produces mechanical stresses. Short duration heating (~10 nsec) produces very high frequencies (Krautkramer and Krautkramer 1990). Pulsed lasers are well suited to the sudden heating of a material surface because the energy can be released by means of a Q-switch. Two types of laser ultrasonic sources can be assumed: ablative and thermoelastic (Scruby and Drain 1990). The ablation regime assumes that loading is purely due to normal stresses while the thermoelastic regime implies that the loading is pure shear stress. For low-power lasers the acoustic source is thermoelastic and the expansion is induced in the direction parallel to the surface of the medium. In presence of high-power lasers the stresses perpendicular to the surface become dominant (ablation regime). The advantages of laser wave generation are the broadband frequency range achievable (DC-20 MHz) and the fact that the sample can be located at a large distance from the laser source.

Three elastic waves are generated by a laser event: longitudinal, shear and surface waves. When the test piece has a waveguide shape, the longitudinal and shear waves combine to form a guided wave.

Air-coupled transducers were used in the rail monitoring system as ultrasound receivers. These transducers use a micro-machined silicon conducting backplate, which contains arrays of small cylindrical holes. These holes act as air springs underneath the membrane, which is a metallized polymer film. A bias voltage is required to obtain consistent charge variations when the membrane moves (Gan *et al.* 2001).

The essential elements of the laser/air-coupled non-contact setup for rail inspection are shown in Fig. 7.1a. The light source was a Q-switched, Nd:YAG pulsed laser operating at 1064 nm with an ~ 8 nsec pulse duration. Through conventional optics the laser beam was focused on the rail head to a 20 mm-long line to excite high frequency surface waves traveling along the rail running (longitudinal) direction (z-axis). The present investigation was focused to detecting small cracks on the order of few millimeters in depth; thus only high frequency (above 100 kHz) waves were considered. Micro-machined, air-coupled capacitive transducers were used for signal detection.

As traditionally done with conventional wedge transducers, the alignment angle of the air-coupled detectors was adjusted to maximize the sensitivity to the guided waves. The optimum air-coupled detection angle from the normal to the rail surface, θ, is given by Snell's law of refraction:

$$\theta = \arcsin(\frac{c_{air}}{c_p} \sin \theta_p) \tag{6.3}$$

where c_p is the phase velocity of the guided wave in the rail, $c_{air} = 330$ m/sec is the wave velocity in air and $\theta_p = 90°$ for a guided wave propagating parallel to the rail surface. Considering that high-frequency nondispersive waves, surface waves, propagate at $c_p = 3,000$ m/sec, Eq. (6.3) gives $\theta = 6.3°$. Thus $\theta = 6.3°$ toward the laser source is the optimum orientation to detect the first arrival of the wave propagating from the laser source (Fig. 7.1b).

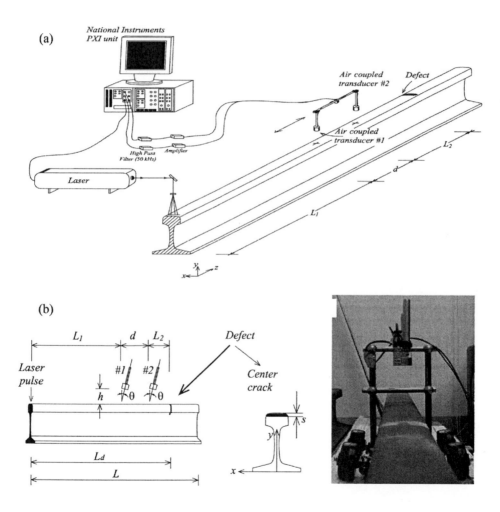

Figure 7.1. (a) Laser/air-coupled system for detecting surface-breaking transverse cracks in rail head; (b) sensor arrangement in the "transmission" mode; (c) photo of the air-coupled sensors and a test rail.

Also in this application, the high-speed data acquisition unit based on National Instruments PXI technology and running under LabVIEW© was for signal acquisition. All ultrasonic signals were acquired at a 5 MHz sampling rate.

A possible inspection deployment is illustrated in Fig. 7.1c. Two different sets of tests were carried out in order to evaluate the pattern recognition approach with unsupervised and supervised algorithm, respectively. Details are given in the next sections.

7.3. Unsupervised Learning Algorithms

7.3.1. Test Protocol

The unsupervised learning test was carried out along one 2440 mm (8 ft) long, 115-lb A.R.E.M.A. rail section donated by San Diego Trolley, Inc. A transverse defect approximately located at the center of the rail head top surface (center crack) was monitored (Fig. 7.2b). This defect was in the form of a notch machined at depths ranging from a minimum of about half millimeter to a maximum of about 4 mm (Fig. 7.2a and b). The width ranged from 2 mm to 3.5 mm, and the length from 22 mm and 45 mm. The largest notch size corresponded to a cross-sectional Head Area (%H.A.) reduction of less than 7%. The seven different notch sizes are schematized in Fig. 7.2b and are hereafter referred to as Defect 1 through Defect 7. Defect 0 represents the no defect case. The corresponding %H.A. reductions are listed in Table 7.1 and plotted in Fig. 7.2c.

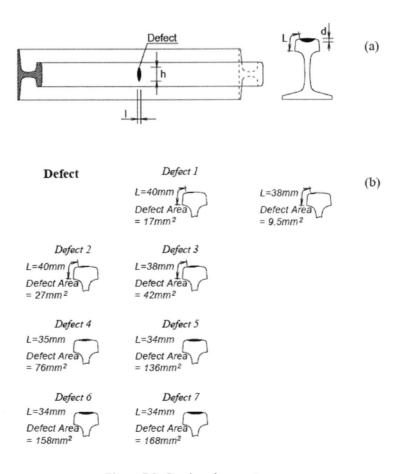

Figure 7.2. Continued on next page.

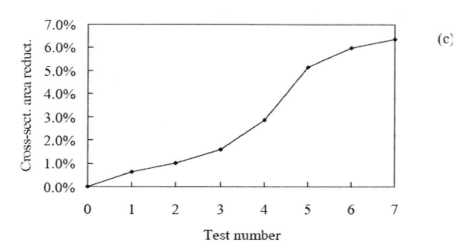

Figure 7.2. (a) The rail notch-like defect monitored; (b) the different notch sizes examined; (c) the cross-sectional head area reduction as a function of notch depth.

Table 7.1. Rail inspection. Definition of defect sizes and number of tests performed for each size.

Def. type	% H.A. reduction	# Acquisition tests
0	0	160
1	0.64	80
2	1.02	60
3	1.59	60
4	2.88	60
5	5.15	60
6	5.98	60
7	6.36	60

Two air-coupled sensors were positioned above the rail head in order to have the crack in between them. Sensor #1 was placed at a 110 mm distance from the laser source while sensor #2 was placed 690 mm away from sensor #1, i.e. 800 mm away from the laser source. The choice of the distance between the sensors depended on the wave damping losses and the de-noising performance of the DWT processing: a short span allows better detection of short wavelength waves which are the most sensitive to the presence of shallow defects; on the contrary a long span increases the inspection speed. Therefore, a compromise between accuracy and inspection speed has to be ultimately achieved.

The air-coupled detectors were oriented at $\theta = 6.3°$, i.e. toward the laser source for optimum implementation of the "transmission" damage detection mode.

Several measurements were taken for each of the seven defect sizes in order to evaluate the statistical repeatability of the inspection. For the baseline (undamaged) rail (Defect 0), 160 data were acquired over two testing days. Defect 1 was monitored 80 times over two different testing days. All other defect sizes (Defect 2 through Defect 7) were monitored 60 times each (Table 7.1).

7.3.2. DWT-Based Feature Extraction

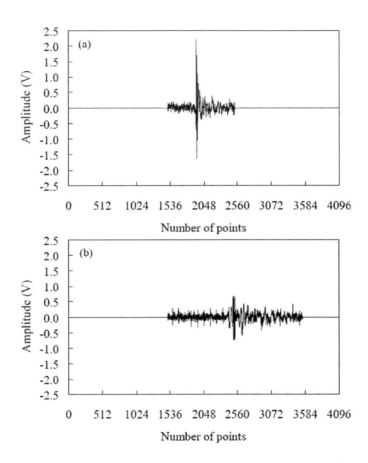

Figure 7.3. Zero-padded ultrasonic signals detected from (a) air-coupled sensor #1 and (b) air-coupled sensor #2.

The ultrasonic signals were gated in order to focus only on the transmitted wave propagating directly from the laser source to the sensors according to the "transmission" defect detection mode. The time-windows covered the expected arrival of the guided wave considering wave speed in rail and in air. The signals were zero padded in order to achieve a total number of 4,096 points. Typical pre-processed ultrasonic signals for sensors #1 and #2 are illustrated in Fig. 7.3.

The db10 mother wavelet was used for the processing. The frequency bandwidths of the wavelet decomposition of a signal sampled at 5 MHz are illustrated in the flowchart of Fig. 7.4. Since the specific experiment designed to test the unsupervised learning approach was targeted to the detection of shallow defects, the low frequency components of the waves were disregarded as they are not affected by the presence of such defects. On the other hand, the propagation of the high frequency components is subjected to high damping and therefore it was also neglected. The focus was on the third wavelet detail level, which included the frequency range between 270 kHz and 590 kHz, corresponding to the semi-wavelength range 2.5÷5.5 mm.

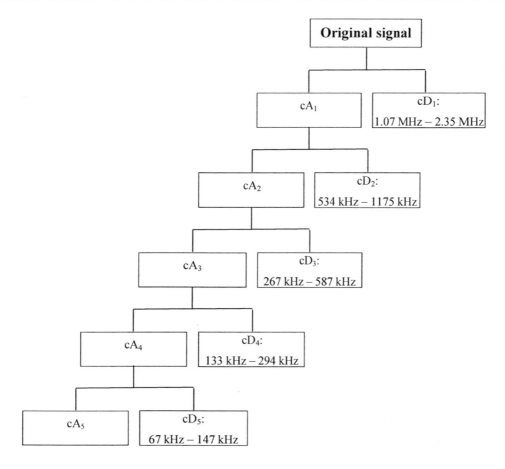

Figure 7.4. Bandwidth of each wavelet level by using the db 10 and a 5 MHz sampling frequency.

The third level was then thresholded in order to retain the eight largest wavelet coefficients. These were used for signal reconstruction and for the statistical analysis to construct the D.I. vector. Since the difference between the defects examined in this portion of the study was subtle, the difference in terms of transmitted ultrasound energy was also expected to be small. Hence the selection of the largest coefficients, rather than a fixed threshold, seemed to be more effective in resolving the small variations.

The D.I. was computed by considering the expression in Eq. 4.2, where F_{sens2} is the feature related to the further sensor from the laser source and F_{sens1} is the feature related to the sensor closer to the laser source.

The features selected for the rail study are listed in Table 7.2 and they are associated to a progressive number; these were the Var, the RMS, the peak amplitude and the Ppk amplitude of the wavelet coefficient vector. Features calculated after reconstructing the ultrasonic signals included the variance (Var Rcstr), the Root Mean Square (RMS Rcstr), the peak amplitude (Max Rcstr) and the peak-to-peak amplitude (Ppk Rcstr), the peak amplitude (Max FFT) and the Area (Area FFT) of the FFT, the peak amplitude (Max HT) and the Area (Area HT) of the HT.

Table 7.2. Rail inspection. List of features considered in the unsupervised learning defect detection

Feat #	Features
1	Variance of the wavelet coefficient vector
2	RMS of the wavelet coefficient vector
3	Max amplitude of the wavelet coefficient vector
4	Peak-to-peak amplitude of the wavelet coefficient vector
5	Variance of the DWT reconstructed time signal
6	RMS of the DWT reconstructed time signal
7	Max amplitude of the DWT reconstructed time signal
8	Peak-to-peak amplitude of the DWT reconstructed time signal
9	Max amplitude of the FFT of the DWT reconstructed time signal
10	Area of the FFT of the DWT reconstructed time signal
11	Max amplitude of the Hilbert Transform of the DWT reconstructed time signal
12	Area of the Hilbert Transform of the DWT reconstructed time signal

7.3.3. Noise Corruption

The baseline distribution of the D.I.s was constructed by corrupting a GUW measurement with three different levels of white Gaussian noise. The noisy signals were created by the MATLAB *randn* function, pre-multiplied by the following factors: 0.002, 0.01 and 0.0625 that are hereafter indicated as "low noise", "medium noise" and "high noise", respectively. For each noise level, and for each signal, 10 samples were created. Since the baseline originally consisted of 160 ultrasonic signals, the total number of baseline samples was increased to 1,600.

The same approach was taken to generate a large number of data for the damaged conditions. The signals acquired for each of the defect sizes were corrupted by the low-, medium- and high-noise level generating a total of 4,400 testing data samples for each noise level. By employing Eq. (6.1) the SNR between the sensor #1 signal and the two 0.002 and 0.0625 noise levels was about 49 dB and 21 dB, respectively. The SNR between the sensor #2 signal and the two 0.002 and 0.0625 noise levels was about 40 dB and 11 dB, respectively. Clearly, the latter two values decreased with increasing cross-sectional H.A. reduction.

Fig. 7.5 shows the D.I. for some of the features listed in Table 7.2 as a function of the sample number. Namely, the D.I. shown were computed from the following features: Var, RMS and Ppk of the wavelet coefficients (Figs. 7.5a, c and e), Max and Area of the Fourier transform of the reconstructed signal (Figs. 7.5b and d) and Area of the HT of the reconstructed signals (Fig. 7.5f). The general trend is similar for all the features presented; the D.I. decreases with increasing defect size. The first 1,600 points refer to the baseline while the subsequent 4,400 points refer to the damage conditions of the structure (testing data). Some of the feature show a liner decrease of D.I. with increasing defect size.

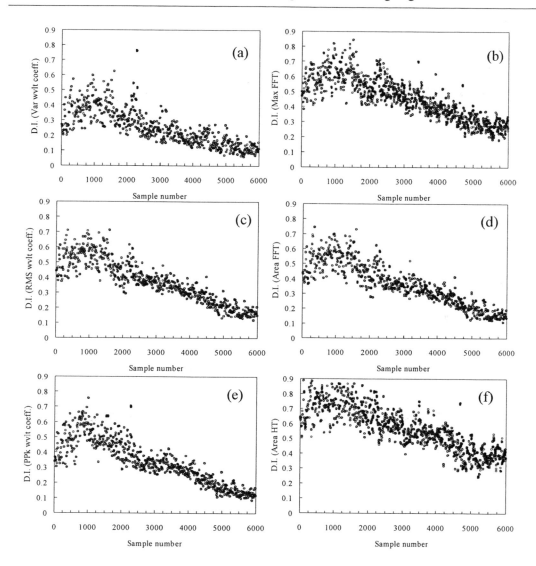

Figure 7.5. Damage Index as a function of the sample number for data corrupted with low-noise level and computed from (a) the variance of the wavelet coefficients, (b) the maximum amplitude of the FFT of the reconstructed signal, (c) the root-mean-square of the wavelet coefficients, (d) the area of the FFT spectrum of the reconstructed signal, (e) the peak-to-peak amplitude of the wavelet coefficients, (f) the area of the HT of the reconstructed signal.

The selection of the third DWT level only was led by the preliminary study of several plots as those shown in Fig. 7.5.

7.3.4. Univariate Analysis for Low Noise

This section presents the outlier analysis results when the selected features were considered separately.

From the values of the D.I. calculated from each feature, the discordancy test was carried out. Fig. 7.6 shows the deviation statistics z_ζ of the feature 2 (Table 7.2), i.e. the RMS of the retained wavelet coefficients, of all sample data for signals corrupted with the low noise level of 0.002. The discordancy values are plotted in semi-logarithmic scale in Fig. 7.6a and in linear scale in Fig. 7.6b. The mean and the standard deviation of all 1,600 D.I.s of the baseline were computed by the inclusive deviation for each of the baseline samples from Eq. (5.1). In Fig. 7.6, the horizontal lines represent the 99.73% confidence threshold. It can be seen that none of the baseline samples exceeds the threshold and thus classifies as an outlier.

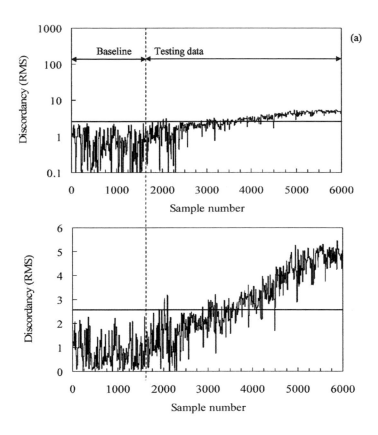

Figure 7.6. Discordancy test for the baseline (undamaged) and damaged rail data corrupted with the low level noise. Feature D.I. (RMS wavelet coefficients) as a function of the sample number: (a) semi-logarithmic scale; (b) linear scale.

For the damaged conditions, the exclusive method was adopted. It can be seen that a progressively increase of the discordancy is visible after sample number 3,000, corresponding to an H.A. reduction of only 2%. The majority of the discordancy values are above the threshold for sample numbers above 4,200 corresponding to defects larger than 5% H.A. reduction.

Table 7.3. summarizes the results for the univariate analysis applied to the ultrasonic signals corrupted with the low level noise; the table shows the number of outliers for each feature and for each defect sizes. The outliers detected for the undamaged structure (defect 0) are false positives; the outliers detected at the other structure conditions properly indicate a faulty condition. The last two rows of Table 7.3 report the total number of outliers (excluding those from the baseline) and the percentage of outliers properly recognized. It is evident that each feature performs differently. The features that perform better, i.e. provide the larger number of outliers, are the RMS of the wavelet coefficient vector, the RMS of the reconstructed signal, and the areas of the FFT and of the HT of the reconstructed signal. It is well known that these features are related to the energy content of the signal. Defects 6 and 7, equal or larger than 6% H.A. reduction, are detected in 100% of the cases (600 outliers out of 600 samples), which means 100% defect detection success rate. Some features provide 96.7% success for 5% H.A. reductions (defect 5). Unfortunately, many of the features are not able to detect defects below 2% H.A. reduction, reiterating the importance of the proper feature selection.

Table 7.3. Rail inspection. Outliers detected from the univariate discordancy test for each D.I. feature and for each defect type when the ultrasonic signals are corrupted with the low-level noise

Def.	Wavelet Coefficient				Reconstructed Signal				Fourier Transf.		Hilbert Transf.	
	Var	Rms	Max	Ppk	Var	Rms	Max	Ppk	Max	Area	Max	Area
0	10	0	0	10	0	0	10	10	2	10	10	10
1	10	32	0	0	0	32	50	30	0	20	40	26
2	0	10	0	0	0	10	0	0	0	20	0	40
3	0	230	0	20	0	230	0	0	39	242	0	179
4	75	499	0	91	87	499	228	139	225	402	160	217
5	250	580	208	530	440	580	537	498	430	560	500	413
6	329	600	549	590	590	600	570	570	572	600	570	575
7	540	600	580	600	600	600	600	600	600	600	600	595
Tot.	1204	2551	1337	1831	1717	2551	1985	1837	1866	2444	1870	2045
Tot. %	27.6 %	58.0 %	30.4 %	41.6 %	39.0 %	58.0 %	45.1 %	41.8 %	42.4 %	55.6 %	42.5 %	46.5 %

7.3.5. Multivariate Analysis for Low Noise

The features considered separately in the previous section were subsequently used simultaneously to construct a multi-dimensional D.I. vector for the outlier analysis. The deviation statistics is now replaced by the MSD of Eq. (5.3). As for the strand inspection, the purpose of combining features was to increase the sensitivity to the presence and the size of the damage compared to the single-feature analysis.

However the use of all twelve features may not be necessary; in addition, since some feature are less sensitive than others, as shown in Table 7.3, the selection of all twelve features may degrade the defect detection performance. To investigate this aspect, a parametric analysis was carried out. All of the features were considered ranging from all combinations of two-dimensional D.I. vectors to the single combination of a 12-dimensional D.I. vector. A total of 4,083 cases were analyzed.

As expected the performance of the multivariate analysis strongly depended on the selection of the features to construct the D.I.; the performance was quantified based upon the percentage of outliers detected among the testing data. This performance ranged from the best 82.1% success rate of the following feature combination (1, 2, 3, 5, 9) in Table 7.2 to the worst 30.5% success rate of the feature combination (3, 4, 8, 9, 10, 11, 12) in Table 7.2. The MSD from these combinations as a function of sample number of GUW corrupted with the low noise level is presented in Fig. 7.7a and b. In Fig. 7.7a no false negatives arise for defect sizes 6 and 7, while only 23 and 10 false negatives arise for defect sizes 4 and 5, which translate into 96.2% and 98.3% detection success rates. For the same defect sizes the percentage of successful detection (outliers among the testing data) given in the worst performance drops to 20%, 21%, 81.8% and 97.3% for defect sizes 4, 5, 6 and 7, respectively.

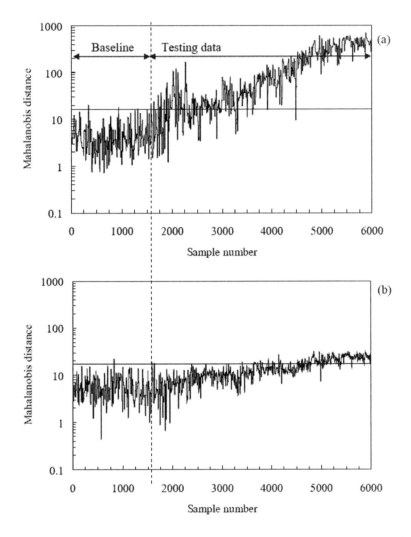

Figure 7.7. Mahalanobis squared distance as a function of the sample number computed from multi-dimensional feature-based Damage Index vector for the baseline (undamaged) and the testing rail data corrupted with the low level noise; (a) best feature combination; (b) worst feature combination.

7.3.6. Multivariate Analysis for Medium and High Noise

Following the same approach discussed in the previous section, Fig. 7.8a and b show the MSD results of the five-dimensional D.I. vectors extracted from the ultrasonic signals corrupted with the medium and high noise levels of 0.01 and 0.0625, respectively. Although not clearly visible in the logarithmic scale adopted in the plots, compared to the low noise results of Fig. 7.7a, the heavier noise corruption compromises the ability to detect small notches and, particularly, increases the number of false negatives. In addition, the sensitivity to the defect dimension is reduced, since the values of the MSD are smaller at the same sample number.

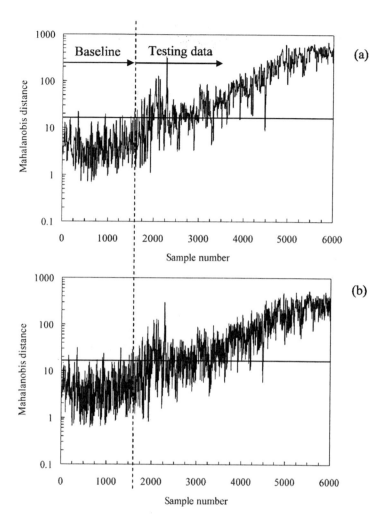

Figure 7.8. Mahalanobis squared distance as a function of the sample number; (a) best feature combination for medium-level noise; (b) best feature combination for high-level noise.

Table 7.4 summarizes the number of outliers detected in the multivariate analyses for the low and the high-noise level considered and for different selections of features. As before, the outliers are false positive indications for the baseline data (Defect 0) and, instead, correct indications of anomalies for the testing data (Defects 1-7). The best combination of features 1, 2, 3, 5, 9 as indicated in Table 7.2 scored the 1st, 2nd, and 1st best performance for the noise levels 0.002, 0.01 and 0.0625, respectively. As it can be seen in Table 7.4, the best combination was able to recognize damage conditions in over 82% of the cases, with 100% or closer success rate for H.A. reductions above 3%. Similar considerations can be carried out by looking at the results from the right column of Table 7.4 referring to the high noise case. As expected and already discussed, the overall performance of the algorithm is lower, even if the percentage of success for defects as small as 5% of H.A. reduction is still encouraging.

Table 7.4. Outliers detected from the multivariate analysis for the best D.I. feature combination and for each defect size when the ultrasonic signals are corrupted with low noise and high noise

Def	Best (Medium level noise)		Best (High level noise)	
	# outl.	% outl.	# outl.	% outl.
0	40	2.5%	31	1.9%
1	348	43.5%	291	36.4%
2	407	67.8%	273	45.5%
3	490	81.7%	453	75.5%
4	577	96.2%	552	92.0%
5	590	98.3%	587	97.8%
6	600	100.0%	600	100.0%
7	600	100.0%	600	100.0%
Tot.	**3612**		**3356**	
Tot. %	**82.1%**		**76.3%**	

7.4. Supervised Learning Algorithm

The supervised learning test was carried out on two 2100 mm (7 ft) long, 115-lb A.R.E.M.A. rail section also donated by San Diego Trolley, Inc.

The experimental setup was the same as before (Fig. 7.1b). The distance between the laser source and the first sensor (L_1) was 700 mm and the distance between the laser source and the second sensor (L_1+d) was 1200 mm; the defect was machined 1000 mm away from the laser source.

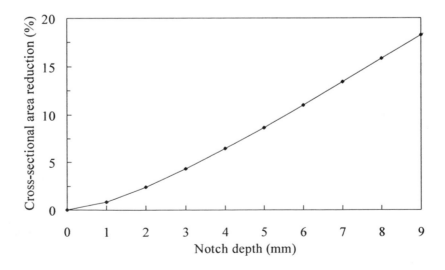

Figure 7.9. Supervised learning algorithm: reduction of the cross-sectional area of the railhead as a function of notch depth.

The transverse crack examined was machined by saw-cutting at increasing depths from a minimum of 0.5 mm to a maximum of 8.5 mm (s in Fig. 7.1b). The relation between the depth s (extended up to 9 mm) and the corresponding cross-sectional H.A. reduction is illustrated in Fig. 7.9; the largest crack depth resulted in a cross-sectional H.A. reduction of about 20%.

More than one wavelet level was considered here. Particularly, the following thresholds were imposed on the DWT decompositions of sensor # 1 and sensor # 2: 80% for level 3, 70% for level 4 and 80% for level 5; detail levels 1 and 2 were ignored. The following features were selected: Var, RMS, kurtosis, peak amplitude and Ppk amplitude of the thresholded wavelet coefficient vectors. The signals were then reconstructed from the corresponding wavelet coefficient vectors and the peak and the Ppk amplitudes were calculated. Additional features included the peak amplitude and the area below the FFT frequency spectrum within the frequency range 100 kHz – 900 kHz, the area below the HT and the peak amplitude of the HT, all of the reconstructed signals.

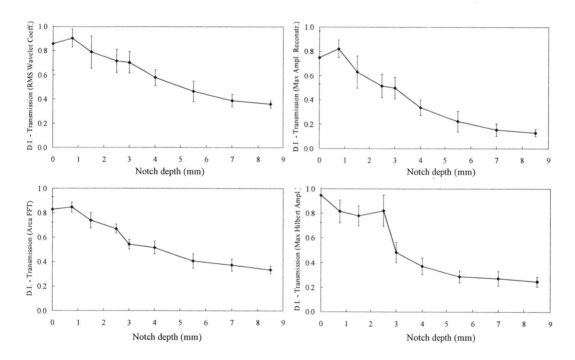

Figure 7.10. Damage Index as a function of the center crack depth for the supervised learning algorithm. (a) RMS of the wavelet coefficients; (b) peak amplitude of the wavelet coefficients; (c) area of the FFT spectrum of the reconstructed signal; (d) peak amplitude of the HT of the reconstructed signal.

Fig. 7.10 shows the "transmission" D.I. calculated from Eq. (4.2) using some of the selected features, namely the RMS of the wavelet coefficient vector (Fig. 7.9a), the peak amplitude of the reconstructed signal (Fig. 7.9b), the area of the Fourier transform of the reconstructed signal (Fig. 7.9c) and the peak amplitude of the HT of the reconstructed signal (Fig. 7.9d). The mean value of ten measurements is plotted as a function of crack depth and the vertical line is equal to $2\sigma_{dev}$. By observing the slope of the plots in Fig. 7.9, it can be seen that some features show a larger sensitivity to the crack depth (slope of the D.I. curves) than others.

In the pattern recognition algorithm the defect sizes were subdivided into three classes (Classes 1, 2 and 3), corresponding to % H.A. reduction in the ranges 0% - 1.1% (to be considered the pristine condition), 1.5% - 9.9%, and 10% - 20%. Each class was coded with a 2-digit binary number, namely 00 for class 1, 01 for class 2 and 10 for class 3.

Five of ten acquisitions for each damage condition were used as training data while the remaining data were used to test the network.

A parametric analysis was conducted in order to find the network design that optimizes network performance, i.e. obtaining the largest percentage of testing data correctly classified. In order to ensure the same initial conditions, the initial biases and weights of the network were kept constant. Table 7.5 summarizes the features selected and the network parameters considered for the optimization study.

Table 7.5. Rail inspection. Features and network parameters considered for the study of the optimal network design.

Feature #	D.I. feature	# of features	# of hidden neurons
1	Variance (WCV)	1 to 7	6 to 22
2	Root mean square (WCV)	step 1	step 2
3	Kurtosis (WCV)		
4	Peak amplitude (WCV)		
5	Peak amplitude (Rcstr)		
6	Area (FFT)		
7	Peak amplitude (HT)		

WCV: wavelet coefficient vector
Rcstr: Reconstructed signal
HT: Hilbert Transform of the reconstructed signal
FFT: Fast-Fourier Transform of the reconstructed signal

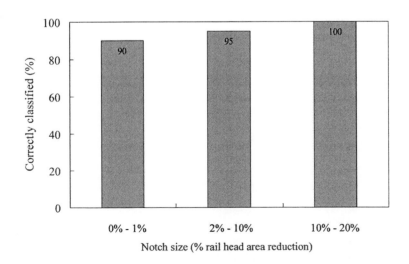

Figure 7.11. Rail defect classification performance of the optimized network design as a function of the size classes.

The learning rate and additional momentum were equal to 0.2 and 0.5, respectively. The number of hidden neurons varied between 6 and 22; all possible combinations of the selected features, varying from one feature only (mono-dimensional input vector) to all 7 features together, were considered. The percentage of correct classification varied between 95.6% (10 hidden neurons and four features) and 8.9% (18 hidden neurons and one feature only). The best success rate (95.6%) was accomplished by selecting the following four features: the Var, the RMS and the peak amplitude of the wavelet coefficient vector, and the area below the FFT of the reconstructed signal.

To visualize the performance of the *optimized* network design, the histograms in Fig. 7.11 are presented. The percentage of correct classification is plotted as a function of the three defect size classes.

Testing data belonging to Class 1 were properly classified in the 90% of the cases and the remaining 10% were associated with Class 2, which clusters the % H.A. reductions between 1.5% and 9.9%. Properly classifying the defect-free case means avoiding false positives. Data in Class 2 were properly classified in 95% of the cases; the remaining 5% of the cases was associated with Class 1, which means a false-negative classification that is the most critical since it does not recognize the presence of a defect. Finally, data from the largest defect Class 3 were properly classified in 100% of the cases.

8. Conclusion

This chapter presented a procedure for the health monitoring of structural components by means of guided ultrasonic waves. The procedure can be applied to many types of waveguides, such as rods, plates, shells, pipes, rails, which can all benefit from the advantages of guided wave probing. These advantages include medium to large inspection ranges and sensitivity to small defects. Structural monitoring is performed by detecting the interaction of the waves with the structural defects. This interaction can be detected either in a reflection mode or in a transmission mode, requiring different arrangements of the inspection probes relative to the defect. The key aspect of this research is the elaboration of the guided wave signals in order to extract robust damage-sensitive features, and the subsequent use of statistical pattern recognition algorithms able to detect and classify damage in the presence of environmental variability. Feature extraction is performed by the Discrete Wavelet Transform decomposition that isolates the relevant information on the defects and rejects noise. Specific features considered include various statistical parameters of the wavelet decompositions and of the de-noised reconstructions of the guided wave signals. These features are then assembled into a D.I. vector that takes a form specific to the defect detection mode adopted ("reflection" or "transmission"). The D.I. vector is used in an outlier analysis for defect detection and in an artificial neural network for defect classification. By increasing the dimension of the D.I. vector, the sensitivity to defect presence and size increases. The wavelet-based features confer large defect sensitivity while keeping the dimension of the vector small for computational efficiency.

The general procedure is demonstrated on two applications: the detection of notch-like defects in multi-wire strands used in civil structures and in railroad tracks. The first application uses permanently attached magnetostrictive coils to excite and detect the waves.

The second application uses a non-contact, hybrid system based on a pulsed laser and air-coupled sensors for high-speed, in motion ultrasonic inspection.

Acknowledgments

The research presented in this chapter was supported by the U.S. National Science Foundation grants CMS-0221707 and CMS-84249, by the U.S. Federal Railroad Administration grant# DTFR53-02-G-00011 and by a 2004 grant from the Von Liebig Center for Technology Advancement of the University of California, San Diego. The in-kind support of San Diego Trolley, Inc. is gratefully acknowledged for providing the rail sections used in the experiments. The contribution of Mrs Elisa Sorrivi at the University of Bologna, Italy to some of the research reported in section 6.5 is much appreciated. Mr. Ivan Bartoli at UC San Diego is acknowledged for preparing Figs. 6.2, 7.1, 7.2a and 7.2b.

List of Abbreviations

ANN	Artificial Neural Network
b.w.	Broken wire
c_g	Wave group velocity
c_L	Bulk longitudinal wave velocity
c_L^*, c_S^*	Complex bulk longitudinal and shear velocities
c_p	Wave phase velocity
c_T	Bulk shear wave velocity
coif 1	Coiflet mother wavelet of order 1
D.I.	Damage Index
D_ζ	Mahalanobis Squared Distance
db10, db40	Daubechies mother wavelet of order 10, 40
DWLF, DWHF	Analysis Discrete Wavelet Lowpass and Highpass Filter
DWLF', DWHF'	Synthesis Discrete Wavelet Lowpass and Highpass Filter
DWT	Discrete Wavelet Transform
E	Network error
EMAT	Electromagnetic Transducer
f	Wave frequency
F	Mother wavelet center frequency
FF BP	Feed Forward Back Propagation
FFT	Fast Fourier Transform
GUWs	Guided Ultrasonic Waves
HT	Hilbert Transform
[**K**]	Covariance Matrix
k	Wavenumber
J_β	Bessel function of order β
Max	Maximum
MSD	Mahalanobis Square Distance
MsS	Magnetostrictive sensor

n	DWT translation parameter
N_j	Number of wavelet coefficients at a detail level j
N_k	Shape functions
NDE	Non-destructive evaluation
P	Poynting vector
Ppk	Peak-to-peak
RMS	Root mean square
s	DWT scaling parameter
SHM	Structural Health Monitoring
SNR	Signal-to-Noise Ratio
U(r), V(r), W(r)	Wave amplitudes associated with the radial, angular and axial displacement along a solid cylinder
u_1, u_2, u_3	Displacement components
U_k, V_k, W_k	Amplitudes of the nodal displacement components
Var	Variance
V_e	Energy velocity
$\bar{x}$	Baseline mean
x_ζ	Potential outlier
$y_{kj}, \hat{y}_{kj}$	Network target and Network output
z_ζ	Deviation statistics
α_L, α_T	Longitudinal wave and shear wave attenuation
ε_{ij}	Strain components
η	ANN Learning rate
η_m	ANN Additional momentum
λ	Wavelength
λ, μ	Lame' constants
σ	Standard deviation
σ_{ij}	Stress components
ω	Circular frequency

References

Aalami, B. "Waves in Prismatic Guides of Arbitrary Cross Section" *J. Appl. Mech.* 1973, 40, 1067-1072.

Abbate, A.; Koay J.; Frankel, J.; Schroeder, S. C.; Das, P. "Application of Wavelet Transform Signal Processor to Ultrasound" *Proc. 1994 Ultrasonic Symposium* 1994, 94CH3468-6, 1147-1152.

Abbate, A.; Koay, J.; Frankel, J.; Schroeder, S. C.; Das, P. "Signal Detection and Noise Suppression Using a Wavelet Transform Signal Processor: Application to Ultrasonic Flaw Detection" *IEEE Trans. Ultrason. Ferroelectr. Freq. Control.* 1997, 44, 14-26.

Achenbach, J. D. Wave Propagation in Elastic Solids; North-Holland Pub. Co.: Amsterdam, The Netherlands, 1973.

Alleyne, D. N.; Pavlakovic, B.; Lowe, M. J. S.; Cawley, P. "Rapid Long-range Inspection of Chemical Plant Paperwork Using Guided Waves" *Insight* 2002, 43, 93-96.

Auld, B. A. Acoustic Fields and Waves in Solids; R.E. Krieger: Malabar, FL, 1990.

Barnett, V.; Lewis, T. Outliers in statistical data; John Wiley and Sons: New York, NY, 1994.

Bartoli, I.; Lanza di Scalea, F.; Fateh, M.; Viola, E. "Modeling Guided Wave Propagation with Application to the Long-Range Defect Detection in Railroad Tracks" *NDT&E Int.* 2005, 38, 325-334.

Beard, M. D. Guided Wave Inspection of Embedded Cylindrical Structures; PhD Dissertation, Imperial College: London, UK, 2002

Beard, M. D.; Lowe, M. J. S.; Cawley, P. "Ultrasonic Guided Waves for Inspection of Grouted Tendons and Bolts" *ASCE J. Mat. Civ. Engr.* 2003, 15, 212-218.

Bernard, A.; Lowe, M. J. S.; Deschamps, M. "Guided Waves Energy Velocity in Absorbing and Non-Absorbing Plates" *J. Acoust. Soc. Am.* 2001, 110, 186-196.

Cawley, P.; Lowe, M. J. S.; Alleyne, D.; Pavlakovic, B.; Wilcox, P. "Practical Long Range Guided Wave Testing: Applications to Pipes and Rail" *Mat. Eval.* 2003, 61, 66-74.

Chase, S. B. "Smarter Bridges, Why and How?" *Smart. Mat. Bull.* 2001, 2, 9-13.

Chen, H.-L.; Wissawapaisal, K. "Measurement of Tensile Forces in a Seven-wire Prestressing Strand Using Stress Waves" *ASCE J. Eng. Mech.* 2001, 127, 599-606.

Chen, H.-L.; Wissawapaisal, K. "Application of Wigner-Ville Transform to Evaluate Tensile Forces in Seven-wire Prestressing Strands" *ASCE J. Eng. Mech.* 2002, 128, 1206-1214.

Cho, S. H.; Kim, Y. K.; Kim, Y. Y. "The Optimal Design and Experimental Verification of the Bias Magnet Configuration of a Magnetostrictive Sensor for Bending Wave Measurement" *Sens. Actuators A Phys.* 2003, 107 (3), 225-232.

Chree, C. "The Equation of an Isotropic Elastic Solid in Polar and Cylindrical Coordinates, their Solutions and Application" *Trans. Cambridge Philos. Soc.* 1889, 14, 250-369.

Clark, R. "Rail Flaw Detection: Overview and Needs for Future Developments" *NDT&E Int.* 2004, 37, 111-118.

Davies, R. M. "A Critical Study of the Hopkinson Pressure Bar" *Philos. Trans. R. Soc. Lond.* 1948, 240, 375-457.

Federal Railroad Administration, *Safety Statistics Data*: 1992-2002, U.S. Department of Transportation, 2005.

Gan, T. H.; Hutchins, D. A.; Billson, D. R.; Schindel, D. W. "The Use of Boradband Acoustic Transducers and Pulse-Compression Techniques for Air-Coupled Ultrasonic Imaging" *Ultrasonics* 2001, 39, 181-194.

Gavrić, L. "Computation of Propagating Waves in Free Rail Using a Finite Element Technique" *J. Sound Vib.* 1995, 185(3), 531-543.

Godin, N.; Huguet, S.; Gaertner, R.; Salmon, L. "Clustering of Acoustic Emission Signals During Tensile Tests on Unidirectional Glass/Polyester Composite Using Supervised and Unsupervised Classifiers" *NDT&E Int.* 2004, 37, 253-264.

Graff, K. F. Wave Motion in Elastic Solids; *Dover Publications*, Inc.: New York, NY., 1991.

Guttormsson, S. E.; Marks II, R. J.; El-Sharkawi, M. A. "Elliptical Novelty Grouping for On-Line Short-Turn Detection Of Excited Running Rotors" *IEEE Trans. Energy Convers.* 1999, 14, 16-22.

Hayashi, T.; Song, W.-J.; Rose, J. L. "Guided Wave Dispersion Curves for a Bar With an Arbitrary Cross-Section, a Rod and Rail Example" *Ultrasonics* 2003, 41, 175-183.

Jensen, A.; La Cour-Harbo, A. Ripples in Mathematics: *The Discrete Wavelet Transform*; Springer: Berlin, Germany, 2001.

Joule, J. P. "On the Effect of Magnetism upon the Dimension of Iron and Steel Bars" *Philosophical Magazine* 1847, 30, 76-92.

Karlsson, G.; Vetterli, M. "Extension of Finite Length Signals for Sub-band Coding" *Signal Process.* 1989, 17, 161-168.

Kenderian, S.; Cerniglia, D.; Djordjevic, B. B.; Garcia, G.; Sun, J.; Snell, M. "Rail Track Field Testing Using Laser/Air Hybrid Ultrasonic Technique" *Mat. Eval.* 2003, 61, 1129-1133.

Kenderian, S.; Djordjevic, B. B.; Green, R. E.; Cerniglia, D. "Laser-Air, Hybrid, Ultrasonic Testing of Railroad Tracks" U.S. *Patent Application Publication* 2004/0003662, 2004.

Kim, K.-B.; Yoon, D.-J.; Jeong, J.-C.; Lee, S.-S. "Determining the Stress Intensity Factor of a Material With an Artificial Neural Network from Acoustic Emission Measurements" *NDT&E Int.* 2004, 37, 423-429.

Kim, H.; Melhem, H. "Damage Detection of Structures by Wavelet Analysis" *Eng. Struct.* 2004, 26 , 347-362.

Kim, Y. Y.; Park, C. Il; Cho, S. H.; Han, S. W. "Torsional Wave Experiments with a New Magnetostrictive Transducer Configuration" *J. Acoust. Soc. Am.* 2005, 117, 3459-3468.

Kolsky, H. Stress Waves in Solids; *Dover Publications* Inc.: New York, NY, 1963.

Krautkrämer, J.; Krautkrämer, H. Ultrasonic Testing of Materials; Springer-Verlag: Berlin, Germany, 1990.

Kwun, H.; Teller, C. M. "Detection of Fractured Wires in Steel Cables Using Magnetostrictive Sensors" *Mat. Eval.* 1994, 503-507.

Kwun, H.; Teller, C.M. "Nondestructive Evaluation of Steel Cables and Ropes Using Magnetostrictively Induced Ultrasonic Waves and Magnetostrictively Detected Acoustic Emissions", U.S. Patent No. 5,456,113, 1995

Kwun, H.; Bartels, K. A. "Magnetostrictive Sensor Technology and its Application" *Ultrasonics* 1998, 36, 171-178.

Kwun, H.; Bartels, K. A.; Hanley, J. J. "Effect of Tensile Loading on the Properties of Elastic-Wave in a Strand" *J. Acoust. Soc. Am.* 1998, 103(6), 3370-3375.

Kwun, H.; Kim, S. Y.; Light, G. M. "The Magnetostrictive Sensor Technology for Long Range Guided Wave Testing and Monitoring of Structures" *Mat. Eval.* 2003, 80-84.

Laguerre, L.; Aime, J.-C.; Brissaud, M. "Magnetostrictive Pulse-Echo Device for Non-Destructive Evaluation of Cylinder Steel Materials using Longitudinal Guided Waves" *Ultrasonics* 2002, 39, 503-514.

Lanza di Scalea, F.; Rizzo, P.; Seible, F. "Stress Measurement and Defect Detection in Steel Strands by Guided Stress Waves" *ASCE J. Mater. Civ. Eng.* 2003, 15, 219-227.

Lanza di Scalea, F.; McNamara, J. "Ultrasonic NDE of Railroad Tracks: Air-Coupled Cross-Sectional Inspection and Long-Range Inspection" *Insight,* 45, 394-401, 2003.

Lanza di Scalea, F.; Bartoli, I.; Rizzo, P.; Fateh, M. "High-Speed Defect Detection in Rails by Non-contact Guided Ultrasonic Testing" *Transp. Res. Record* no. 1961, *J. Transp. Res. Board* 2005, 66-77.

Lee, H.; Kim, Y. Y. "Wave Selection Using a Magnetomechanical Sensor in a Solid Cylinder" *J. Acoust. Soc. Am.* 2002, 112, 953-960.

Love, A. E. *Treatise on the Mathematical Theory of Elasticity*; Dover Publications: New York, NY, 1944.

Mallat, S. G. "A Theory for Multiresolution Signal Decompostion: The Wavelet Representation" *IEEE Trans. Pattern Anal. Mach. Intell.* 1989, 11, 674-93

Mallat, S. G. A Wavelet Tour of Signal Processing; Academic Press: New York, NY, 1999.

Manson, G.; Worden, K. "Experimental Validation of a Structural Health Monitoring Methodology. Part II. Novelty Detection on a Gnat Aircraft" *J. Sound Vib.* 2003, 259, 345-363.

McNamara, J.; Lanza di Scalea, F. "Improvements in Non-Contact Ultrasonic Testing of Rails by the Discrete Wavelet Transform" *Mat. Eval.* **62**, 2004, 365-372.

McNamara, J.; Lanza di Scalea, F.; Fateh, M. "Automatic Defect Classification in Long-Range Ultrasonic Rail Inspection Using a Support Vector Machine-based Smart System" *Insight* 2004, 46, 331-337.

Meitzler, A. H. "Mode Coupling Occurring in the Propagation of Elastic Pulses in Wires" *J. Acoust. Soc. Am.* 1961, 33, 435-445.

Mirsky, I. "Wave Propagation in Transversely Isotropic Cylinders Part 1: Theory" *J. Acoust. Soc. Am.* 1964a, 37, 1016-1021.

Mirsky, I. "Wave Propagation in Transversely Isotropic Cylinders Part 2: Numerical results" *J. Acoust. Soc. Am.* 1964b, 37, 1022-1026.

Morse, R. W. "Compressional Waves Along an Anisotropic Circular Cylinder having Haxagonal Symmetry" *J. Acoust. Soc. Am.* 1954, 26, 1018-1021.

Nagy, P. B. "Longitudinal Guided Wave Propagation in a Trasversely Isotropic Rod Immersed in Fluid" *J. Acoust. Soc. Am.* 1995, 98, 454-457.

Ogden, R. T. Essential Wavelets for Statistical Applications and Data Analysis; Birkhäuser: Boston, MA, 1997.

Omenzetter, P.; Brownjohn, J. M. W.; Moyo, P. "Identification of Unusual Events in Multi-Channel Bridge Monitoring Data" *Mech. Syst. Signal Process.* 2004, 18, 409-430.

Onoe, M.; McNiven, H. D.; Mindlin, R. D. "Dispersion of Axially Symmetric Waves in Elastic Solids" *J. Appl. Mech.* 1962, 29, 729-734.

Oukheloou, L.; Aknin, P.; Perrin, J.-P. "Dedicated Sensor and Classifier of Rail Head Defects" *Control Eng. Practice* 1999, 7, 57-61.

Paget, C. A.; Grondel, S.; Levin, K.; Delebarre, C. "Damage Assessment in Composites by Lamb Waves and Wavelet Coefficients" *Smart Mat. Struct.* 2003, 12, 393-402.

Pao, Y. H.; Mindlin, R. D. "Dispersion of Flexural Waves in an Elastic, Circular Cylinder" *J. Appl. Mech.* 1960, 27, 513-520.

Pao, Y. H. "The Dispersion of Flexural Waves in an Elastic, Circular Cylinder – Part 2" *J. Appl. Mech.* 1962, 29, 61-64.

Parker, D. "Pacific Bridge Collapse Throws up Doubt on Repair Method" *New Civil Engineer* 1996a, 3-4.

Parker, D. "Tropical Overload", *New Civil Engineer* 1996b, 18-21.

Pavlakovic, B.; Lowe, M.; Alleyne, D.; Cawley, P. "Disperse: a General Purpose Program for Creating Dispersion Curves" *Rev. Prog. Quant. Nondestruct. Eval.* 1997, 16a, 185-192.

Pavlakovic, B. N. "Leaky Guided Ultrasonic Waves in NDT" PhD Dissertation, Imperial College: London, UK, 1998.

Pavlakovic, B. N.; Lowe, M. J. S.; Cawley, P. "The Inspection of Tendons in Post-tensioned Concrete Using Guided Ultrasonic Waves" *Insight* 1999, 41, 446-452.

Pavlakovic, B. N.; Lowe, M. J. S.; Cawley, P. "High-frequency Low-loss Ultrasonic Modes in Imbedded Bars" *J. Appl. Mech.* 2001, 68, 67-75.

Pochhammer, J. "Ueber die Fortpflanzungsgeschwindigkeiten kleiner Schwingungen in einem unbegrenzten isotropen Kreiszylinder", *Journal fuer reine und angewandte Math* 1876, 81, 324-376.

Pohl, R.; Erhard, A.; Montag, H.-J.; Thomas, H.-M.; Wüstenberg, H. "NDT Techniques for Railroad Wheel and Gauge Corner Inspection" *NDT&E Int.* 2004, 37, 89-94.

Reis, H.; Ervin, B. L.; Kuchma, D. A.; Bernhard, J. T. "Estimation of Corrosion Damage in Steel Reinforced Mortar Using Guided Waves" *J. Press. Vess. Technol* 2005, Special Issue on the Nondestructive Evaluation of Pipeline and Vessel Structures, 127, 255-261.

Rizzo, P.; Lanza di Scalea, F. "Acoustic Emission Monitoring of Carbon-Fiber-Reinforced-Polymer Bridge Stay Cables in Large-Scale Testing" *Exp. Mech.* 2001, 41, 282-290.

Rizzo, P.; F. Lanza di Scalea, F. "Load Measurement and Health Monitoring in Cable Stays via Guided Wave Magnetostrictive Ultrasonics" *Mat. Eval.* 2004a, 62, 1057-1065.

Rizzo, P.; Lanza di Scalea, F. "Wave Propagation in Multi-Wire Strands by Wavelet-Based Laser Ultrasound" *Exp. Mech.* 2004b, 44, 407-415.

Rizzo, P.; Lanza di Scalea, F. "Ultrasonic Inspection of Multi-Wire Steel Strands with the Aid of the Wavelet Transform" *Smart Mat. Struct.* 2005, 14, 685-695.

Rizzo, P.; Lanza di Scalea, F. "Discrete Wavelet Transform for Enhancing Defect Detection in Strands by Guided Ultrasonic Waves" *Int. J. Struct. Health Monit.,* 2006, 5, 297-308.

Rizzo, P.; Bartoli, I.; Marzani, A.; Lanza di Scalea, F. "Defect Classification in Pipes by Neural Networks using Multiple Guided Ultrasonic Wave Features" *J. Press. Vess. Tech.* 2005, Special Issue on the Nondestructive Evaluation of Pipeline and Vessel Structures, 127, 294-303.

Rose, J. L. Ultrasonic Waves in Solid Media; Cambridge University Press: Cambridge, UK, 1999

Rose, J. L. "Standing on the Shoulders of Giants: An Example of Guided Wave Inspection" *Mat. Eval.* 2002, 60, 53-56.

Rose, J. L.; Avioli, M. J.; Mudge, P.; Sanderson, R. "Guided Wave Inspection Potential of Defects in Rail" *NDTandE Int.,* 37, 153-161, 2004.

Scruby, C.; Drain, L. Laser Ultrasonics: Techniques and Applications; Adam Hilger: New York, NY, 1990.

Staszewski, W.; Pierce, G.; Worden, K.; Philp, W.; Tomlinson, G.; Culshaw, B. "Wavelet Signal Processing for Enhanced Lamb-Wave Defect Detection in Composite Plates Using Optical Fiber Detection" *Opt. Eng.* 1997, 36, 1877-1888.

Staszewski, W. "Wavelet Based Compression and Feature Selection for Vibration Analysis" *J. Sound Vib.* 1998, 211, 735-760.

Staszewski, W. "Intelligent Signal Processing for Damage Detection in Composite Materials" *Comp. Sci. Tech.* 2002, 62, 941-950.

Staszewski, W.; Boller, C.; Tomlinson, G. Health Monitoring of Aerospace Structures; John Wiley and Sons: Munich, Germany, 2004.

Strang, G.; Nguyen, T. Wavelets and Filter Banks; *Wellesley-Cambridge Press*: Wellesley, MA, 1996.

Thurston, R. "Elastic Waves in Rods and Clad Rods" *J. Acoust. Soc. Am.* 1978, 64, 1-37.

Van Nevel, A.; DeFacio, B.; Neal, S. P. "An Application of Wavelet Signal Processing to Ultrasonic Nondestructive Evaluation" *Rev. Prog. Quant. Nondestruct. Eval.* 1996, 15, 733-740.

Villari, E. "Change of Magnetization by Tension and by Electric Current", *Ann. Physik* (Leipzig) 1865, 126, 87-122.

Vogt, T.; Lowe, M.; Cawley, P. "Cure Monitoring Using Ultrasonic Guided Waves in Wires" *J. Acoust. Soc. Am.* 2003, 114, 1303–1313.

Wakiwaka, H.; Mitamura, M. "New Magnetostrictive Type Torque Sensor for Steering Shaft", *Sens. Actuators A Phys.* 2001, 91, 103-106.

Washer, G. "The Acoustoelastic Effect in Prestressing Tendons", PhD Dissertation, Johns Hopkins University: Baltimore, MD, 2001.

Washer, G.; Green, R. E.; Pond, R. B. "Velocity Constants for Ultrasonic Stress Measurements in Prestressing Tendons" *Res. Nondes. Eval.* 2002, 14, 81-94.

Watson, S. C.; Stafford, D. "Cables in Trouble", *Civil Engineering* 1988, 58, 38-41.

Wilcox, P.; Evans, M.; Pavlakovic, B.; Alleyne, D.; Vine, K.; Cawley, P.; Lowe, M. J. S. "Guided Wave Testing of Rail" *Insight* 2003, 45, 413-420.

Woodward, R. J. "Collapse of Ynys-y-Gwas Bridge, West Glamorgan" Proc. - Institution of Civil Engineers 1988, 84(1), 635-669.

Worden, K. "Structural Fault Detection Using a Novelty Measure" *J. Sound Vib.* 1997, 201, 85-101.

Worden, K.; Manson, G.; Fieller, N. R. J. "Damage Detection Using Outlier Analysis" *J. Sound Vib.* 2000a, 229, 647-667.

Worden, K.; Pierce, S. G.; Manson, G.; Philp, W. R.; Staszewski, W.; Culshaw, B. "Detection of Defects in Composite Plates Using Lamb Waves and Novelty Detection", *Int. J. Syst. Sci.* 2000b, 31, 1397-1409.

Worden, K.; Sohn, H.; Farrar, C. R. "Novelty Detection In a Changing Environment: Regression and Interpolation Approaches" *J. Sound Vib.* 2002, 258, 741-761.

Xu, P. C.; Datta, S. K. "Characterization of Fibre-Matrix Interface by Guided Waves: Axisymmetric Case" *J. Acoust. Soc. Am.* 1991, 89, 2573-2583.

Yam, L. H.; Yan, Y. J.; Cheng, L.; Jiang, J. S. "Identification of Complex Crack Damage for Honeycomb Sandwich Plate Using Wavelet Analysis and Neural Networks," *Smart Mater. Struct.* 2003, 12, 661-671.

Zang, C.; Imregun, M. "Structural Damage Detection Using Artificial Neural Networks and Measured FRF Data Reduced via Principal Component Projection" *J. Sound Vib.* 2001, 242, 813-827.

Zemanek, J. "An Experimental and Theoretical Investigation of Elastic Wave Propagation in a Cylinder" *J. Acoust. Soc. Am.* 1971, 51(1), 265-283.

In: Progress in Smart Materials and Structures
Editor: Peter L. Reece, pp. 291-361

ISBN : 1-60021-106-2
© 2007 Nova Science Publishers, Inc.

Chapter 9

MICROMECHANICAL ANALYSES OF SMART COMPOSITE MATERIALS

Jacob Aboudi[*]
Department of Solid Mechanics, Materials & Systems,
Faculty of Engineering, Tel-Aviv University,
Ramat-Aviv 69978, Israel

Abstract

By embedding a smart material in a polymeric and/or metallic constituent, a smart composite is obtained which can be utilized in various technical applications. In order to predict the behavior of this composite, a micromechanical analysis is performed which provides the overall (global) response from the known constitutive relations of the individual phases, their material properties, volume ratios, and by considering their detailed interaction. In this review chapter, micromechanical analyses of multiphase materials are presented which are capable of establishing the global constitutive relations of various types of smart composites which possess periodic microstructure. These include composites with piezoelectric, piezomagnetic, electrostrictive, magnetostrictive and shape-memory alloy phases. Both linear, nonlinear and large deformation analyses are presented. In addition, polymeric and metallic matrix composites are considered, where in the latter case, the inelastic behavior of the elastoplastic or viscoplastic constituent must be taken into account. Furthermore, bounded composites with arbitrarily distributed embedded smart phases (piezoelectric, shape-memory alloy, electrorheological, magnetorheological and fiber optic) are analyzed. Finally, the resulting micromechanically established global constitutive relations are employed to predict the dynamic behavior and buckling of smart composite plates which form a type of smart composite structure.

1. Introduction

Smart, intelligent and adaptive materials and structures are capable of sensing external conditions and, as a result, can respond in a way that improves the performance of the system in

---

[*]E-mail address:aboudi@eng.tau.ac.il

which they are incorporated. Thus, smart materials are capable to sense and actuate according to the desired task that they are supposed to perform. Detailed discussions of various types of smart materials and structures can be found in the books by [1], [2], [3], [4], [5], [6], and recent review articles by [7] and [8], for example.

When a smart material is embedded as a phase in a matrix (e.g., piezoelectric fibers in a polymeric matrix), a smart composite material is obtained, which often has a better performance than the monolithic smart constituent. In order to predict the behavior of the resulting smart composite (e.g., its effective stiffness tensor), a micromechanics analysis has to be performed which can determine the composite's response from the knowledge of the properties of the individual constituents, their relative volume ratios and by considering their detailed interaction. In the present review chapter, a recently developed micromechanics analysis referred to as "High-Fidelity Generalized Method of Cells" (HFGMC) is presented, which is capable of predicting the response of many types of multiphase smart composites by establishing their global (macroscopic) constitutive equations. This analysis is based on the homogenization procedure for composites with periodic microstructure. Once global constitutive relations of a smart multiphase material have been developed, it is possible to utilize these equations for the analysis of a corresponding smart composite structure (e.g., a laminated plate or shell, the layers of which are smart composites). In addition, micromechanics analyses are presented for bounded composites with distinct boundaries that include arbitrarily located smart material phases. Such analyses provide the computed field variables at any desired location. Both HFGMC and its predecessor model, the "Generalized Method of Cells" (GMC), were recently reviewed by [9].

This review chapter starts in Section 2. with a presentation of the micromechanical HFGMC analysis for linear thermoelastic composites. As a result, the effective elastic stiffness and thermal stress tensors are established. The reliability of this prediction is verified by comparison with other approaches. By generalizing this analysis in Section 3., it is readily possible to predict the effective properties of electro-magneto-thermo-elastic composites and the effective thermal expansion, pyroelectric and pyromagnetic coefficients. Piezoelectric and piezomagnetic composites form special cases of the presented general analysis. In Section 4. the hysteresis behavior of ferroelectric composites is predicted. Due to the nonlinear constitutive equations of the monolithic ferroelectric phase, the micromechanical analysis is necessarily nonlinear, yielding incremental macroscopic constitutive relations. As a result, the global constitutive relations of the ferroelectric composite appear in an incremental form. This is followed by Section 5. which deals with electrostrictive composites. The response of the monolithic electrostrictive phase is nonlinear, hence (as in Section 4.), the micromechanical analysis is incremental, as well as the resulting global constitutive equations. Section 6. is concerned with magnetostrictive composites in which the nonlinear constitutive relations of the monolithic magnetostrictive phase are taken to be quite similar to those that represent the electrostrictive phase that was dealt with in Section 5.. Section 7. generalizes the linear electro-magneto-thermo-elastic behavior of Section 3. to a nonlinear one. Here the constitutive equations of the monolithic phase are described by nonlinear electro-magneto-thermo-elastic relations that can be established by expanding the Gibbs potential to a second order. Due to this nonlinearity, the micromechanically established macroscopic constitutive relations of the composite appear, as expected, in an incremental form. In Section 8., the HFGMC analysis is further generalized to model the inelastic

behavior of composite materials that consist of inelastic phases (e.g., graphite fibers reinforcing a metallic matrix). The reliability of the predicted inelastic composite response is
verified by comparisons with analytical solutions that can be established under special circumstances and with a finite element solution. This Section is a prelude to the analysis in
Section 9. of shape-memory alloy fibers embedded in a matrix, thus forming an inelastic
composite material. This is because the transformation strain of the shape-memory alloy
phase plays the same role as the inelastic strain of a conventional metallic phase. Section
10. discusses the modeling of shape-memory alloy fibers undergoing large deformation. By
generalizing the HFGMC model to finite strain analysis, it is possible to predict the large
deformation of a composite that is composed of shape-memory alloy fibers reinforcing a
metallic or a rubber-like matrix.

Thus far, micromechanical analyses were applied on composites with periodic microstructure extending over an infinite domain. It is possible to develop a micromechanics
analysis that is capable of predicting the behavior of bounded composites with arbitrarily
distributed embedded smart materials, see Section 11.. Such an approach was previously
developed for the analysis of functionally graded materials [10] and it is presently generalized to predict the behavior of composites with embedded piezoelectric patches, shape-
memory alloy phases, electrorheological inclusions and optical fibers. It is obvious that
this same approach can be followed to analyze bounded composites with other types of
embedded smart inclusions.

The macroscopic (global) constitutive equations that have been established by a suitable micromechanical analysis (e.g., HFGMC) can be utilized to analyze smart composite
structures (e.g., plates and shells). This is illustrated in Section 12. where the theory of laminated composite plates is employed to predict the dynamic response and thermal buckling
of active plates with shape-memory alloy fibers.

2.   The High-Fidelity Generalized Method of Cells Micromechanical Model

The specific micromechanical model which will be employed in this chapter to predict
the response of smart composite materials with periodic microstructure is the HFGMC. In
the following, this model is presented and illustrated in the simplest case of thermoelastic phases. The presentation follows ref. [11] for the most general case of discontinuous
reinforcement (triply-periodic). Composites with continuous fibers (doubly-periodic) are
obtained as a special case and were discussed by [12]. The HFGMC model assumes that
the multiphase material possesses a periodic microstructure, such that a repeating unit cell
can be identified. It is based on the homogenization technique, which provides the correct boundary conditions that must be applied to the repeating unit cell that represents the
material's periodic microstructure under multiaxial loading.

2.1.   The Homogenization Procedure

Consider a multiphase composite in which the microstructures are distributed periodically
in the space that is given with respect to the global coordinates (x_1, x_2, x_3), see Fig.1(a).

Figure 1(b) shows the repeating unit cell of the periodic composite. In the framework of the homogenization method the displacements are asymptotically expanded as follows:

$$u_i(\mathbf{x}, \mathbf{y}) = u_{0i}(\mathbf{x}, \mathbf{y}) + \delta\, u_{1i}(\mathbf{x}, \mathbf{y}) + \dots \tag{2.1}$$

where $\mathbf{x} = (x_1, x_2, x_3)$ are the macroscopic (global) coordinate system, and $\mathbf{y} = (y_1, y_2, y_3)$ are the microscopic (local) coordinates that are defined with respect to the repeating unit cell. The size of the unit cell is further assumed to be much smaller than the size of the body, so that the relation between the global and local systems is:

$$y_i = \frac{x_i}{\delta} \tag{2.2}$$

where δ is a small scaling parameter characterizing the size of the unit cell. This implies that a movement of order unity on the local scale corresponds to a very small movement on the global scale.

The homogenization method is applied to composites with periodic microstructures. Thus:

$$u_{\alpha i}(\mathbf{x}, \mathbf{y}) = u_{\alpha i}(\mathbf{x}, \mathbf{y} + n_p \mathbf{d}_p) \tag{2.3}$$

with $\alpha = 0, 1, \dots$, where n_p are arbitrary integer numbers and the constant vectors $\mathbf{d}_p$ determine the period of the structure.

Due to the change of coordinates from the global to the local systems the following relation must be employed in evaluating the derivative of a field quantity:

$$\frac{\partial}{\partial x_i} \rightarrow \frac{\partial}{\partial x_i} + \frac{1}{\delta}\frac{\partial}{\partial y_i} \tag{2.4}$$

The quantities u_{0i} are the displacements in the homogenized region and hence they are not functions of y_i.

Let:

$$u_{0i} = u_{0i}(\mathbf{x}) \equiv \bar{u}_i \tag{2.5}$$

and

$$u_{1i} \equiv \tilde{u}_i(\mathbf{x}, \mathbf{y}) \tag{2.6}$$

where the latter are the fluctuating displacements which are unknown periodic functions. These displacements arise due to the heterogeneity of the medium.

The strain components are determined from the displacement expansion (2.1) yielding, in conjunction with Eq. (2.4), the following expression:

$$\epsilon_{ij} = \bar{\epsilon}_{ij}(\mathbf{x}) + \tilde{\epsilon}_{ij}(\mathbf{x}, \mathbf{y}) + O(\delta) \tag{2.7}$$

where

$$\bar{\epsilon}_{ij}(\mathbf{x}) = \frac{1}{2}\left(\frac{\partial \bar{u}_i}{\partial x_j} + \frac{\partial \bar{u}_j}{\partial x_i}\right) \tag{2.8}$$

and

$$\tilde{\epsilon}_{ij}(\mathbf{x}, \mathbf{y}) = \frac{1}{2}\left(\frac{\partial \tilde{u}_i}{\partial y_j} + \frac{\partial \tilde{u}_j}{\partial y_i}\right) \tag{2.9}$$

This shows that the strain components can be represented as a sum of the average strain $\bar{\epsilon}_{ij}(\mathbf{x})$ in the composite and a fluctuating strain $\tilde{\epsilon}_{ij}(\mathbf{x}, \mathbf{y})$. It can be easily shown that:

$$\frac{1}{V_y} \int \epsilon_{ij} dV_y = \frac{1}{V_y} \int (\bar{\epsilon}_{ij} + \tilde{\epsilon}_{ij}) dV_y = \bar{\epsilon}_{ij}$$

where V_y is the volume of the repeating unit cell. This follows directly from the periodicity of the fluctuating strain, implying that the average of the fluctuating strain taken over the unit repeating cell vanishes. For a homogeneous material, it is obvious that the fluctuating displacements and strains identically vanish.

Using Eq. (2.7), one can readily represent the displacements in the form:

$$u_i(\mathbf{x}, \mathbf{y}) = \bar{\epsilon}_{ij} x_j + \tilde{u}_i + O(\delta^2) \tag{2.10}$$

For an elastic material, the stresses are related to the strains according to the Hooke's law as follows:

$$\sigma_{ij} = C_{ijkl} \, \epsilon_{kl} \tag{2.11}$$

where $C_{ijkl}(\mathbf{x})$ are the components of the stiffness tensor of the composite's phases. The stiffness tensor forms a periodic function that is defined in the unit repeating cell in terms of the local coordinates $\mathbf{y}$ such that:

$$C_{ijkl}(\mathbf{x}) = C_{ijkl}(\mathbf{y}) \tag{2.12}$$

Substituting Eq. (2.7) into (2.11) and differentiating with respect to the microvariable coordinates y_i leads to:

$$\frac{\partial}{\partial y_j} \left\{ C_{ijkl}(\mathbf{y}) \left[\bar{\epsilon}_{kl}(\mathbf{x}) + \tilde{\epsilon}_{kl}(\mathbf{x}, \mathbf{y}) \right] \right\} = 0 \tag{2.13}$$

Let us define the following stress quantities

$$\sigma_{ij}^0 = C_{ijkl}(\mathbf{y}) \, \bar{\epsilon}_{kl}(\mathbf{x}) \tag{2.14}$$

$$\sigma_{ij}^1 = C_{ijkl}(\mathbf{y}) \, \tilde{\epsilon}_{kl}(\mathbf{x}, \mathbf{y}) \tag{2.15}$$

with the latter being the fluctuating stresses.
It follows that:

$$\frac{\partial \sigma_{ij}^1}{\partial y_j} + \frac{\partial \sigma_{ij}^0}{\partial y_j} = 0 \tag{2.16}$$

which are the strong form of the equilibrium equations. It is readily seen that the first term in Eq. (2.16) involves the unknown fluctuating periodic displacements $\tilde{u}_i$, while the second term produces pseudo-body forces whose derivatives are actually zero everywhere, except at the interfaces between the phases.

For given values of the average strains $\bar{\epsilon}_{kl}$, the unknown fluctuating displacement are governed by Eqs. (2.16), subject to periodic boundary conditions that are prescribed at the boundaries of the repeating unit cell. In addition to these boundary conditions, one needs to impose the continuity of displacements and tractions at the internal interfaces between the phases that fill the repeating unit cell.

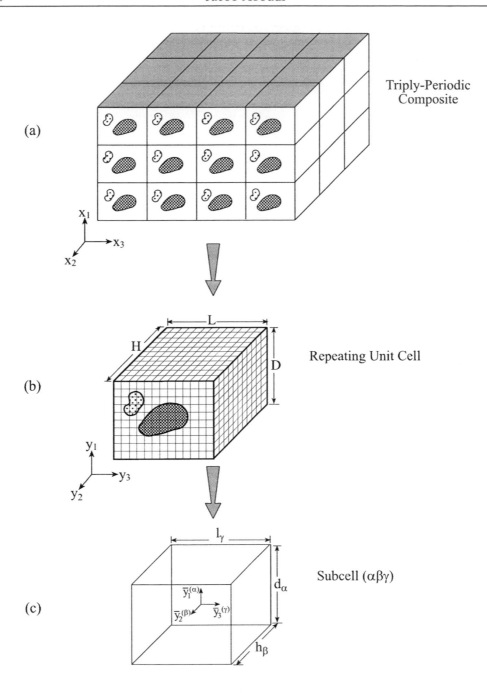

Figure 1. (a) A multiphase composite with triply-periodic microstructures defined with respect to global coordinates (x_1, x_2, x_3). (b) The repeating unit cell is represented with respect to local coordinates (y_1, y_2, y_3). It is divided into N_α, N_β and N_γ subcells, in the y_1, y_2 and y_3 directions, respectively. (c) A characteristic subcell $(\alpha\beta\gamma)$ with local coordinates $\bar{y}_1^{(\alpha)}, \bar{y}_2^{(\beta)}$ and $\bar{y}_3^{(\gamma)}$ whose origin is located at its center.

Referring to Fig. 1(b), the repeating unit cell is given by a parallelepiped defined with respect to the local coordinates by $0 \leq y_1 \leq D, 0 \leq y_2 \leq H, 0 \leq y_3 \leq L$. Consequently, the periodic boundary conditions are given by

$$\tilde{u}_i(y_1 = 0) = \tilde{u}_i(y_1 = D) \tag{2.17}$$

$$\sigma_{1i}(y_1 = 0) = \sigma_{1i}(y_1 = D) \tag{2.18}$$

$$\tilde{u}_i(y_2 = 0) = \tilde{u}_i(y_2 = H) \tag{2.19}$$

$$\sigma_{2i}(y_2 = 0) = \sigma_{2i}(y_2 = H) \tag{2.20}$$

$$\tilde{u}_i(y_3 = 0) = \tilde{u}_i(y_3 = L) \tag{2.21}$$

$$\sigma_{3i}(y_3 = 0) = \sigma_{3i}(y_3 = L) \tag{2.22}$$

where the total stress, which is given by Eq. (2.11), is expressed as:

$$\sigma_{ij} = \sigma_{ij}^0 + \sigma_{ij}^1 \tag{2.23}$$

It is also necessary to fix the displacement field at a point in the repeating unit cell (e.g. at a corner).

Once a solution of Eqs. (2.16), subject to the internal interfacial conditions and periodic boundary conditions (2.17)-(2.22) has been established, one can proceed and determine the strain concentrations tensor associated with the defined repeating unit cell. This tensor expresses the local strain in the cell in terms of the global applied external strain (localization). To this end, let us define the 4th-order tensor $\tilde{\mathbf{A}}$ as follows:

$$\tilde{\epsilon} = \tilde{\mathbf{A}}(\mathbf{y}) \, \bar{\epsilon} \tag{2.24}$$

It relates the fluctuating strain to the applied average strain. By using Eq. (2.7), we readily obtain the requested mechanical strain concentration tensor $\mathbf{A}^M(\mathbf{y})$ as follows

$$\epsilon = \bar{\epsilon} + \tilde{\mathbf{A}}(\mathbf{y}) \, \bar{\epsilon} = [\mathbf{I}_4 + \tilde{\mathbf{A}}(\mathbf{y})] \, \bar{\epsilon} \equiv \mathbf{A}^M(\mathbf{y}) \, \bar{\epsilon} \tag{2.25}$$

where $\mathbf{I}_4$ is the 4th-order identity tensor.

To construct the mechanical strain concentration tensor $\mathbf{A}^M(\mathbf{y})$, a series of problems must be solved as follows. Solve Eqs. (2.16) in conjunction with the internal interfacial and periodic boundary conditions with $\bar{\epsilon}_{11} = 1$ and all other components being equal to zero. The solution of Eqs. (2.16) readily provides A_{ij11}^M for $i, j = 1, 2, 3$. This procedure is repeated with $\bar{\epsilon}_{22} = 1$ and all other components equal to zero, which provides A_{ij22}^M, and so on.

Once the mechanical strain concentration tensor $\mathbf{A}^M(\mathbf{y})$ has been determined, it is possible to compute the effective stiffness tensor of the multiphase composite as follows. Substitution of ϵ given by Eq. (2.25) in (2.11) yields:

$$\sigma = \mathbf{C}(\mathbf{y}) \, \mathbf{A}^M(\mathbf{y}) \, \bar{\epsilon} \tag{2.26}$$

Taking the average of both sides of Eq. (2.26) over the repeating unit cell yields the average stress in the composite in terms of the average strain via the effective elastic stiffness tensor $\mathbf{C}^*$, namely:

$$\bar{\sigma} = \mathbf{C}^* \bar{\epsilon} \tag{2.27}$$

where

$$\mathbf{C}^* = \frac{1}{V_y} \int \mathbf{C}(\mathbf{y}) \, \mathbf{A}^M(\mathbf{y}) \, dV_y \tag{2.28}$$

2.2. Method of Solution of the Repeating Unit Cell

The previous analysis has been presented for a multiphase composite with elastic phases in which HFGMC has been developed to analyze thermoelastic composites. An approximate solution for the displacement field is constructed based on volumetric averaging of the field equations, together with the imposition of the periodic boundary conditions and continuity conditions in an average sense between the subcells used to characterize the materials' microstructure. This is accomplished by dividing the repeating unit cell of Fig. 1(b) into N_α, N_β and N_γ subcells in the y_1, y_2 and y_3 directions, respectively. Each subcell is labeled by the indices $(\alpha\beta\gamma)$ with $\alpha = 1, ..., N_\alpha$; $\beta = 1, ..., N_\beta$ and $\gamma = 1, ..., N_\gamma$, and may contain a distinct homogeneous material. The dimensions of subcell $(\alpha\beta\gamma)$, Fig. 1(c), in the y_1, y_2 and y_3 directions are denoted by d_α, h_β and l_γ, respectively. A local coordinate system $(\bar{y}_1^{(\alpha)}, \bar{y}_2^{(\beta)}, \bar{y}_3^{(\gamma)})$ is introduced in each subcell whose origin is located at its center. The approximate fluctuating displacement field in each subcell of the repeating unit cell of Fig. (1c) is represented as a quadratic expansion in terms of local coordinates $(\bar{y}_1^{(\alpha)}, \bar{y}_2^{(\beta)}, \bar{y}_3^{(\gamma)})$ centered at the subcell's midpoint. A higher order representation of the fluctuating field is necessary in order to capture the local effects created by the field gradients and the microstructure of the composite. Thus, the 2nd-order expansion of the displacement vector $\mathbf{u}^{(\alpha\beta\gamma)}$ in the subcell is given by

$$
\begin{aligned}
\mathbf{u}^{(\alpha\beta\gamma)} \;=\;& \bar{\boldsymbol{\epsilon}} \cdot \mathbf{x} + \mathbf{W}_{(000)}^{(\alpha\beta\gamma)} + \bar{y}_1^{(\alpha)}\mathbf{W}_{(100)}^{(\alpha\beta\gamma)} + \bar{y}_2^{(\beta)}\mathbf{W}_{(010)}^{(\alpha\beta\gamma)} + \bar{y}_3^{(\gamma)}\mathbf{W}_{(001)}^{(\alpha\beta\gamma)} \\
&+ \frac{1}{2}\left(3\bar{y}_1^{(\alpha)2} - \frac{d_\alpha^2}{4}\right)\mathbf{W}_{(200)}^{(\alpha\beta\gamma)} + \frac{1}{2}\left(3\bar{y}_2^{(\beta)2} - \frac{h_\beta^2}{4}\right)\mathbf{W}_{(020)}^{(\alpha\beta\gamma)} \\
&+ \frac{1}{2}\left(3\bar{y}_3^{(\gamma)2} - \frac{l_\gamma^2}{4}\right)\mathbf{W}_{(002)}^{(\alpha\beta\gamma)}
\end{aligned}
\tag{2.29}
$$

where $\mathbf{W}_{(000)}^{(\alpha\beta\gamma)}$, which is the fluctuating volume-averaged displacement vector. $\mathbf{W}_{(000)}^{(\alpha\beta\gamma)}$ and the higher-order terms $\mathbf{W}_{(lmn)}^{(\alpha\beta\gamma)}$ must be determined from the coupled governing equations (2.16), as well as the periodic boundary conditions (2.17)-(2.22), that the fluctuating field must fulfill, in conjunction with the interfacial continuity conditions between all subcells of the repeating unit cell. It should be emphasized that all these conditions are imposed in the average (integral) sense. For example, Eq. (2.17) is imposed in the framework of HFGMC as follows:

$$
\begin{aligned}
&\int_{-h_\beta/2}^{h_\beta/2}\int_{-l_\gamma/2}^{l_\gamma/2}\left[\tilde{u}_i^{(1\beta\gamma)}\Big|_{\bar{y}_1^{(1)}=-d_1/2}\right]d\bar{y}_2^{(\beta)}d\bar{y}_3^{(\gamma)} = \\
&\int_{-h_\beta/2}^{h_\beta/2}\int_{-l_\gamma/2}^{l_\gamma/2}\left[\tilde{u}_i^{(N_\alpha\beta\gamma)}\Big|_{\bar{y}_1^{(N_\alpha)}=d_{N_\alpha}/2}\right]d\bar{y}_2^{(\beta)}d\bar{y}_3^{(\gamma)}
\end{aligned}
$$

for $\beta = 1, ..., N_\beta$ and $\gamma = 1, ..., N_\gamma$.

The total number of unknowns that describe the fluctuating field in the subcell $(\alpha\beta\gamma)$ is 21. Consequently, the governing equations for the interior and boundary cells form a system of $21N_\alpha N_\beta N_\gamma$ algebraic equations in the unknown field coefficients that appear in the quadratic expansions (2.29).

The final form of this system of equations can be symbolically represented by

$$\mathbf{KU} = \mathbf{f} \tag{2.30}$$

where the structural stiffness matrix $\mathbf{K}$ contains information on the geometry and mechanical properties of the materials within the individual subcells $(\alpha\beta\gamma)$ within the cells comprising the multiphase periodic composite. The displacement vector $\mathbf{U}$ contains the unknown displacement coefficients in each subcell that appear on the right-hand-side of Eq. (2.29). The force $\mathbf{f}$ contains information on the thermomechanical properties of the materials filling the subcells, the applied average strains $\bar{\epsilon}_{ij}$ and the imposed temperature deviation ΔT.

The solution of Eq. (2.30) enables the establishment of the following localization relation which expresses the average strain $\bar{\epsilon}^{(\alpha\beta\gamma)}$ in the subcell $(\alpha\beta\gamma)$ in terms of the external applied strain $\bar{\epsilon}$ in the form:

$$\bar{\epsilon}^{(\alpha\beta\gamma)} = \mathbf{A}^{M(\alpha\beta\gamma)}\bar{\epsilon} + \mathbf{A}^{T(\alpha\beta\gamma)} \tag{2.31}$$

where $\mathbf{A}^{M(\alpha\beta\gamma)}$ is the mechanical strain concentration matrix of the subcell $(\alpha\beta\gamma)$, and $\mathbf{A}^{T(\alpha\beta\gamma)}$ is a vector that involves the current thermoelastic effects in the subcell.

The final form of the effective constitutive law of the multiphase thermo-inelastic composite, which relates the average stress $\bar{\sigma}$ and strain $\bar{\epsilon}$, is established as follows:

$$\bar{\sigma} = \mathbf{C}^* \bar{\epsilon} - \mathbf{\Gamma}^* \Delta T \tag{2.32}$$

In this equation, $\mathbf{C}^*$ is the effective elastic stiffness matrix of the composite which is given by the closed-form expression:

$$\mathbf{C}^* = \frac{1}{DHL} \sum_{\alpha=1}^{N_\alpha} \sum_{\beta=1}^{N_\beta} \sum_{\gamma=1}^{N_\gamma} d_\alpha h_\beta l_\gamma \, \mathbf{C}^{(\alpha\beta\gamma)} \, \mathbf{A}^{M(\alpha\beta\gamma)} \tag{2.33}$$

where $\mathbf{C}^{(\alpha\beta\gamma)}$ is the stiffness tensor of the material filling subcell $(\alpha\beta\gamma)$. In addition, $\mathbf{\Gamma}^*$ and ΔT denote the effective thermal stress and the temperature deviation from a reference temperature, respectively.

Let $\mathbf{\Gamma}^{(\alpha\beta\gamma)}$ denote the thermal stress vector of the material filling the subcell $(\alpha\beta\gamma)$. It is given by the product of the stiffness tensor $\mathbf{C}^{(\alpha\beta\gamma)}$ of the material by its coefficient of thermal expansion. The effective thermal stress vector in Eq. (2.32) $\mathbf{\Gamma}^*$ can be determined from Levin's theorem [13], which directly provides the effective thermal stress vector $\mathbf{\Gamma}^*$ in terms of the individual thermal stress vectors $\mathbf{\Gamma}^{(\alpha\beta\gamma)}$ of the phases and the mechanical concentrations matrices $\mathbf{A}^{M(\alpha\beta\gamma)}$, as follows:

$$\mathbf{\Gamma}^* = \frac{1}{DHL} \sum_{\alpha=1}^{N_\alpha} \sum_{\beta=1}^{N_\beta} \sum_{\gamma=1}^{N_\gamma} d_\alpha h_\beta l_\gamma [\mathbf{A}^{M(\alpha\beta\gamma)}]^{tr} \mathbf{\Gamma}^{(\alpha\beta\gamma)} \tag{2.34}$$

where $[\mathbf{A}^{M(\alpha\beta\gamma)}]^{tr}$ denotes the transpose of $\mathbf{A}^{M(\alpha\beta\gamma)}$. The effective coefficients of thermal expansion can be readily obtained from $\mathbf{\Gamma}^*$ according to:

$$\alpha^* = \mathbf{C}^{*-1}\mathbf{\Gamma}^* \tag{2.35}$$

Alternatively, it is possible to establish Γ^* without utilizing Levin's result. This can be accomplished by employing again representation (2.31), while utilizing the thermoelastic concentration vector $\mathbf{A}^{T(\alpha\beta\gamma)}$, which can be determined by applying a temperature deviation in the absence of mechanical loadings. The final form of the global constitutive relation is given again by Eq. (2.32), but with Γ^* expressed by

$$\Gamma^* = \frac{-1}{DHL} \sum_{\alpha=1}^{N_\alpha} \sum_{\beta=1}^{N_\beta} \sum_{\gamma=1}^{N_\gamma} d_\alpha h_\beta l_\gamma \left[\mathbf{C}^{(\alpha\beta\gamma)} \mathbf{A}^{T(\alpha\beta\gamma)} - \Gamma^{(\alpha\beta\gamma)} \right] \qquad (2.36)$$

Both expressions (2.34) and (2.36) provide identical results.

2.3. Applications: Effective Constants

In ref. [12], the authors verified the prediction veracity of HFGMC by considering boron, graphite and glass fibers reinforcing aluminum and epoxy matrices, whose properties are given in Table 1. In Table 2-4, comparisons between the effective elastic moduli and coefficients of thermal expansions of two types of unidirectional fiber composites as predicted by the present HFGMC model, as well as its predecessor GMC model and two finite element solutions that were presented by [14] and [15].

Table 1. Elastic moduli of constituent fiber and matrix materials.

Material	E_{11} (GPa)	E_{22} (GPa)	G_{12} (GPa)	ν_{12}	ν_{23}	α_{11} $(10^{-6}/C)$	α_{22} $(10^{-6}/C)$
Boron fiber	379.3	379.3	172.41	0.10	0.10	8.1	8.1
Aluminum matrix	68.3	68.3	26.3	0.30	0.30	23.0	23.0
Graphite fiber	235.0	14.0	28.0	0.20	0.25	–	–
Epoxy matrix	4.8	4.8	1.8	0.34	0.34	–	–
Glass fiber	69.0	69.0	28.75	0.20	0.20	–	–

It can be readily observed that the present HFGMC predicts reliable results. It should be mentioned that in addition to the verification of the effective moduli prediction capability of HFGMC, the veracity of internal elastic field distribution can be also verified, see [12] for details.

2.4. Reformulation of HFGMC

As stated before, the system of equations (2.30) consists of the $21N_\alpha N_\beta N_\gamma$ unknowns that appear on the right-hand-side of Eq. (2.29). However, the size of this matrix (and thus the solution time) can be significantly reduced by utilizing the continuity of displacements across the interfaces between the phases. This reformulation of HFGMC for computational speed was originally put forth by [16] for the special case of continuously reinforced composites (i.e., doubly-periodic case) and linear elasticity. Therein, the number of unknowns is shown to be reduced (by approximately half when compared with the original formulation presented above) by treating the average displacements at the surfaces as the only unknowns. The reformulation of HFGMC in the case of discontinuous fibers (i.e., the triply periodic case) was similarly performed [17] by defining the average displacement at the surface. Thus, in the present case of discontinuous fibers (i.e., the three-dimensional case), the average displacement at the surface $\bar{y}_1^{(\alpha)} = -d_\alpha/2$ of subcell $(\alpha\beta\gamma)$, see Fig. 1(c), is given by·

$$\overset{(1)(\alpha\beta\gamma)}{\bar{\mathbf{u}}} = \frac{1}{h_\beta l_\gamma} \int_{-h_\beta/2}^{h_\beta/2} \int_{-l_\gamma/2}^{l_\gamma/2} \left[\mathbf{u}^{(\alpha\beta\gamma)} \Big|_{\bar{y}_1^{(\alpha)}=-\frac{d_\alpha}{2}} \right] d\bar{y}_2^{(\beta)} d\bar{y}_3^{(\gamma)} \tag{2.37}$$

with similar definitions of the average surface displacements $\overset{(2)(\alpha\beta\gamma)}{\bar{\mathbf{u}}}$ and $\overset{(3)(\alpha\beta\gamma)}{\bar{\mathbf{u}}}$ evaluated at the surfaces $\bar{y}_2^{(\beta)} = -h_\beta/2$ and $\bar{y}_3^{(\gamma)} = -l_\gamma/2$, respectively, of this subcell. It should be emphasized that the introduction of the average surface displacements at $\bar{y}_1^{(\alpha)} = d_\alpha/2$, $\bar{y}_2^{(\beta)} = h_\beta/2$ and $\bar{y}_3^{(\gamma)} = l_\gamma/2$ is not necessary because these displacements are equal, due to the interfacial continuity, to $\overset{(1)(\alpha+1,\beta,\gamma)}{\bar{\mathbf{u}}}$, $\overset{(2)(\alpha,\beta+1,\gamma)}{\bar{\mathbf{u}}}$ and $\overset{(3)(\alpha,\beta,\gamma+1)}{\bar{\mathbf{u}}}$, respectively. Consequently, in terms of these average surface displacements in all subcells, the new system of equations that replaces Eq. (2.30) in the HFGMC reformulation is given by

$$\mathbf{K}'\bar{\mathbf{U}} = \mathbf{f}' \tag{2.38}$$

where as in Eq. (2.30) the matrix $\mathbf{K}'$ contains information on the geometry and thermomechanical properties of the materials within the individual subcells $(\alpha\beta\gamma)$, and the vector $\bar{\mathbf{U}}$ contains the unknown average surface displacements of all subcells. The latter can be related to the unknown field quantities $\mathbf{W}_{(lmn)}^{(\alpha\beta\gamma)}$ which appear on the right-hand side of Eq. (2.29). Here too, the mechanical vector $\mathbf{f}'$ contains information on the applied average strains $\bar{\epsilon}$ and the imposed temperature deviation ΔT. As a result of this reformulation in terms of the average surface displacements, the size of the system of algebraic equations (2.30) becomes: $9N_\alpha N_\beta N_\gamma + 3(N_\alpha N_\beta + N_\alpha N_\gamma + N_\beta N_\gamma)$, as compared to the previous original formulation with $21N_\alpha N_\beta N_\gamma$. Extensive comparisons of the computational efficiency of the original versus the reformulated HFGMC in the two-dimensional and three-dimensional cases, in the presence of inelasticity and imperfect bonding between the phases, have been presented by [17].

3. Electro-Magneto-Thermo-Elastic Composites

3.1. Effective Behavior of Unidirectional Electro-Magneto-Thermo-Elastic Composites

The constitutive equations that govern the interaction of elastic, electric, magnetic and thermal fields in a electro-magneto-thermo-elastic medium relate the stresses σ_{ij}, strains ϵ_{ij}, electric field E_i, magnetic field H_i and temperature deviation ΔT from a reference temperature as follows:

$$\sigma_{ij} = C_{ijkl}\epsilon_{kl} - e_{kij}E_k - q_{kij}H_k - \Lambda_{ij}\Delta T \quad i,j,k,l = 1,...,3 \tag{3.1}$$

where C_{ijkl}, e_{ijk}, q_{ijk} and Λ_{ij} denote the fourth order elastic stiffness tensor, the third order piezoelectric tensor, the third order piezomagnetic tensor, and the second order thermal stress tensor of the material, respectively.

In addition, the electric displacement vector D_i is also expressed in terms of the strain, electric field, magnetic field and temperature in the form:

$$D_i = e_{ikl}\epsilon_{kl} + \kappa_{ik}E_k + a_{ik}H_k + p_i\Delta T \tag{3.2}$$

where κ_{ik}, a_{ik}, and p_i are the second order dielectric tensor, the second order magnetoelectric coefficient tensor, and the pyroelectric vector, respectively.

Finally, the magnetic flux density vector B_i is given in terms of the mechanical, electric, magnetic fields and temperature by

$$B_i = q_{ikl}\epsilon_{kl} + a_{ik}E_k + \mu_{ik}H_k + m_i\Delta T \tag{3.3}$$

where μ_{ik} and m_i are the second order magnetic permeability tensor and the pyromagnetic vector, respectively.

Let the vectors $\mathbf{X}$ and $\mathbf{Y}$ be defined as follows:

$$\mathbf{X} = [\epsilon_{11}, \epsilon_{22}, \epsilon_{33}, 2\epsilon_{23}, 2\epsilon_{13}, 2\epsilon_{12}, -E_1, -E_2, -E_3, -H_1, -H_2, -H_3] \tag{3.4}$$

$$\mathbf{Y} = [\sigma_{11}, \sigma_{22}, \sigma_{33}, \sigma_{23}, \sigma_{13}, \sigma_{12}, D_1, D_2, D_3, B_1, B_2, B_3] \tag{3.5}$$

Consequently, Eqs. (3.1)-(3.3) can be written in the following compact matrix representation:

$$\mathbf{Y} = \mathbf{Z}\,\mathbf{X} - \mathbf{\Gamma}\Delta T \tag{3.6}$$

where the square $12th$-order symmetric matrix of coefficients $\mathbf{Z}$ has the following form:

$$\mathbf{Z} = \begin{bmatrix} \mathbf{C} & \mathbf{e^t} & \mathbf{q^t} \\ \mathbf{e} & \text{-}\kappa & \text{-}\mathbf{a} \\ \mathbf{q} & \text{-}\mathbf{a} & \text{-}\mu \end{bmatrix} \tag{3.7}$$

and

$$\mathbf{\Gamma} = \left\{ \begin{array}{c} \mathbf{\Lambda} \\ \mathbf{p} \\ \mathbf{m} \end{array} \right\} \tag{3.8}$$

In Eq. (3.7), the square matrix $\mathbf{C}$ of the 6th-order represents the 4th-order stiffness tensor written in contracted notation, $\mathbf{e^t}$ denotes the transpose of the rectangular 3 by 6 matrix $\mathbf{e}$ that represents the corresponding third order piezoelectric tensor, $\mathbf{q^t}$ denotes the transpose of the rectangular 3 by 6 matrix $\mathbf{q}$ that represents the corresponding third order piezomagnetic tensor, κ is a square matrix of order 3 that corresponds to the dielectric tensor, $\mathbf{a}$ is a square matrix of the 3rd-order that represents the magnetoelectric coefficients, and μ is a square matrix of the 3rd-order that represents the magnetic permeability tensor. Finally, the 6th-order vector $\mathbf{\Lambda}$, and the two 3rd-order vectors $\mathbf{p}$ and $\mathbf{m}$ in Eq. (3.8) represent the thermal stresses, pyroelectric and pyromagnetic coefficients, respectively, which altogether form the thermal stress-pyroelectric-pyromagnetic vector $\mathbf{\Gamma}$ in (3.6).

In the framework of the HFGMC analysis, the volume average of the stresses $\bar{\sigma}_{ij}$, electric displacements $\bar{D}_i$ and the magnetic flux density $\bar{B}_i$ in the entire repeating cell (namely in the composite) is given by

$$\bar{\mathbf{Y}} = \frac{1}{DHL}\sum_{\alpha=1}^{N_\alpha}\sum_{\beta=1}^{N_\beta}\sum_{\gamma=1}^{N_\gamma} d_\alpha h_\beta l_\gamma \mathbf{Y}^{(\alpha\beta\gamma)} \tag{3.9}$$

Similarly, the volume average of the strains $\bar{\epsilon}_{ij}$, electric field components $\bar{E}_i$ and magnetic field components $\bar{H}_i$ in the composite is given by

$$\bar{\mathbf{X}} = \frac{1}{DHL} \sum_{\alpha=1}^{N_\alpha} \sum_{\beta=1}^{N_\beta} \sum_{\gamma=1}^{N_\gamma} d_\alpha h_\beta l_\gamma \mathbf{X}^{(\alpha\beta\gamma)} \qquad (3.10)$$

where $\mathbf{X}^{(\alpha\beta\gamma)}$ and $\mathbf{Y}^{(\alpha\beta\gamma)}$ are the field variables at subcell $(\alpha\beta\gamma)$ that are given by Eq. (3.4) and (3.5), respectively.

Let $\xi^{(\alpha\beta\gamma)}$ and $\eta^{(\alpha\beta\gamma)}$ denote the electric and magnetic potentials in the subcell, respectively, such that $\mathbf{E}^{(\alpha\beta\gamma)} = -\nabla \xi^{(\alpha\beta\gamma)}$ and $\mathbf{H}^{(\alpha\beta\gamma)} = -\nabla \eta^{(\alpha\beta\gamma)}$. Following the HFGMC methodology, quadratic expansion of the mechanical displacements $\mathbf{u}^{(\alpha\beta\gamma)}$ (that includes $\bar{\epsilon}$, see Eq. (2.29)), of the electric potential $\xi^{(\alpha\beta\gamma)}$ (that includes $\bar{E}$) and of the magnetic potential $\eta^{(\alpha\beta\gamma)}$ (that includes $\bar{H}$) are introduced. The application of the HFGMC analysis consists of imposing in the present electro-magneto-thermo-elastic micromechanical analysis the following conditions (imposed in the average sense):
(1) continuity of displacements at the interfaces between the subcells;
(2) continuity of tractions at the interfaces between the subcells;
(3) continuity of the electric potentials at the interfaces between the subcells;
(4) continuity of normal electric displacements at the interfaces between the subcells;
(5) continuity of the magnetic potentials at the interfaces between the subcells;
(6) continuity of the normal magnetic flux densities at the interfaces between the subcells;
(7) imposition of the periodic continuity conditions of the displacements, tractions, electric potentials, normal electric displacements, magnetic potentials, normal magnetic flux densities.

Consequently, one can establish the following localization relation between the local electro-magneto-elastic field in the subcell $\mathbf{X}^{(\alpha\beta\gamma)}$ and the average external macro field $\bar{\mathbf{X}}$ in the form [11]:

$$\mathbf{X}^{(\alpha\beta\gamma)} = \mathbf{A}^{EME(\alpha\beta\gamma)} \, \bar{\mathbf{X}} \qquad (3.11)$$

where $\mathbf{A}^{EME(\alpha\beta\gamma)}$ is the electro-magneto-elastic concentration matrix associated with subcell $(\alpha\beta\gamma)$. Hence, in conjunction with the averaging procedure given by Eq. (3.9), the following effective isothermal constitutive relations of the electro-magneto-elastic composite can be established:

$$\bar{\mathbf{Y}} = \mathbf{Z}^* \, \bar{\mathbf{X}} \qquad (3.12)$$

where the effective elastic stiffness, piezoelectric, piezomagnetic, dielectric, magnetic permeability and electromagnetic coefficients matrix $\mathbf{Z}^*$ of the composite is given by

$$\mathbf{Z}^* = \frac{1}{DHL} \sum_{\alpha=1}^{N_\alpha} \sum_{\beta=1}^{N_\beta} \sum_{\gamma=1}^{N_\gamma} d_\alpha h_\beta l_\gamma \mathbf{Z}^{(\alpha\beta\gamma)} \, \mathbf{A}^{EME(\alpha\beta\gamma)} \qquad (3.13)$$

where $\mathbf{Z}^{(\alpha\beta\gamma)}$ is the matrix given by Eq. (3.7) which characterizes the monolithic electro-magneto-elastic material that fills subcell $(\alpha\beta\gamma)$.

The structure of the square 12th-order symmetric matrix $\mathbf{Z}^*$ is of the form

$$\mathbf{Z}^* = \begin{bmatrix} \mathbf{C}^* & \mathbf{e}^{*t} & \mathbf{q}^{*t} \\ \mathbf{e}^* & -\kappa^* & -\mathbf{a}^* \\ \mathbf{q}^* & -\mathbf{a}^* & -\mu^* \end{bmatrix} \qquad (3.14)$$

where $\mathbf{C}^*$, $\mathbf{e}^*$, $\mathbf{q}^*$, $\boldsymbol{\kappa}^*$, $\mathbf{a}^*$ and $\boldsymbol{\mu}^*$ are the effective elastic stiffness, piezoelectric, piezomagnetic, dielectric, magnetic permeability and electromagnetic coefficients, respectively.

In order to incorporate the thermal effects in the composite, we utilize Levin's result [13] to establish the effective thermal stress Λ_{ij}^* tensor, pyroelectric p_i^*, and pyromagnetic m_i^* coefficients. To this end, let us define the following vector of thermal stresses, pyroelectric and pyromagnetic coefficients material constants in the subcell $(\alpha\beta\gamma)$:

$$\boldsymbol{\Gamma}^{(\alpha\beta\gamma)} = [\Lambda_1, \Lambda_2, \Lambda_3, \Lambda_4, \Lambda_5, \Lambda_6, p_1, p_2, p_3, m_1, m_2, m_3]^{(\alpha\beta\gamma)} \qquad (3.15)$$

The corresponding vector of effective values is defined by

$$\boldsymbol{\Gamma}^* = [\Lambda_1^*, \Lambda_2^*, \Lambda_3^*\Lambda_4^*, \Lambda_5^*, \Lambda_6^*, p_1^*, p_2^*, p_3^*, m_1^*, m_2^*, m_3^*] \qquad (3.16)$$

According to Levin's result, the relation between $\boldsymbol{\Gamma}^{(\alpha\beta\gamma)}$ and $\boldsymbol{\Gamma}^*$ can be expressed in terms of the electro-magneto-elastic concentration matrices $\mathbf{A}^{EME(\alpha\beta\gamma)}$ which appear in Eq. (3.11). Thus, with the established electro-magneto-elastic concentration matrix $\mathbf{A}^{EME(\alpha\beta\gamma)}$, Eq. (2.34) (in which $\mathbf{A}^{M(\alpha\beta\gamma)}$ is replaced by $\mathbf{A}^{EME(\alpha\beta\gamma)}$) can be employed to provide the effective thermal stress $\boldsymbol{\Lambda}^*$, pyroelectric $\mathbf{p}^*$ and pyromagnetic $\mathbf{m}^*$ vectors of the composite.

Consequently, the final anisothermal micromechanically established constitutive law of the electro-magneto-thermo-elastic multiphase composite is given by

$$\bar{\mathbf{Y}} = \mathbf{Z}^* \bar{\mathbf{X}} - \boldsymbol{\Gamma}^* \Delta T \qquad (3.17)$$

The coefficients of thermal expansion α_i and the associated pyroelectric constants P_i and the pyromagnetic constants M_i of the monolithic material can be assembled to form the vector:

$$\boldsymbol{\Omega} = [\alpha_1, \alpha_2, \alpha_3, \alpha_4, \alpha_5, \alpha_6, P_1, P_2, P_3, M_1, M_2, M_3]$$

This vector is given by

$$\boldsymbol{\Omega} = \mathbf{Z}^{-1}\boldsymbol{\Gamma} \qquad (3.18)$$

where $\mathbf{Z}$ and $\boldsymbol{\Gamma}$ are given by Eqs. (3.7) and (3.8), respectively.
In the same manner, the effective coefficients of thermal expansion α_i^* and the associated pyroelectric constants P_i^* and the pyromagnetic constants M_i^* of the composite can be also assembled into the vector:

$$\boldsymbol{\Omega}^* = [\alpha_1^*, \alpha_2^*, \alpha_3^*, \alpha_4^*, \alpha_5^*, \alpha_6^*, P_1^*, P_2^*, P_3^*, M_1^*, M_2^*, M_3^*] \qquad (3.19)$$

Once $\mathbf{Z}^*$ and $\boldsymbol{\Gamma}^*$ have been established according to Eqs. (3.14) and (3.15), this vector can be readily determined from:

$$\boldsymbol{\Omega}^* = \mathbf{Z}^{*-1}\boldsymbol{\Gamma}^* \qquad (3.20)$$

In a recent publication [18], the predecessor GMC model was extended to analyze thermo-electro-magneto-elastic laminated composites of this type, and to include possible inelastic deformation effects in the presence of a metallic matrix.

3.2. Applications: Effective Constants

Following ref. [19], let us consider a composite consisting of $CoFe_2O_4$ piezomagnetic matrix reinforced by $BaTiO_3$ piezoelectric material. Both phases are transversely isotropic with the axis of symmetry oriented in the 3-direction. The independent material constants of these constituents are given in Tables 1-3 [19].

Table 2. Comparison of predicted effective elastic moduli of a boron/aluminum unidirectional composite ($v_f = 0.47$).

Effective elastic moduli	E_{11} (GPa)	E_{22} (GPa)	G_{12} (GPa)	G_{23} (GPa)	ν_{12}	ν_{23}
HFGMC	215.4	144.0	54.34	45.83	0.195	0.255
Sun and Vaidya [14]	215.0	144.0	57.20	45.90	0.190	0.290
Tamma and Avila [15]	214.7	144.7	54.30	45.60	0.195	0.249
GMC	215.0	141.0	51.20	43.70	0.197	0.261

It should be noted that in both materials the electromagnetic coefficients are zero, i.e., $a_{ij} = 0$. Consider a fibrous composite in which the $BaTiO_3$ continuous fibers are oriented in the 3-direction. The volume fraction of the fibers is denoted by v_f. In this case the HFGMC model can be employed to generate the effective elastic, dielectric, magnetic permeability, piezoelectric, piezomagnetic, and electromagnetic coupling moduli of the fibrous composite for $0 \leq v_f \leq 1$. These moduli were generated by [19], who employed the Mori-Tanaka micromechanical model [20].

Figures 2-3 show these effective moduli as predicted by the HFGMC model, its predecessor GMC [21] and by the Mori-Tanaka (MT) scheme. It is readily seen that in some cases the predictions of the three methods coincide, while in other cases slight deviations are observed. It is seen however that the HFGMC and Mori-Tanaka predictions are very close. More results are given in [22] and [11].

3.3. Laminated Electro-Magneto-Thermo-Elastic Composites

The micromechanically established effective constitutive relations which relate the average stresses to the average strains, electric, magnetic field and temperature

$$\bar{\sigma}_{ij} = C^*_{ijkl}\bar{\epsilon}_{kl} - e^*_{kij}\bar{E}_k - q^*_{kij}\bar{H}_k - \Lambda^*_{ij}\Delta T \qquad (3.21)$$

can be utilized to derive the basic equations for electro-magneto-thermo-elastic laminated plates in conjunction with the classical plate theory. To this end, let us consider N layers of electro-magneto-thermo-elastic materials. For this type of material, the constitutive equations (3.21), that describe the behavior of a single transversely isotropic layer in which the

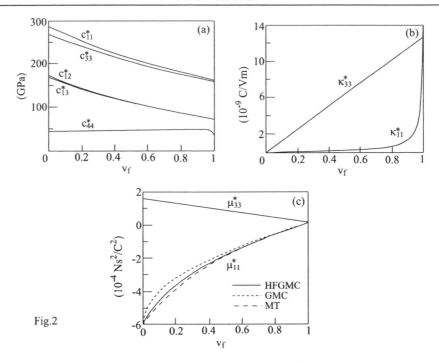

Fig.2

Figure 2. (a) Effective elastic moduli of fibrous composite against the volume fraction of $BaTiO_3$. The predictions of the three micromechanical models coincide. (b) Effective dielectric moduli of fibrous composite against the volume fraction of $BaTiO_3$. The predictions of the three micromechanical models coincide. (c) Comparison between HFGMC, GMC and MT predictions of the effective magnetic permeability moduli of fibrous composite against the volume fraction of $BaTiO_3$.

axis of symmetry is oriented in the 3-direction, can be written in the following form:

$$
\begin{Bmatrix} \bar{\sigma}_{11} \\ \bar{\sigma}_{22} \\ \bar{\sigma}_{33} \\ \bar{\sigma}_{23} \\ \bar{\sigma}_{13} \\ \bar{\sigma}_{12} \end{Bmatrix} = \begin{bmatrix} c_{11}^* & c_{12}^* & c_{13}^* & 0 & 0 & 0 \\ & c_{11}^* & c_{13}^* & 0 & 0 & 0 \\ & & c_{33}^* & 0 & 0 & 0 \\ & & & c_{44}^* & 0 & 0 \\ & & & & c_{44}^* & 0 \\ \text{sym.} & & & & & \frac{1}{2}(c_{11}^* - c_{12}^*) \end{bmatrix} \begin{Bmatrix} \bar{\epsilon}_{11} \\ \bar{\epsilon}_{22} \\ \bar{\epsilon}_{33} \\ 2\bar{\epsilon}_{23} \\ 2\bar{\epsilon}_{13} \\ 2\bar{\epsilon}_{12} \end{Bmatrix}
$$

$$
- \begin{bmatrix} 0 & 0 & e_{31}^* \\ 0 & 0 & e_{31}^* \\ 0 & 0 & e_{33}^* \\ 0 & e_{15}^* & 0 \\ e_{15}^* & 0 & 0 \\ 0 & 0 & 0 \end{bmatrix} \begin{Bmatrix} \bar{E}_1 \\ \bar{E}_2 \\ \bar{E}_3 \end{Bmatrix} - \begin{bmatrix} 0 & 0 & q_{31}^* \\ 0 & 0 & q_{31}^* \\ 0 & 0 & q_{33}^* \\ 0 & q_{15}^* & 0 \\ q_{15}^* & 0 & 0 \\ 0 & 0 & 0 \end{bmatrix} \begin{Bmatrix} \bar{H}_1 \\ \bar{H}_2 \\ \bar{H}_3 \end{Bmatrix} - \begin{Bmatrix} \Lambda_1^* \\ \Lambda_2^* \\ \Lambda_3^* \\ 0 \\ 0 \\ 0 \end{Bmatrix} \Delta T \quad (3.22)
$$

When the behavior of the layer is referred to the plate system of coordinates x, y, z, and by imposing plane stress conditions, the following constitutive relations can be obtained

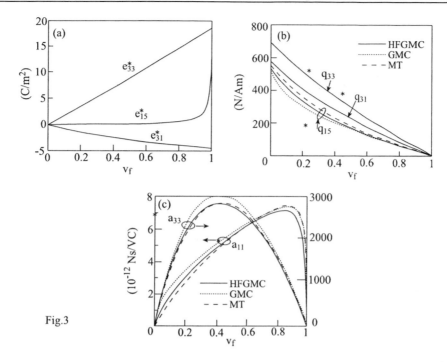

Fig.3

Figure 3. (a) Effective piezoelectric moduli of fibrous composite against the volume fraction of $BaTiO_3$. The predictions of the three micromechanical models coincide. (b) Comparison between HFGMC, GMC and MT predictions of the effective piezomagnetic moduli of fibrous composite against the volume fraction of $BaTiO_3$. (c) Comparison between HFGMC, GMC and MT predictions of the effective electromagnetic coupling moduli of fibrous composite against the volume fraction of $BaTiO_3$.

from (3.22):

$$
\left\{
\begin{array}{c}
\sigma_{xx} \\
\sigma_{yy} \\
\sigma_{xy}
\end{array}
\right\}
= [\bar{Q}]
\left\{
\begin{array}{c}
\epsilon_{xx} \\
\epsilon_{yy} \\
2\epsilon_{xy}
\end{array}
\right\}
-
\left[
\begin{array}{ccc}
0 & 0 & \bar{e}_{11}^* \\
0 & 0 & \bar{e}_{12}^* \\
0 & 0 & \bar{e}_{16}^*
\end{array}
\right]
\left\{
\begin{array}{c}
E_x \\
E_y \\
E_z
\end{array}
\right\}
$$

$$
-
\left[
\begin{array}{ccc}
0 & 0 & \bar{q}_{11}^* \\
0 & 0 & \bar{q}_{12}^* \\
0 & 0 & \bar{q}_{16}^*
\end{array}
\right]
\left\{
\begin{array}{c}
H_x \\
H_y \\
H_z
\end{array}
\right\}
-
\left\{
\begin{array}{c}
\Lambda_x \\
\Lambda_y \\
\Lambda_{xy}
\end{array}
\right\}
\Delta T
\qquad (3.23)
$$

In this equation $[\bar{Q}_{ij}]$ are the reduced stiffness coefficients which are given in terms of the transformed effective stiffnesses $[\bar{c}_{ij}^*]$ by the standard relations, see [23] for example. Similarly, the electric constants $\bar{e}_{ij}$, magnetic constants $\bar{q}_{ij}$ and thermal stress coefficients Λ_x, Λ_y and Λ_{xy} are the corresponding transformed effective coefficients that appear in Eq. (3.22). It is of interest to note that Eq. (3.23) clearly shows that in the rotated system E_z and H_z contribute to the shear stress through the constant $\bar{e}_{16}^*$ and $\bar{q}_{16}^*$ in the present case of transversely isotropic material.

The stress resultants **N** and moments **M** of the laminate can be readily obtained in the form:

$$\left\{ \begin{array}{c} \mathbf{N} \\ \mathbf{M} \end{array} \right\} = \left[\begin{array}{cc} \mathbf{A} & \mathbf{B} \\ \mathbf{B} & \mathbf{D} \end{array} \right] \left\{ \begin{array}{c} \boldsymbol{\epsilon}^0 \\ \boldsymbol{\kappa} \end{array} \right\} - \left\{ \begin{array}{c} \mathbf{N}^E \\ \mathbf{M}^E \end{array} \right\} - \left\{ \begin{array}{c} \mathbf{N}^M \\ \mathbf{M}^M \end{array} \right\} - \left\{ \begin{array}{c} \mathbf{N}^T \\ \mathbf{M}^T \end{array} \right\} \qquad (3.24)$$

where $\mathbf{A}$, $\mathbf{B}$ and $\mathbf{D}$ denote the standard extensional, coupling and bending stiffnesses, and $\boldsymbol{\epsilon}^0$, $\boldsymbol{\kappa}$ represent the middle surface strains and curvatures, respectively. In Eq. (3.24), the electric resultants $\mathbf{N}^E$ and moments $\mathbf{M}^E$ and magnetic resultants $\mathbf{N}^M$ and moments $\mathbf{M}^M$ are defined by

$$\mathbf{N}^E = \left\{ \begin{array}{c} N_x^E \\ N_y^E \\ N_{xy}^E \end{array} \right\} = \sum_{k=1}^{N} \left\{ \begin{array}{c} \bar{e}_{31}^* \\ \bar{e}_{32}^* \\ \bar{e}_{36}^* \end{array} \right\}_k (E_z)_k (h_k - h_{k-1}) \qquad (3.25)$$

$$\mathbf{M}^E = \left\{ \begin{array}{c} M_x^E \\ M_y^E \\ M_{xy}^E \end{array} \right\} = \frac{1}{2} \sum_{k=1}^{N} \left\{ \begin{array}{c} \bar{e}_{31}^* \\ \bar{e}_{32}^* \\ \bar{e}_{36}^* \end{array} \right\}_k (E_z)_k (h_k^2 - h_{k-1}^2) \qquad (3.26)$$

$$\mathbf{N}^M = \left\{ \begin{array}{c} N_x^M \\ N_y^M \\ N_{xy}^M \end{array} \right\} = \sum_{k=1}^{N} \left\{ \begin{array}{c} \bar{q}_{31}^* \\ \bar{q}_{32}^* \\ \bar{q}_{36}^* \end{array} \right\}_k (H_z)_k (h_k - h_{k-1}) \qquad (3.27)$$

$$\mathbf{M}^M = \left\{ \begin{array}{c} M_x^M \\ M_y^M \\ M_{xy}^M \end{array} \right\} = \frac{1}{2} \sum_{k=1}^{N} \left\{ \begin{array}{c} \bar{q}_{31}^* \\ \bar{q}_{32}^* \\ \bar{q}_{36}^* \end{array} \right\}_k (H_z)_k (h_k^2 - h_{k-1}^2) \qquad (3.28)$$

where h_k denotes the thickness of the kth layer.

The thermal resultants $\mathbf{N}^T$ and moments $\mathbf{M}^T$ are given by

$$\mathbf{N}^T = \left\{ \begin{array}{c} N_x^T \\ N_y^T \\ N_{xy}^T \end{array} \right\} = \Delta T \sum_{k=1}^{N} \left\{ \begin{array}{c} \Lambda_x \\ \Lambda_y \\ \Lambda_{xy} \end{array} \right\}_k (h_k - h_{k-1}) \qquad (3.29)$$

$$\mathbf{M}^T = \left\{ \begin{array}{c} M_x^T \\ M_y^T \\ M_{xy}^T \end{array} \right\} = \frac{\Delta T}{2} \sum_{k=1}^{N} \left\{ \begin{array}{c} \Lambda_x \\ \Lambda_y \\ \Lambda_{xy} \end{array} \right\}_k (h_k^2 - h_{k-1}^2) \qquad (3.30)$$

For a detailed analysis of composite plates that consist of piezoelectric materials, see ref. [24].

4.   Hysteresis Behavior of Ferroelectric Fiber Composites

Piezoelectric ceramic materials exhibit linear response as long as the applied electrome-chanical loading is low. At a stress of $10 - 100 MPa$ and electric field of $0.1 - 10 MV/m$, nonlinear effects become appreciable [25]. The nonlinearity is exhibited, in particular, by a hysteresis loop in the relation between the polarization (or electric displacement) and electric field. For the ferroelectric PVDF piezopolymer, on the other hand, higher values of electric field are required to exhibit its hysteresis response [26]. The hysteresis loop results from the delayed responses of polarization reversal and domain switching. The domain switching reflects the energy needed to switch the polarization during each cycle of the field. A qualitative explanation of the hysteretic phenomena in ferroelectric materials is given by [27]. For a recent review article that discusses the utilization of piezoelectric materials as actuators and sensors, see [7]. An overview with special emphasis on experimental evidence of the nonlinear behavior of piezoelectric ceramics has been recently presented by [28].

A recent review article that discusses the behavior and modeling of the hysteresis phenomena of piezoceramics materials is given by [29]. This review paper also provides an extensive list of references to various approaches that have been adopted by several investigators for the modeling of the hysteresis loop. It appears that there are two approaches for the modeling of this behavior. In the first one, a switching criterion is applied in order to generate the hysteresis loop, as discussed in the publications by [30], [31], [32], and [33] for example. In the second approach, phenomenological nonlinear constitutive equations have been proposed for the representation of the hysteresis loop in ferroelectric materials, see [34], [35], [36] and [37], for example.

In a recent paper by [38], a phenomenological model for the hysteresis behavior of piezoceramics was offered. The piezoceramic was assumed to be tetragonal with 180 and 90 degree switching. In [38], the authors formulated a Gibbs energy function from which the coupled strain and electric fields are derived, in conjunction with a function that describes the hysteresis relation between the polarization and electric field in the absence of mechanical loading. The resulting constitutive equations are nonlinear, and the required parameters are just the standard constants that appear in the linear piezoelectric constitutive relations, together with the coercive electric field, the remnant polarization and the saturation polarization. By adjusting an additional parameter in the constitutive equations, comparisons between the predicted hysteresis response with measured data and good agreements were obtained [38].

All the aforementioned papers were concerned with the modeling of the hysteresis behavior of the monolithic piezoelectric materials. As discussed by [39], significant improvements in properties can be achieved through composite technology (e.g. piezoceramic-polymer composites). When the response of composite materials with embedded ferroelectric materials are sought, micromechanical analyses that take into account the detailed interaction between the ferroelectric phase and its surrounding matrix constituent have to be employed. In order to study the nonlinear behavior of piezoelectric fiber composites, the authors in ref. [40] employed a simplified micromechanical analysis which is based on a combination of the iso-stress iso-electric displacement and iso-strain iso-electric field. It should be emphasized, however, that the constitutive relations that were employed by these authors are restricted to the modeling of the composite under monotonically increasing applied electric fields only, so that the hysteresis patterns cannot be considered by their approach. In the following, we describe the hysteresis model of ref. [38] and its incorporation with incremental HFGMC micromechanics analysis. More details are given in [41].

4.1. The Modeling of the Monolithic Ferroelectric Material

In ref. [38], the following expression for the Gibbs energy function that is expressed in terms of the stress σ_{ij} and polarization P_i components is assumed:

$$G(\sigma_{ij}, P_i) = -\frac{1}{2}S_{ijkl}\sigma_{ij}\sigma_{kl} - Q_{ijkl}\sigma_{ij}P_kP_l + F(\mathbf{P}) \qquad i,j,k,l = 1,2,3 \qquad (4.1)$$

where S_{ijkl} are the elastic compliance tensor components of the material, Q_{ijkl} are the components of a tensor that represents the piezoelectric effects and $F(\mathbf{P})$ is a function of the polarization vector $\mathbf{P}$, that describes the hysteresis behavior of the material.

The strains ϵ_{ij} and the electric field E_i components are determined from G as follows:

$$\epsilon_{ij} = -\frac{\partial G}{\partial \sigma_{ij}} \qquad (4.2)$$

$$E_i = \frac{\partial G}{\partial P_i} \qquad (4.3)$$

Consequently, the following constitutive equations result from Eqs. (4.1)-(4.3):

$$\epsilon_{ij} = S_{ijkl}\sigma_{kl} + Q_{ijkl}P_kP_l \qquad (4.4)$$

$$E_i = -Q_{klil}\sigma_{kl}P_l + \frac{\partial F(\mathbf{P})}{\partial P_i} \qquad (4.5)$$

For transversely isotropic piezoelectric materials, the compliance tensor S_{ijkl} consists of five independent components, and Q_{ijkl} has two independent elements: Q_{11} and Q_{12}. If the axis of symmetry of the transversely isotropic material is oriented in the 3-direction, the components stiffness tensor C_{ijkl} (which is the inverse of the compliance tensor S_{ijkl}) are given by

$$
\begin{aligned}
C_{ijkl} = \;& C_{12}\delta_{ij}\delta_{kl} + \frac{1}{2}(C_{22} - C_{12})(\delta_{ik}\delta_{jl} + \delta_{il}\delta_{kj}) \\
+ \;& (C_{23} - C_{12})(\delta_{ij}\delta_{3k}\delta_{3l} + \delta_{kl}\delta_{3i}\delta_{3j}) \\
+ \;& (C_{44} - \frac{1}{2}C_{22} + \frac{1}{2}C_{12})(\delta_{ik}\delta_{3j}\delta_{3l} + \delta_{jk}\delta_{3i}\delta_{3l} + \delta_{il}\delta_{3j}\delta_{3k} + \delta_{jl}\delta_{3i}\delta_{3k}) \\
+ \;& (C_{33} + C_{22} - 2C_{23} - 4C_{44})\delta_{3i}\delta_{3j}\delta_{3k}\delta_{3l} \qquad (4.6)
\end{aligned}
$$

where C_{mn} are the five independent elastic constants of the material and δ_{ij} is the Kronecker delta.

Consequently, the first constitutive equation (4.4) can be rewritten in the form:

$$\sigma_{ij} = C_{ijkl}(\epsilon_{ij} - \epsilon_{ij}^E) \qquad (4.7)$$

where ϵ_{ij}^E are the electric strains which are given by

$$\epsilon_{ij}^E = (Q_{11} - Q_{12})P_iP_j + Q_{12}P_kP_k\delta_{ij} \qquad (4.8)$$

The second constitutive equation (4.5) is given by

$$E_i = -2(Q_{11} - Q_{12})P_k\sigma_{ki} - 2Q_{12}P_i\sigma_{kk} + \frac{\partial F(\mathbf{P})}{\partial P_i} \qquad (4.9)$$

In ref. [38] a limiting process was employed in the vicinity of the remnant polarization of the hysteresis loop, where the response of the nonlinear material should coincide with a linear piezoelectric behavior. This process yields the following expressions for Q_{11} and Q_{12} in terms of the piezoelectric coefficients d_{ijk}, coercive electric field E_c (the magnitude of the electric field required to remove the remnant polarization in the hysteresis loop), remnant polarization P_r and saturation polarization P_s:

$$Q_{11} = -\frac{d_{333}E_c\lambda}{2P_r(P_s - P_r)\, ln(1 - P_r/P_s)}$$

$$Q_{12} = -\frac{d_{311}E_c\lambda}{2P_r(P_s - P_r)\, ln(1 - P_r/P_s)} \tag{4.10}$$

with λ being a scalar to be selected to provide the best fit to the entire loop. It is readily observed that the nonlinear ferroelectric material has to be characterized by the five elastic constants C_{mn}, the piezoelectric constants d_{ijk} and by E_c, P_r, P_s and parameter λ.

The components of the strain tensor ϵ_{ij} are expressed in terms of the mechanical displacement components u_i by

$$\epsilon_{ij} = \frac{1}{2}\left(\frac{\partial u_i}{\partial x_j} + \frac{\partial u_j}{\partial x_i}\right) \qquad i,j = 1,2,3 \tag{4.11}$$

The components of the electric field E_i are obtained from the electric potential ξ via:

$$E_i = -\frac{\partial \xi}{\partial x_i} \tag{4.12}$$

The above constitutive relations must be supplemented by the condition that the static equilibrium of the material must be satisfied, namely:

$$\frac{\partial \sigma_{ji}}{\partial x_j} = 0 \tag{4.13}$$

where the Maxwell's electrostatic stress tensor has been omitted from this equation. Under regular conditions this omission is justified, see [29] for details. Furthermore, in the absence of volume charges, the following Maxwell's equation must be satisfied:

$$\frac{\partial D_i}{\partial x_i} = 0 \tag{4.14}$$

where D_i are the components of the electric displacement.

Due to the nonlinearity of the constitutive equations (4.7)-(4.9), the HFGMC micromechanical analysis, which has been previously described, must be formulated in an incremental form (otherwise, a system of nonlinear algebraic equations must be solved at each increment rather than a linear one). To this end, let us define the following two vectors of increments:

$$\Delta \mathbf{Y}_1 = [\Delta\sigma_{11}, \Delta\sigma_{22}, \Delta\sigma_{33}, \Delta\sigma_{23}, \Delta\sigma_{13}, \Delta\sigma_{12}, \Delta E_1, \Delta E_2, \Delta E_3] \tag{4.15}$$

$$\Delta \mathbf{X}_1 = [\Delta\epsilon_{11}, \Delta\epsilon_{22}, \Delta\epsilon_{33}, 2\Delta\epsilon_{23}, 2\Delta\epsilon_{13}, 2\Delta\epsilon_{12}, \Delta P_1, \Delta P_2, \Delta P_3] \tag{4.16}$$

After some lengthy manipulations, constitutive relations (4.7)-(4.9) can be written in the incremental form:

$$\Delta \mathbf{Y}_1 = \mathbf{H}\, \Delta \mathbf{X}_1 \tag{4.17}$$

where the $9th$-order square symmetric matrix $\mathbf{H}$ can be represented by

$$\mathbf{H} = \begin{bmatrix} \mathbf{C} & \mathbf{M} \\ \mathbf{M}^T & \mathbf{N} \end{bmatrix} \tag{4.18}$$

In this equation, the square matrix $\mathbf{C}$ of the 6th-order represents the 4th-order elastic stiffness tensor C_{ijkl} which is given by Eq. (4.6), $\mathbf{M}$ is a rectangular 3 by 6 matrix that expresses the electromechanical coupling (which is absent when $Q_{11} = Q_{12} = 0$), $\mathbf{M}^T$ is the transpose of $\mathbf{M}$, and $\mathbf{N}$ is a third order square matrix that represents the electric behavior of the material. The latter includes (due to the incremental formulation) the second derivatives $\partial^2 F(\mathbf{P})/\partial P_i^2$ which provide the hysteresis patterns that are exhibited by the material response. It turns out that for the particular form of the hysteresis patterns that are given by [38], the terms that represent these loops appear in the three diagonal elements of matrix $\mathbf{N}$ only. These terms will be denoted by $f_1(E)$, $f_2(E)$ and $f_3(E)$. They are given in terms of a function $f(E)$ as follows:

$$f_1(E) = f(E_1), \quad f_2(E) = f(E_2), \quad f_3(E) = f(E_3) \tag{4.19}$$

The function $f(E)$ is defined in the following. Let:

$$k = ln(1 - P_r/P_s)$$

and

$$P_s' = P_s \left[1 - \exp(k|E_m - E_c|/E_c) \right]$$

where $E_m = E_{max}$ which is the maximum value of the applied electric field. Then $f(E)$ is given by

$$f(E) = -\frac{E_c[1 - \exp(kE_m/E_c)]}{P_s' k} \exp[-kE/E_c] \tag{1.4.20a}$$

$$f(E) = \begin{cases} -\frac{E_c}{P_s k} \exp[-k(E - E_c)/E_c] & E_c \le E \le E_m \\ \\ -\frac{E_c}{P_s' k} \exp[k(E - E_c)/E_c] & -E_m \le E \le E_c \end{cases} \quad \textit{rising line} \tag{1.4.20b}$$

$$f(E) = \begin{cases} -\frac{E_c}{P_s' k} \exp[-k(E + E_c)/E_c] & -E_c \le E \le E_m \\ \\ -\frac{E_c}{P_s k} \exp[k(E + E_c)/E_c] & -E_m \le E \le -E_c \end{cases} \quad \textit{decreasing line} \tag{1.4.20c}$$

Equation (1.4.20a) describes the first polarization process of the initially unpoled material in which the electric field rises from zero to E_m, where the polarization is equal to P_s'. For the polarized material, Eqs. (1.4.20b)/(1.4.20c) describe the rising/decreasing lines in the hysteresis loop along which the polarization P increases/decreases as the electric field E increases/decreases.

In order to impose in the next subsection the equilibrium and Maxwell equations and the proper continuity conditions across the various interfaces of the composite, the constitutive relations of the ferroelectric material need to be given in terms of the increments of the mechanical stresses $\Delta\sigma_{ij}$ and electrical displacements ΔD_i as independent variables. The latter are given by

$$\Delta D_i = \epsilon_0 \, \Delta E_i + \Delta P_i \qquad i = 1, 2, 3 \tag{4.21}$$

where ϵ_0 is the permittivity of free space.

Consequently, the tangential constitutive relation (4.17) can be rewritten in the following standard form:

$$\Delta\mathbf{Y} = \mathbf{Z}\,\Delta\mathbf{X} \tag{4.22}$$

where:

$$\Delta\mathbf{X} = [\Delta\epsilon_{11}, \Delta\epsilon_{22}, \Delta\epsilon_{33}, 2\Delta\epsilon_{23}, 2\Delta\epsilon_{13}, 2\Delta\epsilon_{12}, -\Delta E_1, -\Delta E_2, -\Delta E_3] \tag{4.23}$$

$$\Delta\mathbf{Y} = [\Delta\sigma_{11}, \Delta\sigma_{22}, \Delta\sigma_{33}, \Delta\sigma_{23}, \Delta\sigma_{13}, \Delta\sigma_{12}, \Delta D_1, \Delta D_2, \Delta D_3] \tag{4.24}$$

and the 9th-order symmetric tangent electromechanical matrix $\mathbf{Z}$ is given by

$$\mathbf{Z} = \begin{bmatrix} \mathbf{C} - \mathbf{M}\,\mathbf{N}^{-1}\,\mathbf{M}^T & -\mathbf{M}\,\mathbf{N}^{-1} \\ -\mathbf{N}^{-1}\,\mathbf{M}^T & -\mathbf{N}^{-1} - \epsilon_0\,\mathbf{I} \end{bmatrix} \tag{4.25}$$

where $\mathbf{I}$ is the unit matrix. It should be noted that the elements of matrix $\mathbf{Z}$ consist of the current electromechanical field variables (i.e. they are variable field quantities).

4.2. Incremental Micromechanics Analysis

The basic assumption in HFGMC is that the increments of the mechanical displacements $\Delta\mathbf{u}^{(\alpha\beta\gamma)}$ and electric potential $\Delta\xi^{(\alpha\beta\gamma)}$ in each subcell are expanded into quadratic forms in terms of its local coordinates $(\bar{y}_1^{(\alpha)}, \bar{y}_2^{(\beta)}, \bar{y}_3^{(\gamma)})$:

$$\begin{aligned}
\Delta\mathbf{u}^{(\alpha\beta\gamma)} &= \Delta\bar{\epsilon}\cdot\mathbf{x} + \Delta\mathbf{W}_{(000)}^{(\alpha\beta\gamma)} + \bar{y}_1^{(\alpha)}\Delta\mathbf{W}_{(100)}^{(\alpha\beta\gamma)} + \bar{y}_2^{(\beta)}\Delta\mathbf{W}_{(010)}^{(\alpha\beta\gamma)} + \bar{y}_3^{(\gamma)}\Delta\mathbf{W}_{(001)}^{(\alpha\beta\gamma)} \\
&\quad + \frac{1}{2}\left(3\bar{y}_1^{(\alpha)2} - \frac{d_\alpha^2}{4}\right)\Delta\mathbf{W}_{(200)}^{(\alpha\beta\gamma)} + \frac{1}{2}\left(3\bar{y}_2^{(\beta)2} - \frac{h_\beta^2}{4}\right)\Delta\mathbf{W}_{(020)}^{(\alpha\beta\gamma)} \\
&\quad + \frac{1}{2}\left(3\bar{y}_3^{(\gamma)2} - \frac{l_\gamma^2}{4}\right)\Delta\mathbf{W}_{(002)}^{(\alpha\beta\gamma)}
\end{aligned} \tag{4.26}$$

$$\begin{aligned}
\Delta\xi^{(\alpha\beta\gamma)} &= -\Delta\bar{\mathbf{E}}\cdot\mathbf{x} + \Delta\xi_{(000)}^{(\alpha\beta\gamma)} + \bar{y}_1^{(\alpha)}\Delta\xi_{(100)}^{(\alpha\beta\gamma)} + \bar{y}_2^{(\beta)}\Delta\xi_{(010)}^{(\alpha\beta\gamma)} + \bar{y}_3^{(\gamma)}\Delta\xi_{(001)}^{(\alpha\beta\gamma)} \\
&\quad + \frac{1}{2}\left(3\bar{y}_1^{(\alpha)2} - \frac{d_\alpha^2}{4}\right)\Delta\xi_{(200)}^{(\alpha\beta\gamma)} + \frac{1}{2}\left(3\bar{y}_2^{(\beta)2} - \frac{h_\beta^2}{4}\right)\Delta\xi_{(020)}^{(\alpha\beta\gamma)} \\
&\quad + \frac{1}{2}\left(3\bar{y}_3^{(\gamma)2} - \frac{l_\gamma^2}{4}\right)\Delta\xi_{(002)}^{(\alpha\beta\gamma)}
\end{aligned} \tag{4.27}$$

where $\Delta\bar{\epsilon}$ and $\Delta\bar{\mathbf{E}}$ are the applied (external) strain and electric field increments, respectively, and the unknown terms $\Delta\mathbf{W}_{(lmn)}^{(\alpha\beta\gamma)}$, $\Delta\xi_{(lmn)}^{(\alpha\beta\gamma)}$ must be determined from the fulfillment at each increment of the current equilibrium conditions (4.13), Maxwell equation (4.14), the periodic boundary conditions, and the interfacial continuity conditions of mechanical displacements, tractions, electric potential and electric displacements between subcells. As previously discussed, all these conditions are imposed in the average (integral) sense.

As a result of the imposition of these conditions at each increment of loading, a linear system of algebraic equations is obtained which can be represented in the following form:

$$\mathbf{K}\Delta\mathbf{U} = \Delta\mathbf{g} \qquad (4.28)$$

where the matrix $\mathbf{K}$ contains information on the subcell material properties and dimensions, $\Delta\mathbf{U}$ contains the unknown terms $\Delta\mathbf{W}_{(lmn)}^{(\alpha\beta\gamma)}$, $\Delta\xi_{(lmn)}^{(\alpha\beta\gamma)}$ in the displacement and potential expansions, Eq. (4.26) and (4.27), respectively. The vector $\Delta\mathbf{g}$ contains information on the externally currently applied mechanical and electric field.

Once Eq. (4.28) is solved at a given increment of the electromechanical loading, the local strains and electric fields throughout the repeating unit cell can be determined at this increment as follows:

$$\Delta\mathbf{X}^{(\alpha\beta\gamma)} = \mathbf{A}^{EM(\alpha\beta\gamma)}\,\Delta\bar{\mathbf{X}} \qquad (4.29)$$

Equation (4.29) expresses the strain and electric field increments in the subcell $(\alpha\beta\gamma)$ in terms of the uniform overall strain and electric field increments (i.e., the applied macrostrain and macroelectric field increments) via the instantaneous electromechanical concentration tensors $\mathbf{A}^{EM(\alpha\beta\gamma)}$.

Substitution of Eq.(4.29) into (4.22) yields:

$$\Delta\mathbf{Y}^{(\alpha\beta\gamma)} = \mathbf{Z}^{(\alpha\beta\gamma)}\,\mathbf{A}^{EM(\alpha\beta\gamma)}\,\Delta\bar{\mathbf{X}} \qquad (4.30)$$

where $\mathbf{Z}^{(\alpha\beta\gamma)}$ is given by Eq. (4.25) which characterizes the material in subcell $(\alpha\beta\gamma)$.

Consequently, by averaging Eq. (4.30) over all subcells, the micromechanically established constitutive equations that govern the overall (global) behavior of the multiphase material at the present increment can be represented in the form:

$$\Delta\bar{\mathbf{Y}} = \mathbf{Z}^*\,\Delta\bar{\mathbf{X}} \qquad (4.31)$$

In this equation, $\Delta\bar{\mathbf{Y}}$ is the vector of the averages of the stress and electric displacement vector increments, namely:

$$\Delta\bar{\mathbf{Y}} = \frac{1}{DHL}\sum_{\alpha=1}^{N_\alpha}\sum_{\beta=1}^{N_\beta}\sum_{\gamma=1}^{N_\gamma} d_\alpha h_\beta l_\gamma\,\Delta\mathbf{Y}^{(\alpha\beta\gamma)} \qquad (4.32)$$

and $\Delta\bar{\mathbf{X}}$ is the vector of the averages of the strain and electric field increments in the composite. The matrix $\mathbf{Z}^*$ expresses the current effective electroelastic stiffness tensor which is determined from the instantaneous electromechanical concentration tensors $\mathbf{A}^{EM(\alpha\beta\gamma)}$ in the following closed-form manner:

$$\mathbf{Z}^* = \frac{1}{DHL}\sum_{\alpha=1}^{N_\alpha}\sum_{\beta=1}^{N_\beta}\sum_{\gamma=1}^{N_\gamma} d_\alpha h_\beta l_\gamma\,\mathbf{Z}^{(\alpha\beta\gamma)}\,\mathbf{A}^{EM(\alpha\beta\gamma)} \qquad (4.33)$$

The computation of the composite's response starts with constitutive relations Eq. (4.22) in which all field variables in matrices $\mathbf{Z}^{(\alpha\beta\gamma)}$ are assumed to be zero. By employing this matrix in the micromechanical analysis, the effective matrix $\mathbf{Z}^*$ is computed and the average field increments $\Delta\bar{\mathbf{Y}}$ can be readily determined. The field variables are obtained from:

$$\bar{\mathbf{Y}} = \bar{\mathbf{Y}}\,|_{previous} + \Delta\bar{\mathbf{Y}} \tag{4.34}$$

This process is repeated in the next increment in which the nonlinear field variable terms in the matrices $\mathbf{Z}^{(\alpha\beta\gamma)}$ are involved.

4.3. Applications: Macroscopic Hysteresis Response

Table 3. Comparison of predicted effective coefficients of thermal expansion of a boron/aluminum unidirectional composite ($\mathbf{v}_f = 0.47$).

Effective thermal expansion coefficients	α_{11} (10^{-6}/C)	α_{22} (10^{-6}/C)
HFGMC	11.0	16.7
Tamma and Avila [15]	10.77	17.34
GMC	10.91	16.94

Results are given herein in the case of continuous piezoceramic fibers, oriented in the 3-direction, reinforcing a polymeric matrix. In all cases, the fiber volume fraction is chosen as: $v_f = 0.4$. The fibers are transversely isotropic with the axis of symmetry oriented in the 3-direction. The repeating unit cell, Fig. 1(b), has been divided into $N_\alpha = N_\beta = N_\gamma = 4$ subcells. For continuous fibers oriented in the 3-direction, subcells $(\alpha\beta\gamma)$ with $\alpha, \beta = 2, 3$ and $\gamma = 1, ..., 4$, have been filled by the piezoceramic material, while all other subcells are occupied by the polymeric matrix. The mechanical and electric properties of the fibers are given in Table 1 and 2, respectively. The polymeric matrix is a dielectric material whose parameters are given in Table 3. All results presented in this chapter were computed by setting $\lambda = 1$ in Eq. (4.10), and in all figures the dashed and solid lines display the monolithic piezoceramic fiber and the piezoceramic-polymer composite responses, respectively.

Table 4. Comparison of predicted effective elastic moduli of a graphite/epoxy unidirectional composite ($\mathbf{v}_f = 0.60$).

Effective elastic moduli	E_{11} (GPa)	E_{22} (GPa)	G_{12} (GPa)	G_{23} (GPa)	ν_{12}	ν_{23}
HFGMC	142.9	9.61	6.09	3.10	0.252	0.350
Sun and Vaidya [14]	142.6	9.60	6.00	3.10	0.250	0.350
GMC	143.0	9.47	5.68	3.03	0.253	0.358

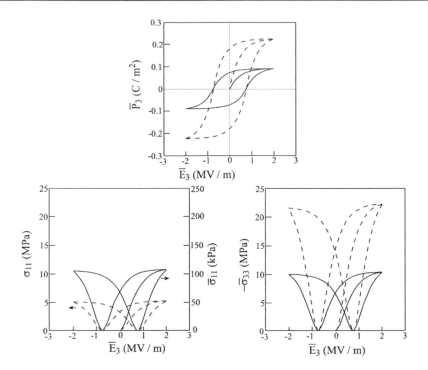

Figure 4.   Comparison between the response of the monolithic piezoceramic material (dashed lines) and the piezoceramic unidirectional composite (solid lines), both of which are subjected to a cyclic electric field ($|E_m| = 2MV/m$) applied in the 3-direction. In both cases the strains are kept equal to zero. The right/left horizontal arrow is directed toward the graph's appropriate ordinate.

Figure 4 exhibits the response of the monolithic piezoceramic fiber and the unidirectional composite to a cyclic electric field, whose amplitude $|E_m| = 2MV/m$, which is applied in the 3-direction (parallel to the fibers) while keeping all strains equal to zero (i.e., $\bar{\epsilon}_{ij} = 0$). The figure shows the resulting induced average polarization $\bar{P}_3$ and average transverse normal stresses $\bar{\sigma}_{11} = \bar{\sigma}_{22}$ and axial stress $\bar{\sigma}_{33}$. It can be observed that the matrix phase has a significant influence on the global electrical and mechanical response. The hysteresis loop of the polarization is reduced by about one half of its size, while the butterfly-like transverse normal stress is reduced by about 50 times (it should be noted that the left and right ordinates are measured in MPa and kPa, respectively). The low values of the transverse normal stress are far beyond the failure stress of the matrix. A much smaller (about one half) reduction in the axial normal stress is obtained.

Table 5. Elastic material properties.

Material	$C_{11}(GPa)$	$C_{12}(GPa)$	$C_{13}(GPa)$	$C_{33}(GPa)$	$C_{44}(GPa)$
$BaTiO_3$	166	77	78	162	43
$CoFe_2O_4$	286	173	170	269.5	45.3

A different type of behavior can be observed when the piezoceramic fiber and the composite are subjected to an electric field which is applied in the 3-direction, in the absence of any applied stress (i.e., $\bar{\sigma}_{ij} = 0$). The corresponding responses are shown in Fig. 5. This figure shows the average polarization $\bar{P}_3$ and the butterfly patterns exhibited by the induced average transverse strains $\bar{\epsilon}_{11} = \bar{\epsilon}_{22}$ and axial strain $\bar{\epsilon}_{33}$. It is readily observed that in the present circumstances the effect of the matrix in which the piezoceramic fibers are embedded on the induced strains appears to be very weak. Further results, which show the behavior of the ferroelectric fiber composite subjected to constant normal tensile and compressive stresses that are applied in the axial and transverse directions, as well as applied transverse shear and axial shear stresses, have been presented in [41].

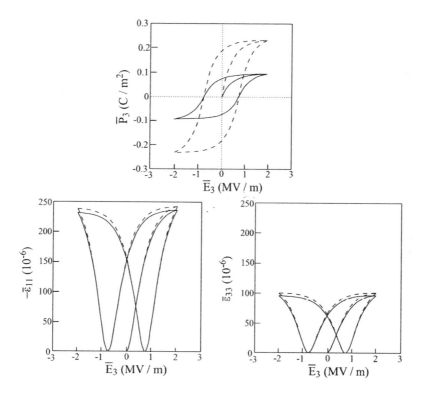

Figure 5.   Comparison between the response of the monolithic piezoceramic material (dashed lines) and the piezoceramic unidirectional composite (solid lines), both of which are subjected to an electric field ($|E_m| = 2MV/m$) applied in the 3-direction. In both cases the stresses are kept equal to zero.

5.   The Response of Electrostrictive Composites

Electrostriction is characterized by the mechanical deformation which occurs in a dielectric material when an electric field is applied. It occurs in any material, but the strains generated in most dielectrics are too small to be utilized for sensing and actuating. However, certain dielectrics (e.g. PMN ceramic) exhibit sufficiently large dielectric permittivities so as to generate appreciable polarization and strains that can be utilized in practical applications.

Table 6. Electric material properties.

Material	$e_{15}(C/m^2)$	$e_{31}(C/m^2)$	$e_{33}(C/m^2)$	$\kappa_{11}(10^{-9}C/Vm)$	$\kappa_{33}(10^{-9}C/Vm)$
$BaTiO_3$	11.6	-4.4	18.6	11.2	12.6
$CoFe_2O_4$	0	0	0	0.08	0.093

Table 7. Magnetic material properties.

Material	q_{15} (N/Am)	q_{31} (N/Am)	q_{33} (N/Am)	μ_{11} $(10^{-6}Ns^2/C^2)$	μ_{33} $(10^{-6}Ns^2/C^2)$
$BaTiO_3$	0	0	0	5	10
$CoFe_2O_4$	550	580.3	699.7	-590	157

Unlike piezoelectric materials, uncharged electrostrictives are isotropic and are not poled. In addition, the relationship between the induced strains and polarization in electrostrictive dielectrics is nonlinear of a second order.

· A brief discussion of electrostrictive materials is given in the book by [3] for example, where several of their advantages and disadvantages are listed. Similarly, numerous advantages of electrostrictive over piezoelectric materials are mentioned in [39], such as the very low hysteresis that is exhibited by the former ones. Several practical applications of electrostrictive materials have been presented by [42] and [43]. A detailed account on piezoelectric and electrostrictive materials, actuators and their practical applications is given in a recent monograph by [44].

Like piezoelectrics, electrostrictives can be utilized by combining them with polymeric materials, thus forming a composite material that can be used in practical applications. For example, in ref. [45] the authors utilized electrostrictive-polymer composite as an ultrasonic probe for medical applications. Electrostrictive particles embedded in composites can be utilized as sensors to detect internal stress status in the composites. This non-distractive tagging method [46], is based on the fact that an application of mechanical stresses to an electrostrictive material generates a measurable voltage under an applied electric field.

A three-dimensional constitutive relation for electrostrictive materials has been recently developed by [47], where it is argued that a constitutive theory based on polarization may

Table 8. Elastic constants of the piezoceramic fibers.

C_{11} (GPa)	C_{12} (GPa)	C_{13} (GPa)	C_{33} (GPa)	C_{44} (GPa)
166	77	78	162	43

Table 9. Electric constants of the piezoceramic fibers.

d_{31} ($10^{-10}C/N$)	d_{33} ($10^{-9}C/N$)	E_c (MV/m)	P_r (C/m^2)	P_s (C/m^2)
-0.795	0.191	0.75	0.2	0.25

model electrostriction better than a theory based on electric field. The material constants that are involved in these relations were obtained by an extensive electromechanical testing of a type of PMN ceramic [48]. These constitutive relations show that the induced strain is proportional to the square of the induced polarization, and displays the saturation effect at high values of electric field. These constitutive equations have been implemented in a finite element analysis to analyze electrostrictive devices, see ref. [49].

5.1.   The Modeling of the Monolithic Electrostrictive Material

In ref. [47], a Helmholtz free energy function ψ is proposed which is expressed in terms of the total strain ϵ_{ij}, polarization P_i and temperature T. From this function the fully coupled constitutive equations of electrostrictive isotropic materials can be derived. This function is given by

$$\psi(\epsilon_{ij}, P_i, T) = \frac{C_{ijkl}}{2}\left[\epsilon_{ij} - \epsilon_{ij}^E - \epsilon_{ij}^T\right]\left[\epsilon_{kl} - \epsilon_{kl}^E - \epsilon_{kl}^T\right] + F(|\mathbf{P}|) \qquad (5.1)$$

where ϵ_{ij}^E and ϵ_{ij}^T are the electric and thermal strains, respectively, $C_{ijkl} = \lambda\delta_{ij}\delta_{kl} + \mu(\delta_{ik}\delta_{jl} + \delta_{il}\delta_{jk})$ are elastic stiffness coefficients of the material (λ and μ are the Lamè constants of the isotropic material, and δ_{ij} is the Kronecker delta), $|\mathbf{P}| = \sqrt{P_1^2 + P_2^2 + P_3^2}$, and the summation convention for repeated Latin indices is employed, with $i, j, k, l = 1, 2, 3$.

The components of the electric strain ϵ_{ij}^E are given by

$$\epsilon_{ij}^E = (Q_{11} - Q_{12})P_iP_j + Q_{12}P_kP_k\delta_{ij} \qquad (5.2)$$

where Q_{11}, Q_{12} are the electrostrictive coefficients of the isotropic material, and $F(|\mathbf{P}|)$ is a function that reflects the symmetric quadratic dependence of the induced strain versus the electric field that is observed in electrostrictive materials, and provides the saturation feature for high values of electric field. This function is given by [47] in the following form:

$$F(|\mathbf{P}|) = \frac{1}{2k_0}\left[|\mathbf{P}|\ln\frac{p_s + |\mathbf{P}|}{p_s - |\mathbf{P}|} + p_s\ln(1 - \frac{|\mathbf{P}|^2}{p_s^2})\right] \qquad (5.3)$$

with k_0 and p_s being two material constants.

The thermal strains ϵ_{ij}^T in Eq. (5.1) are given by

$$\epsilon_{ij}^T = \alpha(T - T_0)\delta_{ij} \qquad (5.4)$$

where α is the coefficient of thermal expansion and T_0 is a reference temperature.

Once the Helmholtz function has been established, one can determine the stress σ_{ij} and electric field E_i components as follows:

$$\sigma_{ij} = \frac{\partial \psi}{\partial \epsilon_{ij}} \tag{5.5}$$

$$E_i = \frac{\partial \psi}{\partial P_i} \tag{5.6}$$

With ψ given by Eq. (5.1), the resulting constitutive relations of the electrostrictive material are given by

$$\sigma_{ij} = C_{ijkl} \left[\epsilon_{kl} - \epsilon_{kl}^E - \epsilon_{kl}^T \right] \tag{5.7}$$

$$E_i = -2(Q_{11} - Q_{12})\sigma_{ij}P_j - 2Q_{12}\sigma_{kk}P_i + f_i(|\mathbf{P}|) \tag{5.8}$$

where

$$f_i(|\mathbf{P}|) = \frac{\partial F}{\partial P_i} = \frac{1}{k_0} \frac{P_i}{|\mathbf{P}|} \, tanh^{-1} \frac{|\mathbf{P}|}{p_s} \tag{5.9}$$

It should be noted that in contrast to a piezoelectric material, Eq. (5.8) shows that an electrostrictive material that is not subjected to an electric field will not polarize under mechanical loading.

Due to the nonlinearity of the constitutive equations, our analysis must be incremental. Furthermore, it is based on formulating the field equations in a tangential form. This formulation will provide a system of linear algebraic equations that need to be solved at each increment (which is of great advantage over solving nonlinear ones in a non-tangential formulation). To this end, the two vectors $\Delta \mathbf{Y}_1$ and $\Delta \mathbf{X}_1$, defined by Eqs. (4.15)-(4.16), are introduced together with the temperature increment ΔT.

After some lengthy manipulations, the constitutive relations (5.7)-(5.8) can be written in the form:

$$\Delta \mathbf{Y}_1 = \mathbf{H} \, \Delta \mathbf{X}_1 + \mathbf{H}_1 \, \Delta T \tag{5.10}$$

where the $9th$-order square symmetric matrix of coefficients $\mathbf{H}$ has the same form as shown by Eq. (4.18), and the $9th$-order vector $\mathbf{H}_1$ can be represented by

$$\mathbf{H}_1 = \begin{bmatrix} \mathbf{M}_1 \\ \mathbf{M}_2 \end{bmatrix} \tag{5.11}$$

where the vectors $\mathbf{M}_1$ and $\mathbf{M}_2$ are of 6 and 3rd- order, respectively. It should be mentioned that in the representation of the ferroelectric response that was discussed in the previous section, the terms that represent the hysteresis loops (i.e., Eqs. (4.20) and their derivatives) appear in the diagonal elements of $\mathbf{N}$ only. In the present electrostrictive model, on the other hand, all the elements of matrix $\mathbf{N}$ involve the functions $f_i(|\mathbf{P}|)$, given by (5.9), and their derivatives.

In the next section, in order to impose the equilibrium and Maxwell equations and the proper continuity conditions across the various interfaces of the composite, the constitutive relations of the electrostrictive ceramic material need to be given in terms of the increments of the mechanical stresses $\Delta \sigma_{ij}$ and electrical displacements ΔD_i as independent variables. The latter are given by Eq. (4.21).

Table 10. Material constants of the isotropic polymeric matrix.

Young's modulus (GPa)	Poisson's ratio	Dielectric constant $(10^{-9}C/Vm)$
1.8	0.4	0.079

Consequently, the tangential constitutive relation (5.10) can be rewritten in the following standard form:

$$\Delta \mathbf{Y} = \mathbf{Z}\,\Delta \mathbf{X} - \boldsymbol{\Gamma}\,\Delta T \tag{5.12}$$

where $\Delta \mathbf{X}$ and $\Delta \mathbf{Y}$ have been defined by Eqs. (4.23) and (4.24), respectively. The 9th-order symmetric tangent electromechanical matrix $\mathbf{Z}$ is given by Eq. (4.25). In Eq. (5.12), the thermal stress-pyroelectric vector $\boldsymbol{\Gamma}$ is given by

$$\boldsymbol{\Gamma} = \left[\begin{array}{c} \mathbf{M}\,\mathbf{M}_1^{-1}\,\mathbf{M}_2 - \mathbf{M}_1 \\ \mathbf{N}^{-1}\,\mathbf{M}_2 \end{array} \right] \tag{5.13}$$

The micromechanical analysis that models multiphase composites with electrostrictive constituents will be carried out using the tangential constitutive relations (5.12).

5.2. Incremental Micromechanical Analysis

Due to the nonlinearity of the electrostrictive material, its constitutive relations have been formulated in the incremental form as shown by Eq. (5.12).

These relations can be incorporated with the HFGMC incremental micromechanical analysis. This analysis is identical to the one discussed for the prediction of the hysteresis behavior of ferroelectric composites. By establishing the instantaneous electromechanical concentration matrix $\mathbf{A}^{EM(\alpha\beta\gamma)}$ and the corresponding current electrothermal concentration vector $\mathbf{A}^{ET(\alpha\beta\gamma)}$ in the subcell, we obtain that:

$$\Delta \mathbf{X}^{(\alpha\beta\gamma)} = \mathbf{A}^{EM(\alpha\beta\gamma)}\,\Delta \bar{\mathbf{X}} + \mathbf{A}^{ET(\alpha\beta\gamma)}\Delta T \tag{5.14}$$

where ΔT is the temperature deviation. The resulting constitutive equation that governs the behavior of the electrostrictive composite is given by

$$\Delta \mathbf{Y} = \mathbf{Z}^*\,\Delta \mathbf{X} - \boldsymbol{\Gamma}^*\Delta T \tag{5.15}$$

where the effective electromechanical tangent tensor $\mathbf{Z}^*$ is given by Eq. (4.33), while the effective thermal and pyroelectric vector is given by (cf. Eq. (2.36)):

$$\boldsymbol{\Gamma}^* = \frac{-1}{DHL}\sum_{\alpha=1}^{N_\alpha}\sum_{\beta=1}^{N_\beta}\sum_{\gamma=1}^{N_\gamma} d_\alpha h_\beta l_\gamma \left[\mathbf{Z}^{(\alpha\beta\gamma)}\mathbf{A}^{ET(\alpha\beta\gamma)} - \boldsymbol{\Gamma}^{(\alpha\beta\gamma)} \right] \tag{5.16}$$

where $\mathbf{Z}^{(\alpha\beta\gamma)}$, given by Eq. (4.25), characterizes the material in subcell $(\alpha\beta\gamma)$.

Table 11. Material properties of PMN at $5°C$ [49] - [50]. The parameters Y, ν and α denote the Young's modulus, Poisson's ratio and the thermal expansion coefficient, respectively, Q_{11} and Q_{12} are coupling coefficients, and p_s, k_0 are parameters that appear in Eq. (5.9).

$Y(GPa)$	ν	$Q_{11}(m^4/C^2)$	$Q_{12}(m^4/C^2)$	$p_s(C/m^2)$	$k_0(m/MV)$	$\alpha(/°C)$
115	0.26	0.0133	-0.00606	0.2589	1.16	1×10^{-6}

5.3. Application: Porous PMN Material Behavior

The derived tangential constitutive relations for multiphase electrostrictive composites will be implemented to investigate the behavior of porous PMN material composite. The properties of the isotropic polycrystalline PMN ceramic material are given in Table 11, taken from ref. [49] and [50].

Consider a porous PMN material that is subjected to an external electric field $\bar{E}_1$ in the 1-direction (that will be referred to as the axial direction), while all external strains $\bar{\epsilon}_{ij}$ and the other two electric components are kept equal to zero, and isothermal conditions are assumed to prevail. Figure 6 shows the resulting overall induced polarization $\bar{P}_1$, and the induced axial and transverse stresses $\bar{\sigma}_{11}$ and $\bar{\sigma}_{22}$, for various values of the pore volume fraction (porosity) v_f ($v_f = 0$ corresponds to the monolithic PMN ceramic with zero porosity whose parameters have been given in Table 11). This figure clearly exhibits the effect of porosity on the electromechanical behavior of the ceramic relaxor. The strong coupling between mechanical and electrical effects is seen to be pronounced even at high values of porosities.

The electric behavior of the electrostrictive material under mechanical loading can be utilized for sensing applications. Thus, by recording the electric response of the material, one can infer about the magnitude of the applied mechanical loading. This is illustrated in Figs. 7 and 8 where, under isothermal conditions, the porous electrostrictive PMN is subjected to pre-stresses $\bar{\sigma}_{11} = 0$ and $\bar{\sigma}_{11} = -80MPa$, respectively, while all other average stress and transverse electric field components are kept equal to zero. Both figures show the average induced polarization $\bar{P}_1$, and the average electric strains $\bar{\epsilon}_{11}^E$ and $\bar{\epsilon}_{22}^E$. The increments of the latter are respectively given by

$$\Delta \bar{\epsilon}_{11}^E = Z_{17}^{*-1} \Delta \bar{D}_1 + Z_{18}^{*-1} \Delta \bar{D}_2 + Z_{19}^{*-1} \Delta \bar{D}_3$$

$$\Delta \bar{\epsilon}_{22}^E = Z_{27}^{*-1} \Delta \bar{D}_1 + Z_{28}^{*-1} \Delta \bar{D}_2 + Z_{29}^{*-1} \Delta \bar{D}_3$$

from which one can determine the current values of these strains. It should be noted that in the absence of an applied pre-stress (as it is the case in Fig. 7), the electric strains are equal to the total strains.

The two figures show the effect of the applied mechanical loading on the behavior of the material with various amounts of porosities. A comparison between the two figures show that the compressive applied stress can be measured by the drop in polarization, the drop in the induced electric strains, and the delay in their saturation. The changes in the

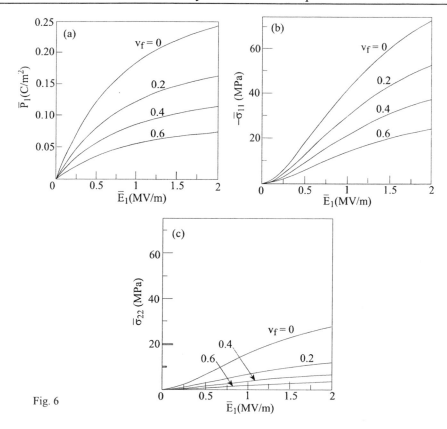

Figure 6. The isothermal effective response of porous PMN ceramic versus applied electric field in the axial direction. All average strains are kept equal to zero. (a) Overall induced polarization, (b) overall induced axial stress, (c) overall induced transverse stress.

electric strain due to the applied pre-stress forms a mean for actuating purposes. Thus, both sensing and actuating can be obtained from the porous PMN ceramic, and the present micromechanical theory provides a mean to quantify these effects. The porous material stress-strain response under constant electric field and its anisothermal behavior are shown in [51], where the GMC model has been employed for the modeling of the electrostrictive composite. Results are also shown therein for a PMN/polymer composite.

6. Analysis of Magnetostrictive Composites

The effect of magnetostriction is similar to that of electrostriction. It refers to the phenomenon in which an applied magnetic field generates elastic strain in ferromagnetic materials (e.g., 2000μ strains at moduli in the order of 100 GPa [4]). Thus, such materials provide actuator capabilities since they can deform in the presence of a magnetic field. In addition, sensing can be also accomplished, since elastic strain applied on a magnetostrictive material produces magnetic polarization. Similarly to electrostriction, magnetostriction is a nonlinear phenomenon because the relation between the applied magnetic field and the resulting elastic strain is nonlinear [52]. In particular, the authors in ref. [53] proposed a set

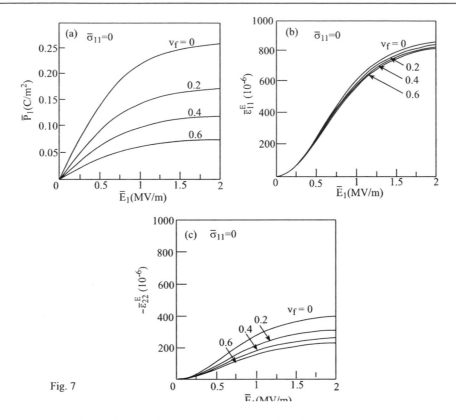

Fig. 7

Figure 7. The isothermal effective response of traction-free porous PMN ceramic versus applied electric field in the axial direction. (a) Overall induced polarization, (b) overall induced axial electric strain, (c) overall induced transverse electric strain.

of constitutive equations that are based on those proposed by [47] which were employed in the previous section for the modeling of electrostrictive materials. In these equations, the electrical field and polarization are replaced by the magnetic field and magnetization, respectively. As a result, the predicted behavior of their magnetostrictive modeling generates the expectedly observed saturation of the strains for a sufficiently large magnetic field.

Consequently, it is possible to utilize the modeling approach of ref. [53] for the characterization of monolithic magnetostrictive materials, and to derive micromechanically established constitutive relations of magnetostrictive composites by employing the HFGMC procedure. This HFGMC modeling approach should be quite similar to the analysis of electrostrictive composites that has been presented in the previous section.

7. Nonlinear Electro-Magneto-Thermo-Elastic Composites

When electro-magneto-thermo-elastic materials (such as piezoelectric) are subjected to strong electric and/or magnetic fields, nonlinear behavior is obtained. As a result, the linear constitutive relations (3.1)-(3.3) are not valid and additional higher-order terms must be included. This can be achieved by expanding the Gibbs potential in a Taylor's series up to a second order (say). It should be mentioned that such an expansion would not provide the

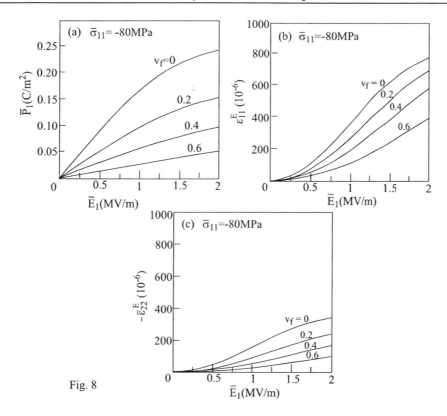

Fig. 8

Figure 8. The isothermal effective response of porous PMN ceramic that is subjected to a compressive pre-stress $\bar{\sigma}_{11} = -80MPa$, versus applied electric field in the axial direction. (a) Overall induced polarization, (b) overall induced axial electric strain, (c) overall induced transverse electric strain.

hysteresis behavior of the material, unless a special switching criterion or a phenomenological model are included.

In order to establish the requested nonlinear constitutive relations, ref. [54] starts with the relation that expresses the balance between the rate of the total energy in a volume V and the power of body forces, intensity of heat sources, work rate of tractions, and the rate of electromagnetic energy:

$$\frac{d}{dt}\int_V \left[\frac{1}{2}\rho \dot{u}_i \dot{u}_i + U\right] dV = \int_V \left[F_i \dot{u}_i + W\right] dV + \int_S \left[t_i \dot{u}_i - q_i n_i - (\mathbf{E} \times \mathbf{H}) \cdot n_i\right] dS \quad (7.1)$$

where U, u_i, F_i, t_i, q_i, W, ρ, E_i, H_i and n_i are the internal energy, mechanical displacements, body forces, tractions, heat flow, heat source intensity, mass density, electric field, magnetic field and the unit vector normal to the surface S, respectively. By using the mechanical equilibrium and Maxwell's equations, the following relation can be established:

$$\dot{U} = \sigma_{ij}\dot{\epsilon}_{ij} + E_i\dot{D}_i + H_i\dot{B}_i + W - \frac{\partial q_i}{\partial x_i} \quad (7.2)$$

where σ_{ij}, ϵ_{ij}, D_i and B_i are the components of the stress, strain, electric displacement and magnetic flux density, respectively.

The internal energy U is a function of ϵ_{ij}, D_i, B_i and the entropy s so that:

$$\dot{U} = \frac{\partial U}{\partial \epsilon_{ij}}\dot{\epsilon}_{ij} + \frac{\partial U}{\partial D_i}\dot{D}_i + \frac{\partial U}{\partial B_i}\dot{B}_i + \frac{\partial U}{\partial s}\dot{s} \qquad (7.3)$$

By employing the Clausius-Duhem inequality:

$$T\dot{s} + T\frac{\partial}{\partial x_i}\left(\frac{q_i}{T}\right) - W \geq 0 \qquad (7.4)$$

where T denotes the temperature, the following relations are obtained:

$$\sigma_{ij} = \frac{\partial U}{\partial \epsilon_{ij}}, \quad E_i = \frac{\partial U}{\partial D_i}, \quad H_i = \frac{\partial U}{\partial B_i}, \quad T = \frac{\partial U}{\partial s} \qquad (7.5)$$

The Gibbs potential is defined by

$$G = U - E_i D_i - H_i B_i - Ts \qquad (7.6)$$

By employing Eq. (7.3) and (7.5) to evaluate the total differential dU and using Eq. (7.6) to determine the differential dG, the following relations can be readily established:

$$\sigma_{ij} = \frac{\partial G}{\partial \epsilon_{ij}}, \quad D_i = -\frac{\partial G}{\partial E_i}, \quad B_i = -\frac{\partial G}{\partial H_i}, \quad s = -\frac{\partial G}{\partial T} \qquad (7.7)$$

By expanding the differential dG to a second order, one obtains (e.g., [55], [56]) three nonlinear constitutive relations for σ_{ij}, D_i and B_i (the entropy is excluded herein) which contain the linear terms in Eqs. (3.1) - (3.3), together with the additional nonlinear terms up to 2nd-order, namely, $\epsilon_{ij}\epsilon_{kl}$, $E_i E_j$, $H_i H_j$, $\epsilon_{ij}E_k$ $\epsilon_{ij}H_k$, $E_i H_j$, $\epsilon_{ij}T$, $E_i T$, $H_i T$ and T^2. These equations can be also represented in a matrix form, see [40] and [57] for the electro-elastic and the fully electro-magneto-thermo-elastic case, respectively.

After some lengthy manipulations, the incremental form of these equations can be represented in the form:

$$\Delta\mathbf{Y} = \mathbf{Z}\Delta\mathbf{X} - \mathbf{\Gamma}\Delta T \qquad (7.8)$$

where $\Delta\mathbf{X}$ and $\Delta\mathbf{Y}$ are the increments of vectors $\mathbf{X}$ and $\mathbf{Y}$ that were defined by Eqs. (3.4) and (3.5), respectively, and $\mathbf{Z}$ and $\mathbf{\Gamma}$ have the same form as Eqs. (3.7) and (3.8).

With the incremental constitutive equations (7.8), the HFGMC micromechanical analysis proceeds as in the linear case that was described in Section 3. by imposing the continuity and periodicity of the increments of the appropriate field variables, yielding the composite's incremental constitutive equation of the form (5.15).

7.1. Application: LiNbO$_3$/PVDF Composite Response

Figure 9 exhibits the behavior of a unidirectional composite that is composed of $LiNbO_3$ fibers (oriented in the 3-direction) embedded in $PVDF$ matrix. The material properties of the nonlinear $LiNbO_3$ piezoelectric fibers and the linear $PVDF$ piezoelectric matrix are given by [40]. This figure is based on [57] micromechanical analysis in which the GMC model has been employed to predict the global nonlinear piezoelectric behavior.

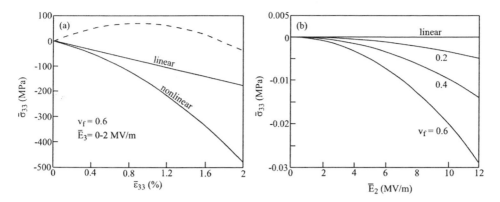

Figure 9. (a) The response of $LiNbO_3/PVDF$ nonlinear piezoelectric composite with fiber volume fraction $v_f = 0.6$ and $E_3 = 0 - 2MV/m$. Also shown are the corresponding linear response where the nonlinear effects have been neglected, and the nonlinear response (dashed line) as predicted by [40]. (b) The $\sigma_{33} - E_2$ response of this composite to an electric field applied in the transverse direction for various values of fiber volume fraction v_f.

Figure 9(a) exhibits a comparison between the nonlinear axial stress-strain response of the composite based on GMC analysis and the corresponding one in which the nonlinearity of the fiber is neglected. It is clearly observed that the effect of nonlinearity is significant, and, as expected, the linear response (whose predicted reliability was demonstrated in Section 3. coincides with the initial tangent of the nonlinear one. Figure 9(a) shows also the nonlinear response which was predicted by [40] by employing a simplifying micromechanics model based on iso-stress iso-electric displacement and iso-strain iso-electric field. Figure 9(b) presents the axial stress σ_{33} in the fiber direction which is induced by the application of electric field E_2 in the transverse direction for various values of fiber volume fraction v_f. Here too, the corresponding linear response (that always predicts zero stress) is included, well exhibiting the strong nonlinear effects under the present circumstances. Further detailed results are given by [57].

8.   HFGMC for Inelastic Composites

In the following section, composites with embedded shape-memory alloy materials are discussed. This type of material involves transformation strains which play the same role as inelastic (plastic or viscoplastic) strains. Consequently, we present in this section the generalization of the HFGMC model for inelastic materials and its applications. The model's reliability is verified by comparisons with exact solutions that can be established under certain circumstances and a finite element procedure, see ref. [58] and [59] for more details.

The constitutive relation that governs the response of the thermo-inelastic isotropic material filling subcell $(\alpha\beta\gamma)$ of Fig. 1(c) can be written in the form (cf. Eq. (2.11)):

$$\sigma^{(\alpha\beta\gamma)} = \mathbf{C}^{(\alpha\beta\gamma)}\epsilon^{(\alpha\beta\gamma)} - 2\mu^{(\alpha\beta\gamma)}\epsilon^{I(\alpha\beta\gamma)} - \mathbf{\Gamma}^{(\alpha\beta\gamma)}\Delta T \tag{8.1}$$

where $\epsilon^{I(\alpha\beta\gamma)}$ and $\mathbf{\Gamma}^{(\alpha\beta\gamma)}$ are the inelastic strain and thermal stress, respectively, ΔT is the temperature deviation from a reference temperature and $\mu^{(\alpha\beta\gamma)}$ is the elastic shear modulus.

By employing the displacement quadratic expansion (2.29), and imposing on an average basis the equilibrium equations, the continuity conditions at the various interfaces between the subcells and the periodicity conditions, the following system of $21N_\alpha N_\beta N_\gamma$ equations is obtained (cf. Eq. (2.30)):

$$\mathbf{KU} = \mathbf{f} + \mathbf{g} \tag{8.2}$$

where the structural stiffness matrix $\mathbf{K}$ contains information on the geometry and mechanical properties of the materials occupying the individual subcells $(\alpha\beta\gamma)$, within the cells comprising the multiphase periodic composite. The displacement vector $\mathbf{U}$ contains the unknown displacement coefficients in each subcell. The force $\mathbf{f}$ contains information on the thermomechanical properties of the materials filling the subcells, the applied average strains $\bar{\epsilon}$ and the imposed temperature deviation ΔT. The inelastic force vector $\mathbf{g}$ appearing on the right-hand side of Eq. (8.2) contains the inelastic effects given in terms of the integrals of the inelastic strain distributions. These integrals depend implicitly on the elements of the displacement coefficient vector $\mathbf{U}$, requiring an incremental procedure for the solution of Eq. (8.2) at each point along the loading path.

The solution of Eq. (8.2) enables the establishment of the following localization relation which expresses the average strain $\bar{\epsilon}^{(\alpha\beta\gamma)}$ in the subcell $(\alpha\beta\gamma)$ to the external applied strain $\bar{\epsilon}$ in the form [59]:

$$\bar{\epsilon}^{(\alpha\beta\gamma)} = \mathbf{A}^{M(\alpha\beta\gamma)}\bar{\epsilon} + \mathbf{A}^{T(\alpha\beta\gamma)}\Delta T + \mathbf{A}^{I(\alpha\beta\gamma)} \tag{8.3}$$

where $\mathbf{A}^{M(\alpha\beta\gamma)}$ is the mechanical strain concentration matrix of the subcell $(\alpha\beta\gamma)$, and the vectors $\mathbf{A}^{T(\alpha\beta\gamma)}$ and $\mathbf{A}^{I(\alpha\beta\gamma)}$ involve the thermoelastic and the current inelastic effects in the subcell.

The final form of the effective constitutive law of the multiphase thermo-inelastic composite, which relates the average stress $\bar{\sigma}$ and strain $\bar{\epsilon}$, is established as follows:

$$\bar{\sigma} = \mathbf{C}^* \bar{\epsilon} - (\bar{\sigma}^T + \bar{\sigma}^I) \tag{8.4}$$

In this equation $\mathbf{C}^*$ is the effective elastic stiffness matrix of the composite whose form is identical to that given by Eq. (2.33), $\bar{\sigma}^T = \mathbf{\Gamma}^* \Delta T$ where $\mathbf{\Gamma}^*$ is given by Eqs. (2.34) or (2.36) and the global inelastic stress $\bar{\sigma}^I$ given by

$$\bar{\sigma}^I = \frac{-1}{DHL} \sum_{\alpha=1}^{N_\alpha} \sum_{\beta=1}^{N_\beta} \sum_{\gamma=1}^{N_\gamma} \left[\mathbf{C}^{(\alpha\beta\gamma)} \mathbf{A}^{I(\alpha\beta\gamma)} - \mathbf{R}_{(000)}^{(\alpha\beta\gamma)} \right] \tag{8.5}$$

where the term $\mathbf{R}_{(000)}^{(\alpha\beta\gamma)}$ represents inelastic stress effects in the phase occupying the subcell $(\alpha\beta\gamma)$, see ref. [59] for more details.

8.1. Application: Graphite/Aluminum Macroscopic Response

In ref. [58] and [59], the predictions obtained from the inelastic HFGMC were compared with those of the multiple concentric cylinder model (MCCM), which provides an analytical solution for a continuous reinforced graphite/aluminum inelastic composite, subjected to an axisymmetric loading due to a spatially uniform temperature [60], as well as under axial shear loading [61].

Table 12. Elastic, plastic and thermal parameters of the isotropic aluminum matrix.

$E(GPa)$	ν	$\alpha(10^{-6}/°C)$	$\sigma_y(MPa)$	$E_s(GPa)$
72.4	0.33	22.5	286.67	11.7

Table 13. Elastic and thermal parameters of the transversely isotropic graphite fiber.

$E_A(GPa)$	ν_A	$E_T GPa)$	ν_T	$G_A(GPa)$	$\alpha_A(10^{-6}/°C)$	$\alpha_T(10^{-6}/°C)$
388.2	0.41	7.6	0.45	14.9	-0.68	9.74

E, ν, α, σ_y and E_s denote the Young's modulus, Poisson's ratio, coefficient of thermal expansion, yield stress and secondary modulus, respectively.

E, ν, G and α denote the Young's modulus, Poisson's ratio, rigidity and coefficient of thermal expansion, respectively, in the axial (A) and transverse (T) directions.

In addition, comparisons with a finite element solution are shown. The properties of the elastic graphite fibers and the elastoplastic aluminum are given in Table 12 and 13, respectively. Figures 10-12 present these comparisons which show excellent agreements. Comparisons between the local field, as predicted by HFGMC and MCCM models and by a finite element procedure, are given by [59].

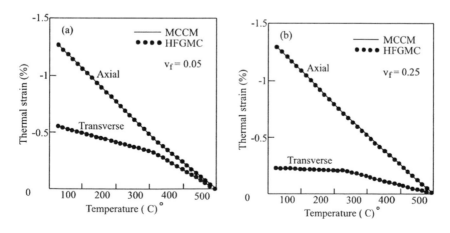

Figure 10. Macroscopic thermal response of a unidirectional graphite/aluminum composite during spatially uniform cooldown from 500 to 25°C: (a) $v_f = 0.05$; (b) $v_f = 0.25$, where v_f denotes the fiber volume fraction.

Extension of the HFGMC model to analyze inelastic composites with imperfect bonding between the phases was performed by [62].

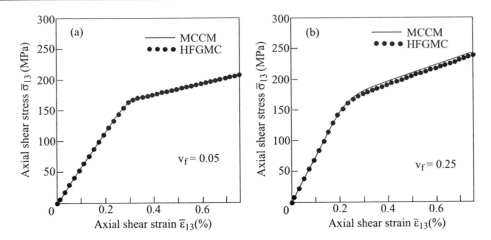

Figure 11.   Macroscopic axial shear stress-strain response of a unidirectional graphite/aluminum composite. (a) $v_f = 0.05$; (b) $v_f = 0.25$, where v_f denotes the fiber volume fraction.

9.   Shape-Memory Alloy Fiber Composites

Shape-memory alloy (SMA) materials form a class of intelligent materials which by the application of stress or temperature they exhibit shape-memory and superelasticity effects. The shape-memory effect results from the application at a constant temperature of mechanical loading and unloading that generates a significant residual strain which can be eliminated by the application of a thermal cycle. The superelasticity effect is obtained by the application, at a higher constant temperature, of mechanical loading and unloading, which generates a nonlinear large deformation, but with zero residual strain. Both effects can be utilized in practical applications.

There are several three-dimensional models which represent the SMA behavior, see [63], [64], [65] for example, and the review articles by [66] and [7] and references cited there. Micromechanical models that can incorporate such three-dimensional SMA constitutive equations would provide the behavior of composites that consist of a matrix with embedded SMA fibers. By employing such micromechanical analyses, one can obtain the composite response under various types of loading, the effect of fiber volume fraction, shape, orientation and several other properties. Examples for such micromechanical investigations are those of [63], [67], [68], and [69].

In the present section, the macroscopic response of composite materials with SMA fibers are predicted by employing the HFGMC in which the constitutive equations of ref. [63] are utilized for the modeling of the fibers. This model has been further extended by [70] to include the effect of compressive loading and, therefore, allowing cyclic loading. To this end, let A_{os}, A_{of}, M_{os}, M_{of} denote the austenitic start, austenitic finish, martensitic start, and martensitic finish temperatures under stress free state, respectively. According to this model the phase transition fraction from austenitic to martensitic $\xi = 0 \rightarrow 1$ is given by

$$\xi = 1 - exp\left[a_M(M_{os} - T) + \frac{b_M}{H}\sigma_{ij}\Lambda_{ij}\right] \quad M_f \leq T \leq M_s \quad (9.1)$$

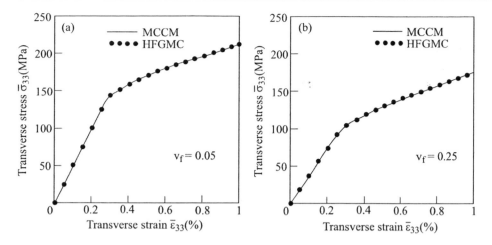

Figure 12. Comparison between HFGMC and finite element (FEM) prediction of the macroscopic transverse stress-strain response of a unidirectional graphite/aluminum composite due to the loading: $\bar{\epsilon}_{22} = -\bar{\epsilon}_{33}$ and $\bar{\epsilon}_{11} = 0$. (a) $v_f = 0.05$; (b) $v_f = 0.25$, where v_f denotes the fiber volume fraction.

whereas the phase transition fraction from martensitic to austenitic $\xi = 1 \rightarrow 0$ is given by

$$\xi = exp\left[a_A(A_{os} - T) + \frac{b_A}{H}\sigma_{ij}\Lambda_{ij}\right] \quad A_s \leq T \leq A_f \tag{9.2}$$

where T is the temperature and H is a material constant. In addition,

$$M_s = M_{os} + \frac{1}{HC_M}\sigma_{ij}\Lambda_{ij} \quad M_f = M_{of} + \frac{1}{HC_M}\sigma_{ij}\Lambda_{ij}$$

$$A_s = A_{os} + \frac{1}{HC_A}\sigma_{ij}\Lambda_{ij} \quad A_f = A_{of} + \frac{1}{HC_A}\sigma_{ij}\Lambda_{ij}$$

where C_M, C_A are material constants, and $a_M = ln(0.01)/(M_s - M_f)$, $a_A = ln(0.01)/(A_s - A_f)$, $b_M = a_M/C_M$, $b_A = a_A/C_A$. Furthermore, the transformation strain rate $\dot{\epsilon}_{ij}^{tr}$ is given by

$$\dot{\epsilon}_{ij}^{tr} = \Lambda_{ij}\dot{\xi} \tag{9.3}$$

with

$$\Lambda_{ij} = \begin{cases} \frac{3}{2}H\frac{s_{ij}}{\sigma_e} & \dot{\xi} > 0 \\ H\frac{\epsilon_{ij}^{tr}}{\epsilon_e^{tr}} & \dot{\xi} < 0 \end{cases} \tag{9.4}$$

where s_{ij} are the stress deviators, $\sigma_e = [(3/2)s_{ij}s_{ij}]^{1/2}$ and $\epsilon_e^{tr} = [(2/3)\epsilon_{ij}^{tr}\epsilon_{ij}^{tr}]^{1/2}$.

The constitutive equation of the SMA fiber is given by

$$\sigma_{ij} = C_{ijkl}[\epsilon_{kl} - \epsilon_{kl}^{tr}] - \alpha_{ij}\Delta T \tag{9.5}$$

where α_{ij} are the thermal expansion tensor.

Table 14. SMA: Material Constants.

	Epoxy	Aluminum	SMA
$E\ (GPa)$	3.45	69.0	21.5
ν	0.35	0.33	0.33
$\alpha\ (10^{-6}/^{\circ}C)$	20.0	23.1	8.8
Inelastic constants		$D_0 = 10000/\ s$ $Z_0 = 52\ MPa$ $Z_1 = 135\ MPa$ $m = 31$ $n = 10$	$M_{0f} = 5^{\circ}C,\ \ M_{0s} = 23^{\circ}C$ $A_{0s} = 29^{\circ}C,\ \ A_{0f} = 51^{\circ}C$ $C_M = 11.3\ MPa/^{\circ}C$ $C_A = 4.5\ MPa/^{\circ}C$ $H = 0.0423$

The above equations provide the state equation for the martensitic volume fraction as follows:

$$\dot{\xi} = -\frac{(R_{ij}\dot{\varepsilon}_{ij} + S\dot{T})}{B} \tag{9.6}$$

where

$$R_{ij} = C_{ijkl}\frac{\partial\Phi}{\partial\sigma_{ij}}, \quad S = \frac{\partial\Phi}{\partial T} - \frac{\partial\Phi}{\partial\sigma_{ij}}C_{ijkl}\alpha_{kl}, \quad B = \frac{\partial\Phi}{\partial\sigma_{ij}}P_{ij} + \frac{\partial\Phi}{\partial\xi}, \quad P_{ij} = \frac{\partial\sigma_{ij}}{\partial\xi},$$

and the function Φ is given by

$$\Phi = \begin{cases} \sigma_{ij}\Lambda_{ij} + HC_M\left(M_{os} - T - \dfrac{\ln(1-\xi)}{a_M}\right), & \dot{\xi} > 0, \ \ M_f \leq T \leq M_s \\[2mm] -\sigma_{ij}\Lambda_{ij} - HC_A\left(A_{os} - T - \dfrac{\ln\xi}{a_A}\right), & \dot{\xi} < 0, \ \ A_s \leq T \leq A_f \end{cases}$$

This SMA constitutive law, Eq.(9.5), which is quite similar to the constitutive law (8.1) of inelastic material, has been recently included in the GMC analysis by [69] for the prediction of the overall behavior of metal and polymer matrix composite plates with embedded SMA fibers. The incorporation of constitutive relations (9.5) into the HFGMC model follows the same procedure that was described in Section 8..

9.1. Application: Unidirectional SMA Fiber Composite Response

The behavior of the SMA material and composites with embedded SMA fibers was investigated by [69]. The material constants were assumed to be temperature independent, and are given in Table 14. The aluminum matrix was characterized by the Bodner-Partom viscoplastic parameters [71].

In all cases, a fiber volume fraction $v_f = 0.3$ was considered.

The great advantage of the use of SMA material stems from the fact that it can be employed as an actuator. This is due to its ability, under certain circumstances, to reduce by a temperature increase the residual strains (generated by a pre-applied mechanical loading). However, if during such heating the reduction of these residual strains is prevented by an external mean, recovery stresses develop. These stresses can be employed to control the behavior of the structure.

In Fig. 13, the uniaxial stress-strain behavior in the 1-direction of the SMA fibers, inelastic aluminum matrix and the SMA/Al unidirectional composite are shown. These behaviors are shown at different temperatures because the SMA constituent exhibits different temperature-dependent responses. At temperature $T = 25^\circ C$, which is lower than the austenitic start temperature, the residual strain obtained after the first unloading to zero stress is rather significant. At a higher temperature $T = 35^\circ C$, which is higher than the austenitic start temperature and lower than the austenitic finish temperature, the residual strain decreases. This residual strain vanishes at $T = 55^\circ C$, which is above the austenitic finish temperature at which the SMA exhibits superelastic behavior. It is clearly seen from Fig. 13 that the inelastic matrix (whose response is exhibited in Fig. 13(d)) has a dominant role on the composite behavior. In particular, in the presence of the inelastic matrix, residual strains are obtained after the first unloading to zero stress at all those temperatures. This has a significant effect on the composite behavior when subjected to a subsequent thermal cycle.

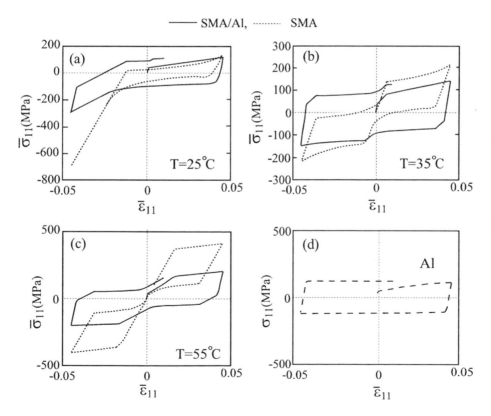

Figure 13. Uniaxial stress-strain response in the 1-direction of SMA, Al and SMA/Al composite.

It is interesting to examine the interaction between the SMA constituent and an elastic resin matrix. Figure 14(a) presents the uniaxial stress-strain response in the 1-direction of the SMA/epoxy composite and the monolithic SMA material at the same temperature values that have been discussed in Fig. 13. Due to the absence of the inelastic mechanism and the lower value of Young's modulus of epoxy, as compared with that of the SMA, the effect of the latter on the overall composite behavior is dominant.

In both Figs. 13 and 14(a), the loading has been applied in the 1-direction in which the SMA fibers are oriented. Figure 14(b) illustrates the behavior of the SMA/Al and SMA/epoxy when the loading is applied in the transverse 2-direction perpendicular to the fibers. It can be readily seen that in the case of metal matrix composite, minor differences are detected when the loading is applied either in the fiber direction or perpendicular to the fibers. In the case of the resin matrix composite, one can observe that the initial elastic slope of the composite, as can be expected, is lower in the transverse loading case. In the inelastic region of the SMA constituent however, the global response of the composite is still dominated by the latter behavior, as in the case of axial loading.

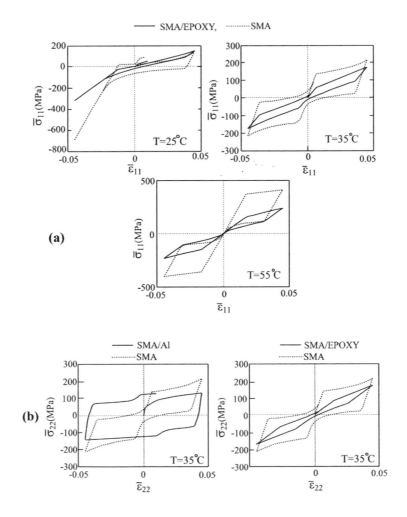

Figure 14. (a) Uniaxial stress-strain response in the 1-direction of SMA, and SMA/epoxy composite; (b) uniaxial stress-strain response in the 2-direction of SMA/Al and SMA/epoxy composite.

In order to demonstrate the ability of the embedded SMA material to control the composite response, it is necessary to examine the behavior due to mechanical loading-unloading, followed by a temperature increase. Figure 15(a) shows the response in the 1-direction of the unidirectional SMA/Al composite and the SMA material both subjected to uniaxial stress loading-unloading at $T = T_R = 35^\circ C$, followed by a temperature increase under traction-free conditions. The shape memory effect in the SMA material, which is exhibited by the decrease of the strain due to the temperature increase, is clearly observed. This effect is caused by the phase transformation which depends, in general, on the temperature and stress state. Under a traction-free situation, this transformation starts together with the temperature increase, as is seen in Fig. 15(a). The transformation process is completed at the austenitic finish temperature $T = A_{of} = 51^\circ C$ where the strain is very small. Further heating induces a slight strain increase due to the thermal expansion of the SMA material. A similar process takes place in the SMA/Al composite. Here however, since the SMA constituent is subjected to residual stresses, the phase transformation which is accompanied by an overall strain reduction starts at a higher temperature (about $80^\circ C$). Further heating decreases the residual overall strain. This strain reduction is more significant than the strain reduction that takes place in the monolithic SMA constituent.

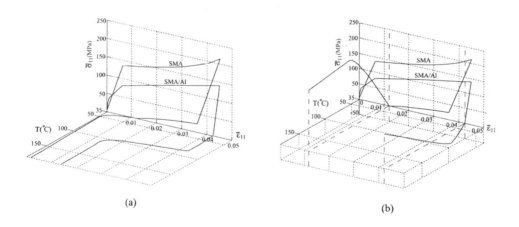

(a)   (b)

Figure 15. (a) Stress-strain-temperature response of SMA and SMA/Al composite exhibiting the strain recovery; (b) stress-strain-temperature response of SMA and SMA/Al composite exhibiting the stress recovery.

As stated before, constraining the composite by preventing the strain reduction, yields significant recovery stresses. Figure 15(b) illustrates the effect of the recovery stresses. It shows the behavior of the monolithic SMA and composite due to axial mechanical loading-unloading, followed by heating under the constraint of constant axial strain. The effect of the phase transformation in the composite is reflected by the slope change in the stress-temperature plane.

10.　Shape-Memory Alloy Fiber Composites Undergoing Large Deformations

The previous investigations have been carried out under the assumption that the deformations are infinitesimal. In ref. [72] and [73], three-dimensional models for the superelastic behavior of SMA materials undergoing finite deformations are presented. Following ref. [73], the finite-strain constitutive law of an SMA material is presented is this section. By incorporating this law with the HFGMC model that is formulated for finite strains, a large strain constitutive law for SMA fiber composites undergoing finite deformations is obtained, which can be employed to investigate their appropriate behavior, see [74] for more details.

10.1.　Finite-Strain Shape-Memory Alloy Constitutive Equations

The constitutive equations of the SMA material undergoing large deformations have been presented by [73]. They are modified herein and adapted to the finite strain HFGMC micromechanical model of [76] and [77].

The present analysis is based on a Lagrangian description according to which the location of a particle in the undeformed state at time $t = 0$ is denoted in the framework of a Cartesian coordinate system by $\mathbf{X} = (X_1, X_2, X_3)$. Let $\mathbf{F}$ denote the deformation gradient. It is decomposed into elastic $\mathbf{F}^e$ and transformation $\mathbf{F}^{tr}$ parts in the following multiplicative form:

$$\mathbf{F} = \mathbf{F}^e\,\mathbf{F}^{tr} \tag{10.1}$$

The elastic right Cauchy-Green deformation tensor is given by

$$\mathbf{C}^e = \hat{\mathbf{F}}^e\,\mathbf{F}^e \tag{10.2}$$

where $\hat{\mathbf{F}}^e$ denotes the transpose of $\mathbf{F}^e$.

In the principal orthonormal directions $\mathbf{N}_A$, $A = 1, 2, 3$, $\mathbf{C}^e$ is represented by

$$\mathbf{C}^e = \text{diag}\left[(\lambda_1^e)^2, (\lambda_2^e)^2, (\lambda_3^e)^2\right] = \sum_{A=1}^{3} \left(\lambda_A^e\right)^2 \mathbf{N}_A \otimes \mathbf{N}_A \tag{10.3}$$

with λ_A^e being the elastic principal stretches.

The free-energy per unit mass of the isotropic material is given in the framework of Ogden's representation [78] by

$$\psi(\lambda_1^e, \lambda_2^e, \lambda_3^e) = \frac{1}{\rho_0}\left[\frac{K}{2}(\theta^e)^2 + G\sum_{A=1}^{3}(e_A^e)^2\right] \tag{10.4}$$

where

$$\theta^e = log\,J^e, \quad J^e = \lambda_1^e\lambda_2^e\lambda_3^e \tag{10.5}$$

and

$$e_A^e = log\,\bar{\lambda}_A^e, \quad \bar{\lambda}_A^e = \frac{\lambda_A^e}{(J^e)^{1/3}} \tag{10.6}$$

with K and G denoting the bulk and shear modulus, respectively, and ρ_0 is the initial density of the material.

The second Piola-Kirchhoff stress tensor $\mathbf{S}$ is given by [79]:

$$\mathbf{S} = 2\rho_0 \frac{\partial \psi}{\partial \mathbf{C}^e} \tag{10.7}$$

Hence in the principal coordinates,

$$S_A = 2\rho_0 \frac{\partial \psi}{\partial (\lambda_A^e)^2} = \frac{\rho_0}{\lambda_A^e} \frac{\partial \psi}{\partial \lambda_A^e}, \quad A = 1, 2, 3 \tag{10.8}$$

It follows from Eq. (10.4) that:

$$S_A = \frac{1}{(\lambda_A^e)^2} \left[K\theta^e + 2Ge_A^e \right] \tag{10.9}$$

In terms of the principal stretches λ_A, $A = 1, 2, 3$, associated with $\mathbf{C} = \hat{\mathbf{F}}\,\mathbf{F}$, let:

$$\theta = log\,J, \quad J = \lambda_1 \lambda_2 \lambda_3 \tag{10.10}$$

$$e_A = log\,\bar{\lambda}_A, \quad \bar{\lambda}_A = \frac{\lambda_A}{J^{1/3}} \tag{10.11}$$

The multiplicative decomposition, Eq. (10.1), yields in the principal directions that:

$$\lambda_A = \lambda_A^e\,\lambda_A^{tr} \tag{10.12}$$

where λ_A^{tr} are the transformation stretches. They are given by [73]:

$$\lambda_A^{tr} = \exp[\epsilon_L \xi_s (d_A + \alpha)] \tag{10.13}$$

where ξ_s is the martensite volume fraction,

$$d_A = \frac{e_A}{\| e \|}$$

with $\| e \| = \sqrt{e_1^2 + e_2^2 + e_3^2}$, and α, ϵ_L are material parameters.

Let:

$$\theta^{tr} = log\,J^{tr}, \quad J^{tr} = \lambda_1^{tr}\lambda_2^{tr}\lambda_3^{tr} \tag{10.14}$$

and

$$e_A^{tr} = log\,\bar{\lambda}_A^{tr}, \quad \bar{\lambda}_A^{tr} = \frac{\lambda_A^{tr}}{(J^{tr})^{1/3}} \tag{10.15}$$

It follows that:

$$\theta^{tr} = 3\epsilon_L \xi_s \alpha \tag{10.16}$$

and

$$e_A^{tr} = \epsilon_L \xi_s d_A \tag{10.17}$$

from which the logarithmic volumetric and deviatoric quantities can be additively written as:

$$\theta = \theta^e + \theta^{tr}, \quad e_A = e_A^e + e_A^{tr} \tag{10.18}$$

Using these relations in Eq. (10.9), the following expression of the second Piola-Kirchhoff component is obtained in the principal directions:

$$S_A = \left(\frac{\lambda_A^{tr}}{\lambda_A}\right)^2 \left[K(\theta - 3\epsilon_L \xi_s \alpha) + 2G(e_A - \epsilon_L \xi_s d_A)\right] \tag{10.19}$$

In finite strain HFGMC micromechanical analysis, it is necessary to represent the constitutive relations of the material in an incremental form, and to establish the associated instantaneous tangent tensor. To this end, the stress increments can be obtained from Eq. (10.19) in the form:

$$\Delta S_A = \sum_{B=1}^{3} \left[\frac{\partial S_A}{\partial \lambda_B^e}\right] \Delta \lambda_B^e - P_A \Delta \xi_s, \quad A = 1, 2, 3 \tag{10.20}$$

where the quantities P_A in the inelastic stress increments $P_A \Delta \xi_s$ are given by

$$P_A = \epsilon_L \left(\frac{\lambda_A^{tr}}{\lambda_A}\right)^2 \left[3K\alpha + 2G d_A\right] \tag{10.21}$$

It can be shown that:

$$\frac{\partial S_A}{\partial \lambda_A^e} = \left(\frac{\lambda_A^{tr}}{\lambda_A}\right)^3 \left[K\left(1 - 2\theta + 6\epsilon_L \xi_s \alpha\right) + \frac{4G}{3}\left(1 - 3e_A + 3\epsilon_L \xi_s d_A\right)\right] \tag{10.22}$$

and

$$\frac{\partial S_A}{\partial \lambda_B^e} = \left(\frac{\lambda_A^{tr}}{\lambda_A}\right)^2 \frac{\lambda_B^{tr}}{\lambda_B} \left[K - \frac{2}{3}G\right], \quad A \neq B \tag{10.23}$$

The fourth order symmetric tangent tensor $\mathbf{D}$ in the principal directions $\mathbf{N}_A$ expresses twice the derivative of the second Piola-Kirchhoff stress tensor with respect to the elastic right Cauchy-Green deformation tensor. It can be determined from the following expression [79]:

$$\mathbf{D} = \sum_{A=1}^{3} \sum_{B=1}^{3} \frac{1}{\lambda_B^e} \frac{\partial S_A}{\partial \lambda_B^e} \mathbf{N}_A \otimes \mathbf{N}_A \otimes \mathbf{N}_B \otimes \mathbf{N}_B \tag{10.24}$$

$$+ \sum_{A=1}^{3} \sum_{B \neq A=1}^{3} \frac{S_B - S_A}{(\lambda_B^e)^2 - (\lambda_A^e)^2} (\mathbf{N}_A \otimes \mathbf{N}_B \otimes \mathbf{N}_A \otimes \mathbf{N}_B + \mathbf{N}_A \otimes \mathbf{N}_B \otimes \mathbf{N}_B \otimes \mathbf{N}_A)$$

where it should be noted that for $\lambda_A^e = \lambda_B^e$, a Taylor expansion shows that:

$$\lim_{\lambda_B^e \to \lambda_A^e} \frac{S_B - S_A}{(\lambda_B^e)^2 - (\lambda_A^e)^2} = \frac{1}{2\lambda_B^e} \left[\frac{\partial S_B}{\partial \lambda_B^e} - \frac{\partial S_A}{\partial \lambda_B^e}\right]$$

So far, the above expressions for the stresses, stress increments, inelastic stress increments and tangent tensors have been referred to the principal directions $\mathbf{N}_A$. A transformation back to the original coordinates in which $\mathbf{F}$ is given would provide the corresponding transformed tensors which will be herein denoted, respectively, by $\mathbf{S}$, $\Delta\mathbf{S}$, $\mathbf{P}'\Delta\xi_s$ and $\mathbf{D}$. In particular, the following expressions can be written in the original coordinates:

$$\Delta\mathbf{S} = \frac{1}{2}\mathbf{D}\,\Delta\mathbf{C}^e - \mathbf{P}'\Delta\xi_s \tag{10.25}$$

which expresses the increment of the second Piola-Kirchhoff stress tensor in terms of the elastic right Cauchy-Green deformation tensor increment and the inelastic stress increment $\mathbf{P}'\Delta\xi_s$.

Since the micromechanical analysis uses the actual stress, let us employ the following relation that provides the first (non-symmetric) Piola-Kirchhoff stress tensor $\mathbf{T}$ in terms of the second Piola-Kirchhoff stress tensor $\mathbf{S}$:

$$\mathbf{T} = \mathbf{S}\,\hat{\mathbf{F}}^e \tag{10.26}$$

Consequently, the following incremental constitutive law is obtained:

$$\Delta\mathbf{T} = \mathbf{R}\,\Delta\mathbf{F}^e - \mathbf{P}\,\Delta\xi_s \tag{10.27}$$

where $\mathbf{R}$ is the current mechanical tangent tensor given by

$$R_{ijkl} = D_{irls}\,F^e_{jr}\,F^e_{ks} + S_{il}\,\delta_{jk} \tag{10.28}$$

and $\mathbf{P} = \mathbf{P}'\hat{\mathbf{F}}^e$ with δ_{jk} being the Kronecker delta.

The phase transformation and activation conditions are expressed by [73] in terms of the Kirchhoff stress components τ_A; $A = 1, 2, 3$ in the principal coordinates. The latter are given in terms of the second Piola-Kirchhoff stresses S_A, Eq. (10.19), as follows:

$$\tau_A = \left(\lambda^e_A\right)^2 S_A \tag{10.29}$$

These conditions are based on the Drucker-Prager loading function which incorporates a possible pressure-dependence of the phase transformation:

$$F(\tau_A) = \sqrt{t_1^2 + t_2^2 + t_3^2} + 3\alpha p \tag{10.30}$$

where $t_A = 2Ge^e_A$ (which are the deviatoric components of the Kirchhoff stress) and $p = K\theta^e$ (which is the pressure given by $(\tau_1 + \tau_2 + \tau_3)/3$), such that $\tau_A = p + t_A$.

For conversion of austenite into martensite $(A \rightarrow S)$, the evolution of ξ_s is given by

$$\dot{\xi}_s = -H^{AS}(1 - \xi_s)\frac{\dot{F}}{F - R_f^{AS}} \tag{10.31}$$

while for conversion of martensite into austenite $(S \rightarrow A)$:

$$\dot{\xi}_s = H^{SA}\xi_s\frac{\dot{F}}{F - R_f^{SA}} \tag{10.32}$$

where the dot denotes a time derivative and:

$$R_f^{AS} = \sigma_f^{AS}\left(\sqrt{\frac{2}{3}} + \alpha\right), \quad R_f^{SA} = \sigma_f^{SA}\left(\sqrt{\frac{2}{3}} + \alpha\right) \tag{10.33}$$

with σ_f^{AS}, σ_f^{SA} being material constants. The quantities H^{AS} and H^{SA} determine the activation conditions that allow to choose between Eqs. (10.31) and (10.32). They are given by

$$H^{AS} = \begin{cases} 1 & R_s^{AS} < F < R_f^{AS}, \quad \dot{F} > 0 \\ 0 & otherwise \end{cases} \tag{10.34}$$

$$H^{SA} = \begin{cases} 1 & R_f^{SA} < F < R_s^{SA}, \quad \dot{F} < 0 \\ 0 & otherwise \end{cases} \tag{10.35}$$

where:

$$R_s^{AS} = \sigma_s^{AS}\left(\sqrt{\frac{2}{3}} + \alpha\right), \quad R_s^{SA} = \sigma_s^{SA}\left(\sqrt{\frac{2}{3}} + \alpha\right) \tag{10.36}$$

with σ_s^{AS}, σ_s^{SA} being material constants.

10.2.   Finite-Strain Constitutive Equations for SMA Fiber Composites

By employing the homogenization technique in the framework of the HFGMC analysis for inelastic materials undergoing finite deformation [77], [74], the following global constitutive law for such composites is obtained:

$$\Delta\bar{\mathbf{T}} = \mathbf{R}^* \, \Delta\bar{\mathbf{F}} - \mathbf{H}^*\Delta\theta - \Delta\bar{\mathbf{T}}_P \tag{10.37}$$

where $\Delta\bar{\mathbf{T}}$, $\Delta\bar{\mathbf{F}}$, and $\Delta\theta$ are the global increments of the first Piola-Kirchhoff stress tensor, global deformation gradient and temperature, respectively. In addition, $\mathbf{R}^*$ and $\mathbf{H}^*$ are the instantaneous effective stiffness and effective thermal stress tangent tensors of the composite and $\Delta\bar{\mathbf{T}}_P$ is the global inelastic stress tensor. This equation governs the behavior of the multiphase composite at finite strains in which any phase is represented by a suitable finite deformation law, including the SMA constitutive equations that have been developed in the previous subsection.

10.3.   Application: Response of SMA/Aluminum Unidirectional Composite Undergoing Large Ceformations

Results are given for unidirectional continuous SMA fiber composites. The repeating unit cell consists of sixteen subcells (see Fig. 1c), four of which are filled with the material properties of the SMA phase which are uniformly surrounded by the other twelve matrix subcells. The material properties of the SMA constituent are given in Table 15.

It should be noted that although the maximum transformation strain is assumed to be 0.07 as indicated in Table 15, the range of deformation on the following graphs has been farther extended in order to get the effects of large deformation behavior.

Figure 16(a) shows the superelastic response of the monolithic SMA material to a uniaxial stress cyclic loading in the 1-direction (say). This curve has been obtained from the

Table 15. Material constants of the SMA fiber [73].

$E(GPa)$	ν	ϵ_L	α	$\sigma_s^{AS}(MPa)$	$\sigma_f^{AS}(MPa)$	$\sigma_s^{SA}(MPa)$	$\sigma_f^{SA}(MPa)$
50	0.3	0.07	0.12	520	600	300	200

micromechanical equations (10.37) with just one-phase material, and by a direct solution of the nonlinear equations (10.19) without invoking the incremental procedure or the micromechanical analysis. Both approaches gave practically the same results. The applied load in a state in which the martensite volume fraction ξ_s is equal to zero (full austenite state). The martensite volume fraction starts to increase at about $F_{11} = 1.01$ and reaches its maximum value $\xi_s = 1$ at $F_{11} = 1.081$ (full martensite state), after which it remains equal to 1 as the loading continues increasing. Upon unloading, ξ_s continues to be equal to 1, but starts to decrease at $F_{11} = 1.07$. This indicates that the variations of ξ_s during loading and unloading are not identical, which can be clearly observed in the inset to this figure. Finally, it should be mentioned that in the absence of any phase transformation (i.e. $\xi_s \equiv 0$), the corresponding response of the SMA material is represented by a nonlinear curve varying from about $4GPa$ to $-5GPa$.

The properties of the elastic-viscoplastic matrix are given in Table 2 of ref. [74] which characterizes an aluminum alloy. The response of the aluminum material to isothermal cyclic uniaxial stress loading applied in the 1-direction (say) at two rates: $1s^{-1}$ and $0.001s^{-1}$ is shown in Fig. 16(b). This figure well exhibits the rate dependence and the Bauschinger effect of the material.

Consider next the elastic-viscoplastic aluminum matrix with embedded SMA fibers oriented in the 1-direction. All results shown in this section are given with an applied rate of $1s^{-1}$ and 30% of SMA volume fraction. The response of the SMA/Al composite to a uniaxial cyclic stress loading applied in the axial 1-direction is shown in Fig. 16(c). The effect of superelasticity that is exhibited by the monolithic SMA constituent is clearly lost in the present situation. In order to exhibit the phase transformation effect, Fig. 16(c) shows also the composite response in the absence of such a transformation in the SMA fiber. In this case, the composite behavior is dominated by the inelastic effects of the matrix and the nonlinearity of the constituents. Further results are given in ref. [74].

Recently, Aboudi and Freed, ref. [75], presented a two-way thermomechanically coupled HFGMC micromechanical analysis. As a result of this generalization of the HFGMC model, the temperature which is coupled to the mechanical effects, is governed by the energy equation and is induced into the composite's constituents as a result of the application of mechanical loadings. This generalized model was applied to predict the behavior of composites that consist of shape memory alloy fibers embedded in metallic and polymeric matrices. The results exhibit the response of the composites to various types of loading, and the effect of the two-way thermomechanical coupling that induces temperature deviations from reference temperatures at which the shape memory and pseudoelasticity effects take place in the fibers.

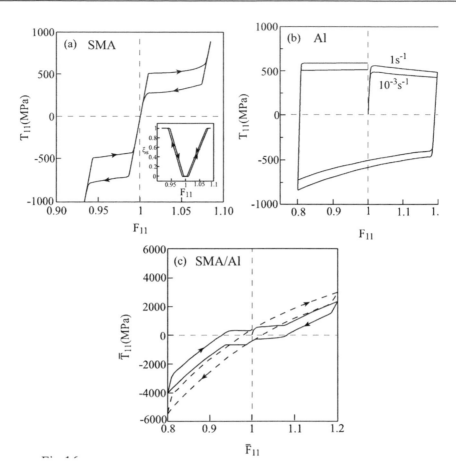

Figure 16. (a) The response of the SMA material subjected to a uniaxial stress loading. The inset to this figure shows the variation of the martensite volume fraction with the applied loading. (b) The response of the elastic-viscoplastic aluminum alloy subjected to a uniaxial stress loading applied at two rates. (c) The response of the SMA/Al composite subjected to a uniaxial stress loading in the fiber direction. The dashed lines show the response in the absence of a phase transformation in the SMA material.

11. Bounded Smart Composites

Thus far, smart composite materials possessing periodic microstructure and extending over an infinite region have been micromechanically analyzed in conjunction with a homogenization procedure by utilizing the HFGMC model. As a result, effective constitutive equations have been established that represent the global behavior of the multiphase material. In the present section a different type of smart composites are analyzed. We consider bounded multiphase composites extending over a finite domain $0 \leq x_1 \leq D$, $0 \leq x_2 \leq H$ and $0 \leq x_3 \leq L$, see Fig. 17. In this type of multiphase materials there are distinct boundaries defined by the surfaces $x_1 = 0, D$; $x_2 = 0, H$ and $x_3 = 0, L$. In addition, no periodicity of microstructure is assumed to exist. We analyze this bounded composite by dividing it into N_α, N_β and N_γ subcells, each one of which can be occupied by a distinct smart or

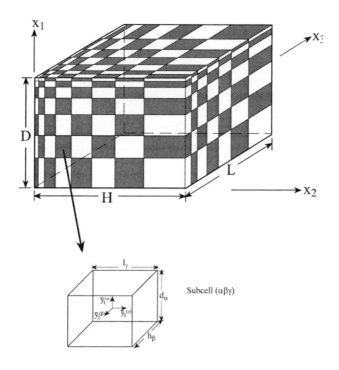

Figure 17. Schematic description of a bounded composite in three directions. The entire region is divided into N_α, N_β and N_γ subcells in the x_1, x_2 and x_3 directions, respectively. Every subcell is labeled by $(\alpha\beta\gamma)$.

conventional material, thus forming a multiphase smart composite. The analysis of this composite is similar to the micro-to-macro model which was designated for the investigation of functionally graded materials, see ref. [10]. It is based on the quadratic expansion of the displacement in each subcell $(\alpha\beta\gamma)$ as follows (cf. Eq. (2.29)):

$$
\begin{aligned}
\mathbf{u}^{(\alpha\beta\gamma)} = \; & \mathbf{W}^{(\alpha\beta\gamma)}_{(000)} + \bar{y}_1^{(\alpha)}\mathbf{W}^{(\alpha\beta\gamma)}_{(100)} + \bar{y}_2^{(\beta)}\mathbf{W}^{(\alpha\beta\gamma)}_{(010)} + \bar{y}_3^{(\gamma)}\mathbf{W}^{(\alpha\beta\gamma)}_{(001)} \\
& + \frac{1}{2}\left(3\bar{y}_1^{(\alpha)2} - \frac{d_\alpha^2}{4}\right)\mathbf{W}^{(\alpha\beta\gamma)}_{(200)} + \frac{1}{2}\left(3\bar{y}_2^{(\beta)2} - \frac{h_\beta^2}{4}\right)\mathbf{W}^{(\alpha\beta\gamma)}_{(020)} \\
& + \frac{1}{2}\left(3\bar{y}_3^{(\gamma)2} - \frac{l_\gamma^2}{4}\right)\mathbf{W}^{(\alpha\beta\gamma)}_{(002)}
\end{aligned}
\tag{11.1}
$$

If one of the constituents is electro-magneto-elastic material, similar expansions of the electric and magnetic potential are assumed. As a result of the implementation of the governing equations and the appropriate continuity conditions at the interfaces between the subcells, as well as the proper boundary conditions at the bounding surfaces $x_1 = 0, D$; $x_2 = 0, H$ and $x_3 = 0, L$, a system of equations is obtained in the form given by Eqs. (2.30), (4.28) and (8.2) for linear, nonlinear and inelastic materials, respectively. It should be emphasized that the governing equations and the interfacial continuity conditions, as well as the boundary conditions that are applied at the composite's surfaces, are imposed in the average (integral) sense. Once this system is solved, any of the requested field variables in the subcells can be readily determined.

It should be noted that two special cases can be obtained from the present general formulation (see ref. [10]):

(1) The bounded composite extends over $0 \leq x_1 \leq D$ in the x_1- direction, but its microstructures are periodic in the x_2 and x_3 directions.

(2) The bounded composite extends over $0 \leq x_1 \leq D$ and $0 \leq x_2 \leq H$, but its microstructures are periodic in the x_3 direction.

In the following, we consider bounded composites with embedded piezoelectric, SMA, electrorheological (or magnetorheological) and fiber optic phases. All these four applications were performed on a composite of the type described by case (1).

11.1. Embedded Piezoelectric Materials

Reference [80] presented the details of the micro-to-macro analysis of a composite that extends along $0 \leq x_1 \leq D$, but it is periodic in the x_2 and x_3 directions, and consists of distributed piezoelectric patches. The effect of different levels of applied voltages at various piezoelectric volume fractions was studied. The actuation process of the piezoelectric patches was examined in a composite which consists of a different number of layers ranging from one layer to a multilayer composite.

11.2. Embedded SMA Fibers

For a bounded composite in the x_1-direction, but periodic in the x_2 and x_3 directions, the effect of embedding SMA fibers has been studied by [81]. The SMA fibers were modeled by the constitutive equations of [82], which resemble the total deformation theory of plasticity (in contrast to the incremental approach of [63] that was discussed in Section 9.).

The uniaxial rate-independent constitutive law of the SMA fibers is given in the form [82]:

$$\dot{\epsilon}_{11} = \frac{\dot{\sigma}_{11}}{E} + |\dot{\epsilon}_{11}| \left| \frac{\sigma_{11} - \beta_{11}}{\sigma_c} \right|^{n-1} \frac{\sigma_{11} - \beta_{11}}{\sigma_c} \tag{11.2}$$

$$\beta_{11} = E \, \alpha \left\{ \epsilon_{11} - \frac{\sigma_{11}}{E} + f_T \, \text{erf} \left[a \left(\epsilon_{11} - \frac{\sigma_{11}}{E} \right) \right] \right\} \tag{11.3}$$

where ϵ_{11}, σ_{11} and β_{11} are the one-dimensional strain, stress and back stress, respectively, and $erf(x)$ denotes the error function. In these equations, the dot denotes the time derivative, E is the Young's modulus, σ_c is a critical stress, n is a constant power controlling the sharpness of transition from the elastic to the inelastic region, f_T is a parameter that controls the type (twinning hysteresis or superelasticity) and the amount of recovery during unloading, a is a constant controlling the smoothness of the curve near the origin, and α is a constant that determines the slope during inelastic deformation.

Let Y denote the value of the threshold stress required to start stress-induced phase transition. It was shown by [82] that the parameter σ_c can be expressed as follows:

$$\sigma_c = Y - f_T \, E \, \alpha (1 + \alpha)^{1/n} \tag{11.4}$$

In ref. [82], a methodology is given which enables the determination of the model parameters from measured data. Furthermore, in [83] the temperature dependence of these parameters for NiTi SMA was established from measurements. It turns out that the parameters a and n can be assumed to be temperature independent.

A multiaxial generalization of Eqs. (11.2) -(11.3) has been presented by [82] and it is given as follows:

$$\dot{\epsilon}_{ij} = \frac{1+\nu}{E}\dot{\sigma}_{ij} - \frac{\nu}{E}\dot{\sigma}_{kk}\delta_{ij} + \sqrt{3\,K_2}\,(3\,J_2)^{\frac{n-1}{2}}\,\frac{s_{ij} - b_{ij}}{\sigma_c} \tag{11.5}$$

$$b_{ij} = \frac{2}{3}E\,\alpha\epsilon_{ij}^I\left\{1 + f_T\left(\frac{2}{3}\sqrt{3\,I_2}\right)^{-1}\,\mathrm{erf}\left(\frac{2a}{3}\sqrt{3\,I_2}\right)\right\} \tag{11.6}$$

where δ_{ij} is the Kronecker delta, and the inelastic strain (transformation strain) ϵ_{ij}^I denotes the difference between the total strain ϵ_{ij} and the elastic strain, namely $\epsilon_{ij}^I = \epsilon_{ij} - (\frac{1+\nu}{E}\sigma_{ij} - \frac{\nu}{E}\sigma_{kk}\delta_{ij})$. In addition,

$$I_2 = \frac{1}{2}\epsilon_{ij}^I\epsilon_{ij}^I, \quad J_2 = \frac{1}{2}\frac{s_{ij} - b_{ij}}{\sigma_c}\frac{s_{ij} - b_{ij}}{\sigma_c}, \quad K_2 = \frac{1}{2}\dot{e}_{ij}\,\dot{e}_{ij}$$

where e_{ij}, s_{ij} and b_{ij} are the deviatoric tensors of ϵ_{ij}, σ_{ij} and β_{ij}, respectively. It can be easily verified that under uniaxial loading conditions, these constitutive relations reduce to Eqs. (11.2)-(11.3) that describe the SMA one-dimensional behavior.

This SMA model was embedded in the bounded composite described above in which the other constituent was taken either as a metallic matrix, which was modeled by the classical plasticity theory, or as an elastic resin matrix. Thus, the detailed interaction of the SMA fibers with the inelastic/elastic matrix in which the effect of the boundaries of the composite are incorporated was studied by the analysis of [81] where numerous results are provided.

This micromechanical approach of bounded composites was recently employed by [104] to analyze a composite with embedded SMA fibers. The composite is subjected to a non-uniform temperature distribution that arises from the process of heating or cooling, and the constitutive equations of the SMA fibres are those presented by [105].

11.3.   Embedded Electrorheological and Magnetorheological Fluids

Electro-Rheological (ER) and Magneto-Rheological (MR) fluids have the property that in the absence of an electric/magnetic field they flow as viscous liquids, but in the presence of an electric/magnetic field they immediately solidify (the response of ER fluids to the excitation voltage is typically less than one millisecond [84]. In the latter situation they remain at rest when they are subjected to shear stresses up to a certain limit (shear yield stress), after which they behave as viscous fluids under the action of shearing. This phenomenon can be utilized in various engineering applications. An extensive discussion of ER fluids and their potential applications can be found in ref. [1] and references therein. A discussion of MR fluids has been recently given by [85].

As stated before, ER and MR fluids can be represented by the viscoplastic Bingham model where the most important parameter is the strong dependence of the shear yield stress on the electric field and magnetic field, respectively. As a result of this dependence, the ER and MR fluids exhibit several orders of magnitude increase in their apparent viscosity.

In order to utilize ER and MR fluids, they must be embedded in the structure whose properties are to be controlled. Furthermore, it might be desirable to incorporate different types of these fluids that are distributed within the structure [1]. In ref. [86], the analysis of bounded composites that include embedded ER or MR fluids is presented. Due to the existence of the viscoplastic mechanism in these fluids, certain adjustments were necessary to accommodate their inelastic behavior.

ER and MR fluids are modeled by the viscoplastic Bingham material (e.g. [87]). Furthermore, it is well established that the behavior of ER and MR fluids can be divided into pre-yield and post-yield regions (e.g. [88], [89], [90], [91] and [85]. Consequently, let us add an elastic strain to the classical Bingham model so that the total strain rate can be written as a sum of the elastic strain rate and the viscous strain rate as follows:

$$\dot{\epsilon}_{ij} = \dot{\epsilon}_{ij}^{E} + \dot{\epsilon}_{ij}^{V} \tag{11.7}$$

where the dot denotes a time derivative with respect to the time t.

The elastic part of the strain rate is related to the stress rate by the Hooke's law of elastic anisotropic materials:

$$\dot{\sigma}_{ij} = C_{ijkl}\, \dot{\epsilon}_{kl}^{E} \tag{11.8}$$

where C_{ijkl} is the elastic stiffness tensor of the material. It is worthwhile to mention however that the present model admits the modeling of viscoelastic materials. However, for such materials Eq.(11.8) needs to be modified accordingly.

For the Bingham material, the rate of the viscous component of the strain in Eq. (11.7) is given by

$$\eta\, \dot{\epsilon}_{ij}^{V} = \begin{cases} 0 & g < 0 \\ g\, s_{ij} & g \geq 0 \end{cases} \tag{11.9}$$

where η is the shear viscosity coefficient and s_{ij} is the stress deviator (i.e. $s_{ij} = \sigma_{ij} - \sigma_{kk}\delta_{ij}$; δ_{ij} being the Kronecker delta). The dimensionless load function g is defined by

$$g = 1 - \frac{Y}{\sqrt{J_2}} \tag{11.10}$$

where Y is the yield stress for pure shear, and $J_2 = s_{ij}s_{ij}/2$ being the second invariant of the stress deviator tensor.

The yield stress of ER and MR fluids depends strongly on the applied electric field and magnetic fields, respectively. For a type of ER fluid, ref. [92] for example used the following formula for the dependence of Y on the electric field E:

$$Y = a\, E + b\, E^2 \tag{11.11}$$

with $a = 0.8867$ and $b = 0.7833$, where E is in MV/m and Y is in kPa. This formula is a least-squares quadratic fit to experimental data. Similarly, a quadratic dependence of the yield stress on the magnetic field strength has been reported by [85].

References [89] and [91] indicate that the yield strain of ER and MR fluids is around 1% for various levels of applied electric and magnetic fields. Using the value of 1% for the yield strain, one can estimate the elastic shear modulus of the fluid for a given value of applied electric field. In Fig. 18, the behavior of ER fluid subjected to a uniaxial stress

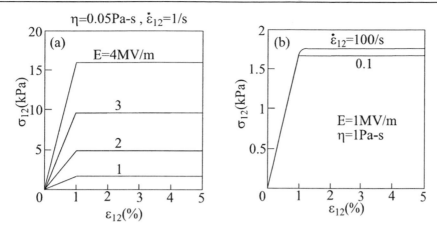

Figure 18. (a) Shear stress-strain response of an ER fluid subjected to various values of electric field, and, (b) the effect of rate of applied loading.

loading is shown for several values of the applied electric field. The dependence of the yield stress of the ER fluid on the electric field is characterized by Eq.(11.11). Also shown in Fig. 18 is the effect of the value of the applied rate of loading on the ER fluid response.

The micro-macro analysis that is presented in ref. [86] for the modeling of electro- and magneto-rheological fluids, which are embedded in multiphase composites, enables the prediction of the overall behavior of ER composites in the presence of several values of the electric field. This includes a layered composite, a single ER inclusion in a solid matrix, and a cluster of ER inclusions embedded in a solid matrix. Furthermore, an extensive investigation of shear wave propagation in a layered ER plate was presented that includes studies of the effect of intensity of the electric field, viscosity and applied pulse time duration.

11.4. Embedded Optical Fibers

This section is based on the analysis of [93] of bounded composites with embedded optical fibers. Fiber-optics are sensors attached to or embedded within the structure may provide real-time information, at the fabrication process or later in service, about temperature, strain, vibration, pressure, etc. Besides being compatible with the composite structure, optical fibers offer many other advantages ([94] and [95]): (i) silica-based optical fibers are mechanically compatible with composite materials both in dimensions and strength; (ii) the fibers are electrically non-conductive and, therefore, insensitive to electromagnetic interferences; (iii) they are easily shaped into almost any geometrical layout of interest; (iv) the same technology supports the intrinsic measurement of many physical phenomena; (v) many sensors can be conveniently multiplexed along the same fiber, and finally; (vi) fiber-optic sensing and data flow between them and their control center could and should benefit from the widespread success of its twin technology of fiber-optic communication. Embedded (or surface-attached) optical fibers can sense the local thermo-mechanical field through a variety of optical processes. Presently, the micro-macro model is applied to those sensors which are based on induced changes in the optical length of the fiber. The treated family of sensors include the Fabry-Perot sensor, as well as its less sensitive and somewhat simpler radio-frequency interferometer. Both sensors are interferometric and their description and predicted performance are discussed by [93].

The authors of ref. [96] were the first to relate the optical length of a fiber to its strain in an interferometric set-up. They derived the phase-strain $\Phi - \epsilon$ relationship for a single-mode fiber and showed that $\Delta\Phi/\Delta\epsilon$ is mainly due to the actual length change, as well as to the strain-induced refractive index change, while its dependence on the effect of strain on the guiding properties of the fiber can be neglected. Later, this theory was extended by [97], such that an arbitrary strain field was considered, together with an arbitrary fiber configuration. Next, ref. [98] introduced the analysis of an embedded optical fiber in which an integral equation modeled the strain induced optical phase shift, such that the entire strain along the fiber was related to the entire phase shift. In [99], the authors have modeled the problem of embedded intrinsic and extrinsic Fabry-Perot optical fiber sensors in an anisotropic medium, bridging over most of the assumptions and approximations offered by previous investigators. Yet, the utilization of Lekhnitskii's theory for inclusions embedded in infinite anisotropic media by the latter, lacked the ability to pin-point local geometrical, thermal and mechanical effects. These are needed when microstructures involving the interactions of the composite fibers and matrix and the optical fiber sensor are examined and their mutual influences are essential in establishing the solution. The choice of the present macro- micromechanical approach enables one to examine the mechanical behavior of the optical fiber and its geometrical changes when subjected to external loading and/or thermal field. In the latter case of temperature effects, the three-dimensional temperature distribution is computed, thereby allowing one to take into account very local influences of different temperature values at the three major constituents: graphite fibers, epoxy matrix and the optical fibers. When micromechanically analyzed, the mutual effects of the host media and the hosted inclusion can be precisely investigated. As a result, local geometrical changes in the optical fiber, such as changes in its cross-section which may introduce optical anisotropy, can be established. Furthermore, phase changes which are the essence of this method of sensing can be accurately determined at any point. The importance of such surveillance lies in the fact that these changes, minor as they seem to be, may affect the optical behavior of the sensor. To this end, the macro-micro analysis of [93] was formulated within the framework of the generalized plane strain assumptions. This approach enables one to examine problems of any type of mechanical and/or thermal fields at any geometrical conditions.

Propagation of light through a cylindrically symmetric optical fiber is completely independent of the polarization of the input light. However, when the fiber is exposed to non-cylindrically symmetric stresses, different input polarizations propagate with different speeds and the analysis becomes quite cumbersome. In our study of a unidirectional tape and a cross-ply laminate with the optical fiber oriented in the x_3−direction and under homogeneous stresses, one can model the fiber as linearly bi-refringent, namely: the two linearly polarized light waves with polarization vectors parallel to the transverse axes do maintain their polarizations while propagating down the embedded fiber, but each travels at its own speed. The electric displacement vector $\mathbf{D}$ of a wave propagating along the fiber, can be expressed [100] as:

$$\mathbf{D}(x_3, t) = A_1 sin(\omega t - k_1 x_3)\hat{\mathbf{X}}_1 + A_2 sin(\omega t - k_2 x_3)\hat{\mathbf{X}}_2 \qquad (11.12)$$

where $\hat{\mathbf{X}}_1$ and $\hat{\mathbf{X}}_2$ are orthogonal unit vectors in the fiber's cross-section plane, A_1 and A_2

are the relevant amplitudes, ω is the angular frequency of the optical radiation, and finally, k_1 and k_2 are the wave numbers.

11.5. The Fabry-Perot Optical Fiber - the Opto-Thermo-Mechanical Relations

In the Fabry-Perot interferometer, multiple-reflections between two parallel mirrors give rise to interference effects for both the transmitted and reflected waves [94].

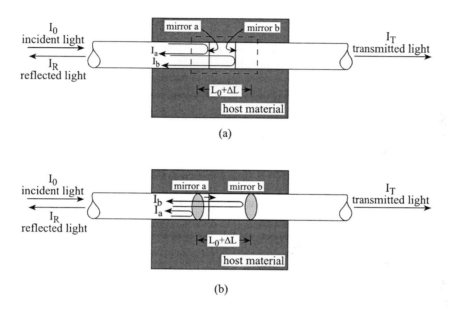

Figure 19. A schematic illustration of a Fabry-Perot optical fiber sensor embedded in a host material. (a) Extrinsic; (b) intrinsic.

In its fiber-optic implementation, the interferometer may be formed by a protected interruption in the fiber, where the two air-glass interfaces form the reflectors, Figure 19(a), or by a short piece of fiber with semi-reflective coatings on both ends, which is fusion spliced to fiber feeders, Figure 19(b). Normally, single-mode fibers are used and the interference which takes place between the mirrors depends on the optical length of the mirror separation (i.e., the product of the length and the refractive index) and affects both the reflected and transmitted intensities. Since the optical fibers in Figure 19(a) are merely used for light transport, this type of sensor is referred to as extrinsic and is predominately used for external applications [101]. The intrinsic Fabry-Perot of Figure 19(b), whose construction does not involve any increase of diameter over that of the fiber itself, serves well as an embedded sensor and has proved its usefulness in quite a few demonstrations [99]. When used in its reflective mode, an incident light whose intensity is I_0 is transmitted through the fiber and upon reaching the first mirror it will be partially reflected at an intensity of I_a and partially transmitted. The transmitted wave will again be partially reflected at an intensity of I_b from the other mirror so that the two waves will propagate back to the transmission station where they are detected. (Since, most often, the semi-reflective coatings have fairly low reflective ($\approx 10\%$), multiple reflections are negligible). For a sensor located a distance $x_3 = x_3^0$, the received electric displacement vector is the sum of the two reflected waves:

$$
\begin{aligned}
\mathbf{D}_{reflected} &= A_1\left[\sqrt{R_a}sin\left(\omega t - 2k_1x_3^0\right) + \sqrt{R_b}sin\left(\omega t - 2k_1(x_3^0 + L)\right)\right]\hat{\mathbf{X}}_1 \\
&+ A_2\left[\sqrt{R_a}sin\left(\omega t - 2k_2x_3^0\right) + \sqrt{R_b}sin\left(\omega t - 2k_2(x_3^0 + L)\right)\right]\hat{\mathbf{X}}_2
\end{aligned}
$$

$$(11.13)$$

where $R_a = I_a/I_0$ and $R_b = I_b/I_0$ are the (assumed polarization-independent) intensity reflection coefficients of the two partial mirrors and L is the distance between the two mirrors measured after the application of the thermomechanical loading. The intensity of the reflected light is given by [99]:

$$
I_R = J\int_0^{2\pi/\omega}|\mathbf{D}|^2dt
$$

$$(11.14)$$

where J is a proportionality constant. After some manipulations (see ref. [93] and references cited there), the following opto-thermomechanical relation can be established for the total reflectivity $R = I_R/I_0$ (the ratio between the reflected and incident intensities):

$$
R = R_a + R_b + 2\sqrt{R_aR_b}cos\left[K_1 + K_2\epsilon_{33} + K_3\epsilon_h + K_4\Delta T\right]cos(K_5\epsilon_s) \qquad (11.15)
$$

where $\epsilon_{33} = \Delta L/L_0$, $\epsilon_h = (\epsilon_{11}+\epsilon_{22})/2$ and $\epsilon_s = \sqrt{(\epsilon_{22} - \epsilon_{11})^2 + \epsilon_{23}^2}$ are the strain in the optical fiber direction, the hydrostatic strain and the maximum shear strain, respectively. In addition,

$$
K_1 = \frac{4\pi n_0 L_0}{\lambda_0}, \quad K_2 = \frac{2\pi n_0 L_0}{\lambda_0}(2 - n_0^2p_{32}), \quad K_3 = -\frac{2\pi n_0^3 L_0}{\lambda_0}(p_{33} + p_{32}),
$$

$$
K_4 = \frac{2\pi L_0}{\lambda_0}\left[n_0^3(p_{33} + 2p_{32})\alpha + 2\frac{dn_0}{dT}\right], \quad K_5 = -\frac{\pi n_0^3 L_0}{\lambda_0}(p_{33} - p_{32})
$$

where n_0, λ_0, T and α are the refraction index of the free fiber, wave length in vacuum, temperature and thermal expansion coefficient of the fiber. The parameters p_{32} and p_{33} are the Pockel's photoelastic constants of the material between the mirrors.

Equation (11.15) provides the relation between the total reflectivity and the mechanical strains $\epsilon_{33}, \epsilon_{33}, \epsilon_{33}$ and thermal strains, as well as the change of the distance between the mirrors.

Figure 20 illustrates the reflected intensity of light and strains that result from the introduction of a Fabry-Perot optical fiber embedded in an epoxy matrix of a graphite/epoxy composite, upon which a transverse external loading, Fig. 20(a), or a temperature variation, Fig. 20(b), have been applied, see [93] for more details and further results.

12. Smart Composite Structures

Once the global (macroscopic) constitutive equations of a smart composite are micromechanically established (e.g., Eq. (3.17), (4.31), (5.15) or (10.37)), one can proceed and analyze the corresponding smart composite structure (e.g., a laminated plate or shell) whose

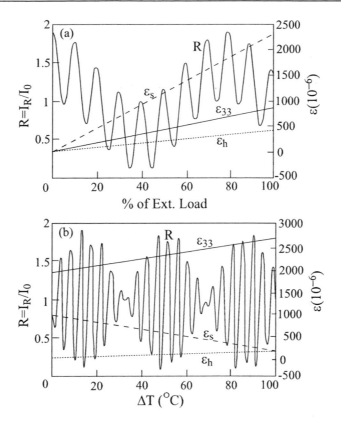

Figure 20. (a) Fabry-Perot normalized reflected light intensity and strains versus increasing external loading, four layers unidirectional tape, zero temperature and applied external transverse loading $\bar{N}_{22}^{ext} = 100MPa$ and $\bar{N}_{33}^{ext} = 100MPa$. (b) Fabry-Perot normalized reflected light intensity and strains versus increasing temperature change, eight layers cross-ply laminates, and applied external transverse loading $\bar{N}_{22}^{ext} = 100MPa$ and $\bar{N}_{33}^{ext} = 100MPa$.

layers form homogeneous smart composite materials, the behavior of which are governed by these equations. The smart composite structure analyses are based on the various structures theories (e.g., the theory of composite plates and shells, see [24] for example). In the following, micromechanically based constitutive equations for SMA fiber composites are utilized to investigate the dynamic response and thermal buckling of SMA laminated plates.

12.1.   Dynamic Response of SMA Composite Plates

In ref. [69], the authors employed the micromechanically established relations of a unidirectional composite with SMA fibers embedded in polymeric or metallic matrices to analyze the nonlinear behavior of infinitely wide composite plates that are subjected to the sudden application of thermal loading. The behavior of the SMA fiber was represented by the model in ref. [63] that was described in Section 9.. In order to illustrate the shape memory effect on the composite structure behavior, two separate situations are considered. In the first one the SMA fibers have been activated (prior to the application of the thermal shock) through a cycle of mechanical loading-unloading applied on the composite, which gives rise to a residual global strain associated with overall traction-free state. In the present investigation this activation procedure involves loading to a strain of 0.045. This pre-loading enables the realization of the shape memory effect during a subsequent thermal loading. The second situation exhibits the behavior of the same structure but without the pre-loading procedure.

Consider a case in which both surfaces of a SMA/epoxy $[0^o]$ plate, being initially at a temperature T_R, are subjected to the following temperature increase with time t:

$$T = T_R + T_m(1 - e^{-\beta t})$$

that is shown in the inset of Fig. 21, where β and T_m are parameters. The resulting dynamic behaviors of the plate in these two situations are shown in Fig. 21, which indicate that the responses of the activated SMA/epoxy plate deviate from the unactivated one immediately after time $t = 0$. This implies that the phase transformation which generates the shape memory effect in the SMA/epoxy plate starts as soon as heating commences, such that under the present circumstances the efficiency of the activation dramatically increases. Other cases where the efficiency of the activation is not so dramatic are discussed in [69].

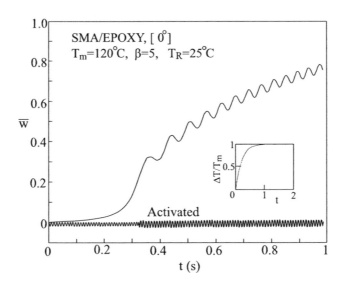

Figure 21. Variation with time of the normalized out-of-plane displacement of SMA/epoxy $[0^o]$ plate subjected to a temperature heating at both surfaces (shown by the inset).

12.2. Thermal Buckling of Activated SMA Composite Plates

The micromechanically established constitutive equations for continuously reinforced SMA metallic or polymeric composites were employed, in conjunction with a plate analysis, by [102] to predict the occurrence of thermal buckling of rectangular laminated plates. The SMA fibers are activated by a mechanical loading-unloading cycle which is applied on the laminated plate, prior to the application of the thermal loading.

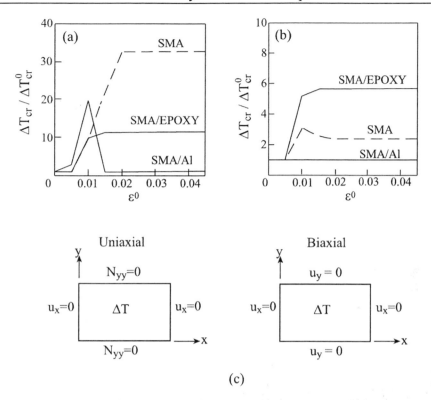

Figure 22. Normalized buckling temperature against activation strain. (a) Uniaxial thermal load; (b) biaxial thermal load; uniaxial and biaxial boundary conditions.

In Fig. 22 the effect of the activation strain on the normalized thermal buckling load of square unidirectional SMA/aluminum and SMA/epoxy composite plates with lamination angle $\theta = 0$ (namely with fibers aligned along the x direction) is presented. Figure 22(a) exhibits the buckling behavior of plates under a uniaxial thermal load, while Fig. 22(b) is for plates subjected to a biaxial thermal load (these two situations are shown in Fig. 22(c)). A value of $\epsilon^0 = 0$ corresponds to an un-activated plate, the buckling temperature of which is denoted by ΔT_{cr}^0. By normalizing the various buckling temperatures with respect to ΔT_{cr}^0, information about the efficiency of the activation can be revealed. The corresponding curve of a monolithic SMA plate is shown for comparison. It can be observed that under uniaxial thermal loading the monolithic SMA plate exhibits the highest buckling temperatures for most values of pre-straining. Under biaxial thermal loading, on the other hand, the buckling temperature of the SMA/epoxy plate is the highest, indicating that the latter exhibits the best performance. A comparison between Fig. 22(a) and Fig. 22(b) shows that the shape memory effect on the buckling temperatures is less pronounced under biaxial boundary conditions. It should be noted that for both SMA and SMA/epoxy plates, under the two types of boundary conditions, a pre-straining to more than about $\epsilon^0 = 0.02$ is not useful. Further results and discussions are given by [102].

13. Conclusion

In the present review chapter, three types of approaches have been presented for the analysis of smart composites. In the first one, micromechanical analyses have been developed for the establishment of macroscopic constitutive equations of smart multiphase composites possessing a periodic microstructure and extending over an infinite domain. In the second type, analyses of bounded composites with distinct boundaries have been presented. These composites contain arbitrarily distributed smart materials. These analyses provide the field variables at any location in the smart composite and its boundaries. Finally, the established macroscopic constitutive equations that govern the behavior of smart composites can be employed to predict the response of corresponding smart composite structures. This has been illustrated in the determination of the dynamic response and thermal buckling of shape-memory alloy composite plates.

Based on the HFGMC and GMC models, NASA Glenn Research Center developed a micromechanics computer code referred to as MAC/GMC, which has many user friendly features and significant flexibility for the analysis of continuous, discontinuous, woven polymer, ceramic, and metal matrix composites with phases that can be represented by arbitrary elastic, viscoelastic, and/or viscoplastic constitutive models. The most recent version of a user guide to this code (version 4) which has been presented by [103], incorporates HFGMC, together with additional material models including smart materials (electromagnetic and shape memory alloys), and yield surface prediction of metal matrix composites.

Acknowledgment

The author gratefully acknowledges the support of the Diane and Arthur Belfer chair of Mechanics and Biomechanics. Special thanks go to Ms. Bette Lewis for her help in formatting this chapter.

References

[1] Gandhi, M. V.; Thompson, B. S. Smart Materials and Structures; Chapman & Hall; London, UK, 1992.

[2] Neelakanta, P. S. Handbook of Electromagnetic Materials: Monolithic and Composite Versions and their Applications; *CRC Press*; Boca Raton, US, 1995.

[3] Banks, H. T.; Smith, R. C.; Wang, Y. Smart Material Structures; Wiley; Chichester, UK, 1996.

[4] Culshaw, B. Smart Structures and Materials; Artech House; Boston, US, 1996.

[5] Janocha, H.; Editor, Adaptronic and Smart Materials: Basics, Materials, Design and Applications; Springer; Berlin, DE, 1999.

[6] Srinivasan, A. V.; McFarland, D. M. Smart Structures; *Cambridge University Press*; Cambridge, UK, 2001.

[7] Chopra, I. Review of state of art of smart structures and integrated systems. *AIAA J.* 2002, vol. 40, 2145-2187.

[8] Tzou, H. S.; Lee, H.-J.; Arnold, S. M. Smart materials, precision sensors/actuators, smart structures, and structronic systems. *Mech. Adv. Mater. Struct.* 2004, vol. 11, 367-393.

[9] Aboudi, J. The generalized method of cells and high-fidelity generalized method of cells micromechanical models - a review. *Mech. Adv. Materl. Struct.* 2004, vol. 11, 329-366.

[10] Aboudi, J.; Pindera, M-J.; Arnold, S. M. Higher-order theory for functionally graded materials. *Composites: Part B* 1999, vol. 30, 777-832.

[11] Aboudi, J. Micromechanical analysis of fully coupled electro-magneto-thermo-elastic multiphase composites, *Smart Mater. Struct.* 2001, vol. 10, 867-877.

[12] Aboudi, J.; Pindera, M.-J.; Arnold, S. M. Linear thermoelastic higher-order theory for periodic multiphase materials. *J. Appl. Mech.* 2001, vol. 68, 697-707.

[13] Levin, V. M. On the coefficients of thermal expansion of heterogeneous materials. *Mech. Solids* 1967, vol. 2, 58-61.

[14] Sun, C. T.; Vaidya, R. S. Prediction of composite properties from a representative volume element. *Compos. Sci. Tech.* 1996, vol. 56, 171-179.

[15] Tamma, K. K.; Avila, A. F. Modeling and computational methodology for high temperature composites. *In Thermal Stresses*; Hetnarski, R. B.; Ed.; Lastran Corporation, Rochester, NY, 1999; Vol. 5, pp. 143-256.

[16] Bansal, Y.; Pindera, M-J. *Testing the predictive capability of the high-fidelity generalized method of cells using an efficient reformulation* . NASA/CR-2004-213043, 2004.

[17] Arnold, S. M.; Bednarcyk, B.; Aboudi, J. *Comparison of the computational efficiency of the original versus reformulated high-fidelity generalized method of cells.* NASA/TM-2004-213438, 2004.

[18] Bednarcyk, B. A. An inelastic micro/macro theory for hybrid smart/metal composites. *Composites Part B* 2003, vol. 34, 175-197.

[19] Li, J. Y.; Dunn, M. L. Micromechanics of magnetoelectroelastic composite materials: average field and effective behavior. *J. Intell. Material Systems & Structures* 1998, vol. 9, 404-416.

[20] Mori, T.; Tanaka, K. Average stresses in matrix and average energy of materials with misfitting inclusions. *Acta Metall.* 1973, vol. 21, 571-574.

[21] Aboudi, J. Micromechanical prediction of the effective behavior of fully coupled electro-magneto-thermo-elastic multiphase composites. NASA/CR-2000-209787, 2000.

[22] Aboudi, J. Micromechanical prediction of the effective coefficients of thermo-piezoelectric multiphase composites, *J.Intell. Mater. Syst. Struct.* 1998, vol. 9, 713-722.

[23] Herakovich, C. T. Mechanics of Fibrous Composites; Wiley; New York, US, 1998.

[24] Reddy, J. N. *Mechanics of Laminated Composite Plates and Shells* ; CRC Press; Boca Raton. US, 2004.

[25] Herbert, J. M. *Ferroelectric Transducers and Sensors; Gordon and Breach Science* ; New York, US, 1982.

[26] Gookin, D. M.; Jacobs, E. W.; Hicks, J. C. Correlation of ferroelectric hysteresis with 33 ferroelastic hysteresis of polyvinylidene fluoride. *Ferroelectrics* 1984, vol. 57, 89-98.

[27] Maugin, G. A.; Pouget, J.; Drouot, R.; Collet, B. Nonlinear Electromechanical Coupling; Wiley; New York, US, 1992.

[28] Hall, D. A. Nonlinearity in piezoelectric ceramics. *J. Mater. Sci.* 2001, vol. 36, 4575-4601.

[29] Kamlah, M. Ferroelectric and ferroelastic piezoceramics - modeling of electromechanical hysteresis phenomena. *Continuum Mech.Thermodyn.* 2001, vol.13, 219-268.

[30] Chan, K. H.; Hagood, N. W. Modeling of nonlinear piezoceramics for structural actuation. *Proc. SPIE* 1994, vol. 2190, 194-205.

[31] Hwang, S. C.; McMeeking, R. M. 1999. A finite element model of ferroelastic polycrystals. *Int. J. Solids Struct.* 1999, vol. 36, 1541-1556.

[32] Kim, S-J.; Jiang, Q. A finite element model for rate-dependent behavior of ferroelectric ceramics. *Int. J. Solids Struct.* 2002, vol. 39, 1015-1030.

[33] Li, F.; Fang, D. Simulations of domain switching in ferroelectrics by a three-dimensional finite element model. *Mech. Mater.* 2004, vol. 36, 959-973.

[34] Bassiouny, E.; Ghaleb, A. F.; Maugin, G. A. Thermodynamical formulation for coupled electromechanical hysteresis effects- I. Basic equations. *Int. J. Engng. Sci.* 1988, vol. 26, 1279-1295.

[35] Zhang, X. D.; Rogers, C. A. A microscopic phenomenological formulation for coupled electromechanical effects in piezoelectricity. *J. Intell. Mater. Sys. Struct.* 1993, vol. 4, 307-316.

[36] Huang L.; Tiersten, H. F. An analytical description of slow hysteresis in polarized ferroelectric ceramic actuators. *J. Intell. Mater. Sys. Struct.* 1998, vol. 9, 417-426.

[37] Kamlah, M.; Tsakmakis, C. Phenomenological modeling of a non-linear electromechanical coupling in ferroelectrics. *Int. J. Solids Struct.* 1999, vol. 36, 669-695.

[38] Zhou, X.; Chattopadhyay, A. Hysteresis behavior and modeling of piezoceramic actuators. *J. Appl. Mech.* 2001, vol. 68, 270-277.

[39] Moulson, A. J.; Herbert, J. M. *Electroceramics*; Chapman & Hall; New York, US, 1990.

[40] Tan, P.; Tong, L. Micromechanics models for non-linear behavior of piezo-electric fiber reinforced composite materials. *Int. J. Solids Struct.* 2001, vol. 38, 8999-9032.

[41] Aboudi, J. Hysteresis behavior of ferroelectric fiber composites. *Smart Mater. Sruct.* 2005, vol. 14, 715-736.

[42] Uchino, K. Electrostrictive actuators: materials and applications. *Ceramic Bull.* 1986, vol. 65, 647-652.

[43] Rittenmyer, K. M. Electrostrictive ceramics for underwater transducer applications. *J. Acoust. Soc. Am.* 1994, vol. 95, 849-856.

[44] Uchino, K. *Piezoelectric Actuators and Ultrasonic Motors*; Kluwer; Boston, US, 1997.

[45] Takeuchi, H.; Masuzawa, H.; Nakaya, C.; Ito, Y. Medical ultrasonic probe using electrostrictive ceramic/polymer composite. *IEEE Ultrasonic Symposium* 1989, pp. 705-708.

[46] Rogers, C. A. Intelligent material systems - the dawn of a new materials age. *J. Intell. Mater. Syst. & Struct.* 1993, vol. 4, 4-12.

[47] Hom, C. L.; Shankar, N. A fully coupled constitutive model for electrostrictive ceramic materials. *J. Intell. Mater. Syst. & Struct.* 1994, vol. 5, 795-801.

[48] Hom, C. L.; Pilgrim, S. M.; Shankar, N.; Bridger, K.; Massuda, M.; Winzer, S. R. Calculation of quasistatic electromechanical coupling coefficients for electrostrictive ceramic materials. IEEE Trans on Ultrason., Ferroelec., *Freq. Contr.* 1994, vol. 41, 542-551.

[49] Hom, C. L.; Shankar, N. A finite element method for electrostrictive ceramic devices. *Int. J. Solids & Struct.* 1996, vol. 33, 1757-1779.

[50] Cross, L. E.; Jang, S. J.; Newnham, R. E.; Nomura, S.; Uchino, K. Large electrostrictive effects in relaxor ferroelectrics. *Ferroelectrics* 1980, vol. 23, 187-192.

[51] Aboudi, J. Micromechanical prediction of the response of electrostrictive multiphase composites. *Smart Mater. Sruct.* 1999, vol. 8, 663-671.

[52] McDonald, P. H. Continuum Mechanics; PWS Publishing Co.; Boston, MA, 1996.

[53] Duenas, T. A.; Hsu, L.; Carman, G. P. Magnetostrictive composite material systems analytical/experimental. In: *Materials for Smart Systems II*; George, E.P.; Gotthardt, R.; Trolier-McKinstry, S.; Wun-Fogle, M. Eds.; MRS; Pittsburg, PA, 1996.

[54] Parton, V. Z.; Kudryavtsev, B. Z. *Electromagnetoelasticity*; Gordon and Breach Science Publisher; New York, US, 1988.

[55] Tan, P.; Tong, L. Prediction of non-linear electromagnetoelastic properties for piezo-electric/piezomagnetic fiber reinforced composites. *Proc. Inst. Mech. Engrs.* vol. 218, part L: *Materials Des. Appl.* 2004, 111-127.

[56] Joshi, S. P. Non-linear constitutive relations for piezoceramic materials. *Smart Mater. Sruct.* 1992, vol. 1, 80-83.

[57] Shvartsman, A. *Micromechanical Analysis of Nonlinear Electro-Magneto-Thermo-Elastic Composites.* M.Sc. Thesis, Tel-Aviv University, 2005.

[58] Aboudi, J.; Pindera, M.-J.; Arnold, S. M. *High-fidelity generalized method of cells for inelastic periodic multiphase materials,* NASA TM-2002-211469, 2002.

[59] Aboudi, J.; Pindera, M.-J.; Arnold, S. M. Higher-order theory for periodic multiphase materials with inelastic phases. *Int. J. Plasticity* 2003, vol. 19, 805-847, 2003

[60] Pindera, M.-J.; Freed, A. D.; Arnold, S. M. Effects of fiber and interfacial layer morphologies on the thermoplastic response of metal matrix composites. *Int. J. Solids Struct.* 1993, vol. 30, 1213-1238.

[61] Williams, T. O.; Pindera, M.-J. An analytical model for the inelastic axial shear response of unidirectional metal matrix composites. *Int. J. Plasticity* 1997, vol. 13, 261-289.

[62] Bednarcyk, B. A.; Arnold, S. M.; Aboudi, J.; Pindera, M.-J. Local field effects in titanium matrix composites subjected to fiber-matrix debonding. *Int. J. Plasticity* 2004, vol. 20, 1707-1737.

[63] Boyd, J. G.; Lagoudas, D. C. Thermomechanical response of shape memory composites. *J. Intell. Mat. Syst. Struct.* 1994, vol. 5, 333-346.

[64] Auricchio, F.; Taylor, R. L.; Lubliner, J. Shape-memory alloys: macromodelling and numerical simulations of the superelastic behavior. Comput. *Methods Appl. Mech. Eng.* 1997, vol. 146, 281-312.

[65] Brocca, M.; Brinson, L. C.; Bazant, Z. P. Three-dimensional constitutive model for shape memory alloy based on microplane model. *J. Mech. Phys. Solids* 2002, vol. 50, 1051-1077.

[66] Birman, V. Review of mechanics of shape memory alloy structures. *Appl. Mech. Rev.* 1997, vol. 50, 629-645.

[67] Kawai, M.; Ogawa, H.; Baburaj, V.; Koga, T. Micromechanical analysis for hysteretic behavior of unidirectional TiNi SMA fiber composite. *J. Intell. Mat. Syst. Struct.* 1996, vol. 10, 14-28.

[68] Song, G. Q.; Sun, Q. P.; Cherkaoui, M. 1999 Role of microstructures in the thermomechanical behavior of SMA composites. *J. Eng. Mater. Tech.* 1999, vol. 121, 86-92.

[69] Gilat, R.; Aboudi, J. Dynamic response of active composite plates: shape memory alloy fibers in polymeric/metallic matrices. *Int. J. Solids & Struct.* 2004, vol. 41, 5717-5731.

[70] Lagoudas, D. C.; Bo, Z.; Qidwai, M. A. A unified thermodynamic constitutive model for SMA and finite element analysis of active metal matrix composites. *Mech. Compos. Materl. Struct.* 1996, vol. 3, 153-179.

[71] Bodner, S. R. Unified Plasticity for Engineering Applications; Kluwer; New York, US, 2002.

[72] Auricchio, F.; Taylor, R. L. Shape-memory alloys: modelling and numerical simulations of the finite-strain superelastic behavior. *Comp. Meth. Appl. Mech. Eng.* 1997, vol. 143, 175-194.

[73] Auricchio, F. A robust integration-algorithm for a finite-strain shape-memory-alloy superelastic model.*Int. J. Plasticity* 2001, vol. 17, 971-990.

[74] Aboudi, J. Micromechanically based constitutive equations for shape-memory fiber composites undergoing large deformations. *Smart Mater. Struct.* 2004, vol. 13, 828-837.

[75] Aboudi, J.; Freed, Y. Two-way thermomechanically coupled micromechanical analysis of shape memory alloy composites. *J. of Mech. Mat. Struct.* 2006, vol. 1, 937-955.

[76] Aboudi, J. Micromechanical analysis of the fully coupled finite thermoelastic response of rubber-like matrix composites. *Int. J. Solids Struct.* 2002, vol. 39, 2587-2612.

[77] Aboudi, J. Micromechanical analysis of the finite elastic-viscoplastic response of multiphase composites. *Int. J. Solids Struct.* 2003, vol. 40, 2793-2817.

[78] Ogden, R. W. Non-linear Elastic Deformations; Ellis Horwood; Chichester, UK, 1984.

[79] Holzapfel, G. A. *Nonlinear Solid Mechanics*; John Wiley; New York, US, 2000.

[80] Shalev, D.; Aboudi, J. Coupled micro to macro analysis of a composite that hosts embedded piezoelectric actuators. *J. Intell. Mater. Sys. Struct.* 1996, vol. 7, 15-24.

[81] Aboudi, J. The response of shape memory alloy composites. *Smart Mater. Sruct.* 1997, vol. 6, 1-9.

[82] Witting, P. R.; Cozzarelli, F. A. Experimental determination of shape memory alloy constitutive model parameters. Active Materials and Smart Structures, *SPIE-The International Society for Optical Engineering* 1995, vol. 2427, 260-275.

[83] Lauermann, M. E. Temperature Dependence of Shape Memory Material *Parameters in Constitutive Law for Nitinol.* M.Sc. Thesis, State University of New York at Buffalo, 1994.

[84] Gandhi M. V.; Thompson, B. S.; Choi, S. B. A new generation of innovative ultra-advanced intelligent composite materials featuring electro-rheological fluids: an experimental investigation. *J. Compos. Materl.* 1989, vol. 23, 1232-1255.

[85] Ashour, O.; Rogers, C. A.; Kordonsky, W. Magnetorheological fluid: materials, characterization, and devices. *J. Intell. Mat. Systm & Struct* 1996, vol. 7, 123-130.

[86] Aboudi, J. Effective behavior and dynamic response modeling of electro-rheological and magneto-rheological fluid composites. *Smart Mater. Struct.* 1999, vol. 8, 106-115.

[87] Malvern, L. E. Introduction to the Mechanics of Continuous Medium; Prentice-Hall; New Jersey, US, 1969.

[88] Conard, H.; Chen, Y.; Sprecher, A. F. 1989, Electrorheology of suspensions of zeolite particles in silicone oil. In *Electrorheological Fluids;* Carlson, J. D.; Sprecher, A. F.; Conard, H.; Eds.; Technomic; Lancaster, US, 1989, pp. 252-264.

[89] Coulter, J. P.; Duclos, T. G. 1989, Applications of electrorheological materials in vibration control. In Electrorheological Fluids; Carlson, J. D.; Sprecher, A. F.; Conard, H.; Eds.; *Technomic; Lancaster*, US, 1989, pp. 300-325.

[90] Weiss, K. D.; Carlson, J. D.; Coulter, J. P. Material aspects of electrorehological systems. *J. Intell. Mat. Systm & Struct.* 1993, vol. 4, 13-34.

[91] Weiss, K. D.; Carlson, J. D.; Nixon, D. A. Viscoelastic properties of magneto- and electro-rheological Fluids Systems. *J. Intell. Mat. Systm & Struct.* 1994, vol. 5, 772-775.

[92] Wang, K. C.; Mclay, R.; Carey, G. F. 1989, ER fluid modelling. In Electrorheological Fluids; Carlson, J. D.; Sprecher, A. F.; Conard, H.; Eds.; Technomic; *Lancaster*, US, 1989, pp. 41-52.

[93] Shalev, D.; Aboudi, J.; Tur, M. A micro-macro model for the effects of thermo-mechanical fields on optical fibers embedded in a laminated composite plate with applications to sensing. *Mech. Compos. Materl. Struct.* 1996, vol. 3, 297-320,

[94] Dakin, J.; Culshaw, B. *Optical Fiber Sensors*: Principles and Components; Artech House; Boston, US, 1988.

[95] Culshaw, B.; Dakin, J. *Optical Fiber Sensors: Systems and Applications* ; Artech House; Boston, US, 1989.

[96] Butter, C. D.; Hocker, G. B. Fiber optics strain gauge. *Appl. Optics* 1978, vol. 17, 2867-2869.

[97] Sirkis, J. S.; Haslach, H. W. Interferometric strain measurement by arbitrary configurated surface mounted optical fiber. *J. Light. Tech.* 1990, vol. 8, 1497-1503.

[98] Sirkis, J. S.; Haslach, H. W. Complete phase-strain model for structurally embedded interferometric optical fiber sensors. *J. Intell. Mater. Syst. & Struct.* 1991, vol. 2, 3-24.

[99] Kim, K. S.; Kollar, L.; Springer, G. S. A model of embedded fiber optic Farby-Perot temperature and strain sensors. *J. Compos. Materl.* 1993, vol. 27, 1618-1662.

[100] Yariv, A.; Yei, P. Optical Waves in Crystals; John Wiley Inc.; New York, US, 1984.

[101] Murphy, K. A.; Gunther, M. F.; Vengsarkar, A. M.; Claus, R. O. Quadrature phase shifted, extrinsic Fabry-Perot optical fiber sensors. *Opt. Lett.* 1991, vol. 16, 273-275.

[102] Gilat, R.; Aboudi, J. Thermal buckling of activated shape memory reinforced laminated plates. *Smart Mater. Struct.* vol. 15,829-838, 2006.

[103] Bednarcyk, B. A.; Arnold, S.M. MAC/GMC 4.0 User's Manual. NASA/TM-2002-212077, 2002

[104] Zhen, Z.; Jian, J.; Yapeng, S. Thermomechanical response of SMA fiber composites with non-uniform temperature distribution. *Acta Mech. Solida Sinica* 2003, vol. 16, 334-344.

[105] Liang, C.; Rogers, C. A. The multi-dimensional constitutive relations for shape memory alloys. *J. Engng. Math.* 1992, vol. 26, 429-443.

INDEX

N

O

S

T